Concepts and Semantics of Programming Languages 1

Series Editor
Jean-Charles Pomerol

Concepts and Semantics of Programming Languages 1

A Semantical Approach with OCaml and Python

Thérèse Hardin
Mathieu Jaume
François Pessaux
Véronique Viguié Donzeau-Gouge

iSTE

WILEY

First published 2021 in Great Britain and the United States by ISTE Ltd and John Wiley & Sons, Inc.

ISTE Ltd
27-37 St George's Road
London SW19 4EU
UK

www.iste.co.uk

John Wiley & Sons, Inc.
111 River Street
Hoboken, NJ 07030
USA

www.wiley.com

Library of Congress Control Number: 2021930488

British Library Cataloguing-in-Publication Data
A CIP record for this book is available from the British Library
ISBN 978-1-78630-530-5

MIX
Paper from responsible sources
FSC
www.fsc.org
FSC® C013604

Contents

Foreword xi

Preface xiii

Chapter 1. From Hardware to Software 1

1.1. Computers: a low-level view 1
1.1.1. Information processing 1
1.1.2. Memories 2
1.1.3. CPUs 3
1.1.4. Peripheral devices 7
1.2. Computers: a high-level view 8
1.2.1. Modeling computations 9
1.2.2. High-level languages 9
1.2.3. From source code to executable programs 10

Chapter 2. Introduction to Semantics of Programming Languages 15

2.1. Environment, memory and state 16
2.1.1. Evaluation environment 16
2.1.2. Memory 18
2.1.3. State 20
2.2. Evaluation of expressions 21
2.2.1. Syntax 21
2.2.2. Values 22
2.2.3. Evaluation semantics 24
2.3. Definition and assignment 26
2.3.1. Defining an identifier 26
2.3.2. Assignment 29
2.4. Exercises 31

Chapter 3. Semantics of Functional Features 35

3.1. Syntactic aspects 35
3.1.1. Syntax of a functional kernel 35
3.1.2. Abstract syntax tree 36
3.1.3. Reasoning by induction over expressions 39
3.1.4. Declaration of variables, bound and free variables 39
3.2. Execution semantics: evaluation functions 42
3.2.1. Evaluation errors 42
3.2.2. Values 43
3.2.3. Interpretation of operators 45
3.2.4. Closures 46
3.2.5. Evaluation of expressions 47
3.3. Execution semantics: operational semantics 54
3.3.1. Simple expressions 55
3.3.2. Call-by-value 56
3.3.3. Recursive and mutually recursive functions 60
3.3.4. Call-by-name 61
3.3.5. Call-by-value versus call-by-name 62
3.4. Evaluation functions versus evaluation relations 64
3.4.1. Status of the evaluation function 64
3.4.2. Induction over evaluation trees 65
3.5. Semantic properties 69
3.5.1. Equivalent expressions 69
3.5.2. Equivalent environments 71
3.6. Exercises 71

Chapter 4. Semantics of Imperative Features 77

4.1. Syntax of a kernel of an imperative language 77
4.2. Evaluation of expressions 81
4.3. Evaluation of definitions 86
4.4. Operational semantics 89
4.4.1. Big-step semantics 89
4.4.2. Small-step semantics 93
4.4.3. Expressiveness of operational semantics 95
4.5. Semantic properties 96
4.5.1. Equivalent programs 96
4.5.2. Program termination 98
4.5.3. Determinism of program execution 100
4.5.4. Big steps versus small steps 103
4.6. Procedures 109
4.6.1. Blocks 109
4.6.2. Procedures 112
4.7. Other approaches 118

4.7.1. Denotational semantics 118
4.7.2. Axiomatic semantics, Hoare logic 129
4.8. Exercises 134

Chapter 5. Types 137

5.1. Type checking: when and how? 139
5.1.1. When to verify types? 139
5.1.2. How to verify types? 140
5.2. Informal typing of a program $\mathbf{Exp}_2$ 141
5.2.1. A first example 141
5.2.2. Typing a conditional expression 142
5.2.3. Typing without type constraints 142
5.2.4. Polymorphism 143
5.3. Typing rules in $\mathbf{Exp}_2$ 143
5.3.1. Types, type schemes and typing environments 143
5.3.2. Generalization, substitution and instantiation 146
5.3.3. Typing rules and typing trees 151
5.4. Type inference algorithm in $\mathbf{Exp}_2$ 154
5.4.1. Principal type 154
5.4.2. Sets of constraints and unification 155
5.4.3. Type inference algorithm 159
5.5. Properties 167
5.5.1. Properties of typechecking 167
5.5.2. Properties of the inference algorithm 167
5.6. Typechecking of imperative constructs 168
5.6.1. Type algebra 168
5.6.2. Typing rules 169
5.6.3. Typing polymorphic definitions 171
5.7. Subtyping and overloading 172
5.7.1. Subtyping 173
5.7.2. Overloading 175

Chapter 6. Data Types 179

6.1. Basic types 179
6.1.1. Booleans 179
6.1.2. Integers 181
6.1.3. Characters 186
6.1.4. Floating point numbers 187
6.2. Arrays 191
6.3. Strings 194
6.4. Type definitions 194
6.4.1. Type abbreviations 195
6.4.2. Records 196

6.4.3. Enumerated types 200
6.4.4. Sum types 202
6.5. Generalized conditional 205
6.5.1. C style `switch/case` 205
6.5.2. Pattern matching 208
6.6. Equality 216
6.6.1. Physical equality 217
6.6.2. Structural equality 218
6.6.3. Equality between functions 220

Chapter 7. Pointers and Memory Management 223

7.1. Addresses and pointers 223
7.2. Endianness 225
7.3. Pointers and arrays 225
7.4. Passing parameters by address 226
7.5. References 229
7.5.1. References in C++ 229
7.5.2. References in Java 233
7.6. Memory management 234
7.6.1. Memory allocation 234
7.6.2. Freeing memory 237
7.6.3. Automatic memory management 239

Chapter 8. Exceptions 243

8.1. Errors: notification and propagation 243
8.1.1. Global variable 245
8.1.2. Record definition 245
8.1.3. Passing by address 245
8.1.4. Introducing exceptions 246
8.2. A simple formalization: ML-style exceptions 247
8.2.1. Abstract syntax 247
8.2.2. Values 248
8.2.3. Type algebra 248
8.2.4. Operational semantics 248
8.2.5. Typing 250
8.3. Exceptions in other languages 250
8.3.1. Exceptions in OCaml 251
8.3.2. Exceptions in Python 251
8.3.3. Exceptions in Java 253
8.3.4. Exceptions in C++ 254

Conclusion 257

Appendix: Solutions to the Exercises 259

List of Notations 287

Index of Programs 289

References 293

Index 295

Foreword

Computer programs have played an increasingly central role in our lives since the 1940s, and the quality of these programs has thus become a crucial question. Writing a high-quality program – a program that performs the required task and is efficient, robust, easy to modify, easy to extend, etc. – is an intellectually challenging task, requiring the use of rigorous development methods. First and foremost, however, the creation of such a program is dependent on an in-depth knowledge of the programming language used, its syntax and, crucially, its semantics, i.e. what happens when a program is executed.

The description of this semantics puts the most fundamental concepts into light, including those of value, reference, exception or object. These concepts are the foundations of programming language theory. Mastering these concepts is what sets experienced programmers apart from beginners. Certain concepts – like that of value – are common to all programming languages; others – such as the notion of functions – operate differently in different languages; finally, other concepts – such as that of objects – only exist in certain languages. Computer scientists often refer to "programming paradigms" to consider sets of concepts shared by a family of languages, which imply a certain programming style: imperative, functional, object-oriented, logical, concurrent, etc. Nevertheless, an understanding of the concepts themselves is essential, as several paradigms may be interwoven within the same language.

Introductory texts on programming in any given language are not difficult to find, and a number of published books address the fundamental concepts of language semantics. Much rarer are those, like the present volume, which establish and examine the links between concepts and their implementation in languages used by programmers on a daily basis, such as C, C++, Ada, Java, OCaml and Python. The authors provide a wealth of examples in these languages, illustrating and giving life to the notions that they present. They propose general models, such as the *kit* presented in Volume 2, permitting a unified view of different notions; this makes it easier for readers to understand the constructs used in popular programming languages and facilitates comparison. This thorough and detailed work provides readers with an understanding of these notions and, above all, an understanding of the ways of using the latter to create high-quality programs, building a safer and more reliable future in computing.

Gilles DOWEK
Research Director, Inria
Professor at the École normale supérieure, Paris-Saclay

Catherine DUBOIS
Professor at the École nationale supérieure
d'informatique pour l'industrie et l'entreprise

January 2021

Preface

This two-volume work relates to the field of programming. First and foremost, it is intended to give readers a solid grounding in the bases of functional or imperative programming, along with a thorough knowledge of the module and class mechanisms involved. In our view, the semantics approach is most appropriate when studying programming, as the impact of interlanguage syntax differences is limited. Practical considerations, determined by the material characteristics of computers and/or "smart" devices, will also be addressed. The same approach will be taken in both volumes, using both mathematical formulas and memory state diagrams. With this book, we hope to help readers understand the meaning of the constructs described in the reference manuals of programming languages and to establish solid foundations for reasoning and assessing the correctness of their own programs through critical review. In short, our aim is to facilitate the development of safe and reliable programs.

Volume 1 begins with a presentation of the computer, in Chapter 1, first at the material level – as an assemblage of components – then as a tool for executing programs. Chapter 2 is an intuitive, step-by-step introduction to language semantics, intended to familiarize readers with this approach to programming. In Chapter 3, we provide a detailed discussion on the subject, with a formal presentation of the execution semantics of functional features. Chapter 4 continues with the same topic, looking at the execution semantics of imperative features. In these two chapters, a clear mathematical framework is used to support our presentation. Also, all of the notions which we introduce in these chapters are implemented in both Python and OCaml to assist readers learning about the semantic concepts in question for the first time. Multiple exercises, with detailed solutions, are provided in both cases. Chapter 5, on the subject of typing, begins by addressing typing rules, which are used to check programs; we then present the algorithm used to infer polymorphic types, along with the associated mathematical notions, all implemented in both languages. Finally, the extension of typing to imperative features is addressed. In Chapter 6, we present the main data types and methods of pattern matching, using a range of examples expressed in different programming languages. Chapter 7 focuses on low-level programming features: endianness, pointers and memory management; these notions are mostly presented using C and C++. Volume 1 ends with a discussion of error processing using exceptions, their semantics is presented in OCaml, and the exception management mechanisms used in Python, Java and C++ are also described (see Chapter 8).

Thus, Volume 1 is intended to give a broad overview of the functional and imperative features of programming, from notions that can be modeled mathematically to notions that are linked to the hardware configuration of computers themselves. Volume 2 focuses on modular and object programming, building on the foundations laid down in Volume 1 since modules, classes and objects are, in essence, the means of organizing functional or imperative constructs. Volume 2 first analyzes the needs of developers in terms of tools for software architecture. Based on this study, an original semantic model, called a *kit*, is drawn up, jointly presenting all the features of the modules and objects that can meet these needs. The semantics of these kits are defined in a rather informal way, as research in this field has not yet led to a mathematical model of this set of features, while remaining relatively simple. From this model, we consider a set of emerging questions, the objective of which is to guide the acquisition of a language. This approach is then exemplified by the study of the module systems of Ada, OCaml and C. Finally, the same approach will be used to deduce a semantic model of class and object features, which will serve to present classes in Java, C++, OCaml and Python from a unified perspective.

This work is aimed at a relatively wide audience, from experienced developers – who will find valuable additional information on language semantics – to beginners who have only written short programs. For beginners, we recommend working on the semantic concepts described in Volume 1 using the implementations in OCaml or Python to ease assimilation. All readers may benefit from studying the reference manual of a programming language, while comparing the presentations of constructs given in the manual with those given here, guided by the questions mentioned in Volume 2.

Note that we do not discuss the algorithmic aspect of data processing here. However, choosing the algorithm and the data representation that fit the requirements of the specification is an essential step in program development. Many excellent works have been published on this subject, and we encourage readers to explore the subject further. We also recommend using the standard libraries provided by the chosen programming language. These libraries include tried and tested implementations for many different algorithms, which may generally be assumed to be correct.

1

From Hardware to Software

This first chapter provides a brief overview of the components found in all computers, from mainframes to the processing chips in tablets, smartphones and smart objects via desktop or laptop computers. Building on this hardware-centric presentation, we shall then give a more abstract description of the actions carried out by computers, leading to a uniform definition of the terms "program" and "execution", above and beyond the various characteristics of so-called electronic devices.

1.1. Computers: a low-level view

Computer science is the science of rational processing of information by computers. Computers have the capacity to carry out a variety of processes, depending on the instructions given to them. Each item of *information* is an element of knowledge that may be transmitted using a signal and encoded using a sequence of symbols in conjunction with a set of rules used to decode them, i.e. to reconstruct the signal from the sequence of symbols. Computers use binary encoding, involving two symbols; these may be referred to as "true"/"false", "0"/"1" or "high"/"low"; these terms are interchangeable, and all represent the two stable states of the electrical potential of digital electronic circuits.

1.1.1. *Information processing*

Schematically, a computer is made up of three families of components as follows:

– memories: store data (information) and executable code (the so-called von Neumann architecture);

– one or more microprocessors, known as CPUs (central processing units), which process information by applying elementary operations;

– peripherals: these enable information to be exchanged between the CPU/memory couple and the outside.

Information processing by a computer – in other terms, the execution of a program – can be summarized as a sequence of three steps: fetching data, computing the results and returning them. Each elementary processing operation corresponds to a configuration of the logical circuits of the CPU, known as a *logic function*. If the result of this function is solely dependent on input, and if no notion of "time" is involved in the computations, then the function is said to be *combinatorial*; otherwise, it is said to be *sequential*.

For example, a binary half-adder, as shown in Figure 1.1, is a circuit that computes the sum of two binary digits (input), along with the possible carry value. It thus implements a combinatorial logic function.

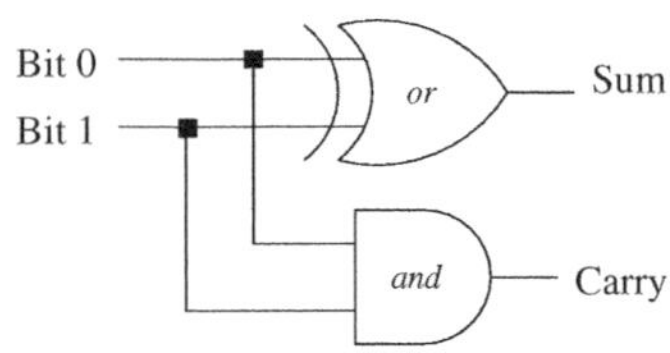

Figure 1.1. *Binary half-adder*

The essential character of a combinatorial function is that, for the same input, the function always produces the same output, no matter what the circumstances. This is not true of sequential logic functions.

For example, a logic function that counts the number of times its input changes relies on a notion of "time" (changes take place in time), and a persistent state between two inputs is required in order to record the previous value of the counter. This state is saved in a *memory*. For sequential functions, a same input value can result in different output values, as every output depends not only on the input, but also on the state of the memory at the moment of reading the new input.

1.1.2. *Memories*

Computers use memory to save programs and data. There are several different technologies used in memory components, and a simplified presentation is as follows:

– RAM (Random Access Memory): RAM memory is both readable and writeable. RAM components are generally fast, but also volatile: if electric power falls down, their content is lost;

– ROM (Read Only Memory): information stored in a ROM is written at the time of manufacturing, and it is read-only. ROM is slower than RAM, but is non-volatile, like, for example, a burned DVD;

– EPROM (Erasable Programmable Read Only Memory): this memory is non-volatile, but can be written using a specific device, through exposure to ultra-violet light, or by modifying the power voltage, etc. It is slower than RAM, for both reading and writing. EPROM may be considered equivalent to a rewritable DVD.

Computers use the memory components of several technologies. Storage size diminishes as access speed increases, as fast-access memory is more costly. A distinction is generally made between four different types of memory:

– mass storage is measured in terabytes and is made either of mechanical disks (with an access time of $\sim$ 10 ms) or – increasingly – of solid-state drive (SSD) blocks. These blocks use an EEPROM variant (electrically erasable) with an access time of $\sim 0.1 - 0.3$ ms, known as *flash memory*. Mass storage is non-volatile and is principally used for the file system;

– RAM, which is external to the microprocessor. Recent home computers and smartphones generally possess large RAM capacities (measured in gigabytes). Embedded systems or consumer development electronic boards may have a much lower RAM capacity. The access time is around 40–50 ηs;

– the *cache* is generally included in the CPU of modern machines. This is a small RAM memory of a few kilobytes (or megabytes), with an access time of around $5 - 10$ ηs. There are often multiple levels of cache, and access time decreases with size. The cache is used to save frequently used and/or consecutive data and/or instructions, reducing the need to access slower RAM by retaining information locally. Cache management is complex: it is important to ensure consistency between the data in the main memory and the cache, between different CPUs or different cores (full, independent processing units within the same CPU) and to decide which data to discard to free up space, etc.;

– *registers* are the fastest memory units and are located in the center of the microprocessor itself. The microprocessor contains a limited number (a few dozen) of these storage zones, used directly by CPU instructions. Access time is around one processor cycle, i.e. around 1 ns.

1.1.3. *CPUs*

The CPU, as its name suggests, is the unit responsible for processing information, via the execution of *elementary instructions*, which can be roughly grouped into five categories:

– data transfer instructions (copy between registers or between memory and registers);

– arithmetic instructions (addition of two integer values contained in two registers, multiplication by a constant, etc.);

– logical instructions (bit-wise and/or/not, shift, rotate, etc.);

– branching operations (conditional, non-conditional, to subroutines, etc.);

– other instructions (halt the processor, reset, interrupt requests, `test-and-set`, `compare-and-swap`, etc.).

Instructions are coded by binary words in a format specific to each microprocessor. A program of a few lines in a high-level programming language is translated into tens or even hundreds of elementary instructions, which would be difficult, error prone and time consuming to write out manually. This is illustrated in Figure 1.2, where a "Hello World!" program written in C is shown alongside its counterpart in x86-64 instructions, generated by the `gcc` compiler.

```
#include <stdio.h>
int main () {
  printf ("Hello world!\n") ;
  return (0) ;
}
```

```
  .section __TEXT
  .globl   _main
  .align   4, 0x90
_main:
  .cfi_startproc
## BB#0:
  pushq    %rbp
Ltmp0:
  .cfi_def_cfa_offset 16
Ltmp1:
  .cfi_offset %rbp, -16
  movq     %rsp, %rbp
Ltmp2:
  .cfi_def_cfa_register %rbp
  subq     $16, %rsp
  leaq     L_.str(%rip), %rdi
  movl     $0, -4(%rbp)
  movb     $0, %al
  callq    _printf
  xorl     %ecx, %ecx
  movl     %eax, -8(%rbp)
  movl     %ecx, %eax
  addq     $16, %rsp
  popq     %rbp
  retq
  .cfi_endproc
  .section __TEXT
L_.str:
  .asciz   "Hello world!\n"
```

Figure 1.2. *"Hello world!" in C and in x86-64 instructions*

Put simply, a microprocessor is split into two parts: a control unit, which decodes and sequences the instructions to execute, and one or more arithmetic and logic units (ALUs) , which carry out the operations stipulated by the instructions. The CPU runs permanently through a three-stage cycle:

1) fetching the next instruction to be executed from the memory: every microprocessor contains a special register, the Program Counter (PC), which records the location (address) of this instruction. The PC is then incremented, i.e. the size of the fetched instruction is added to it;

2) decoding of the fetched instruction;

3) execution of this instruction.

However, the next instruction is not always the one located next to the current instruction. Consider the function `min` in example 1.1, written in C, which returns the smallest of its two arguments.

EXAMPLE 1.1.–

```
C
int min (int a, int b) {
  if (a < b) return (a) ;
  else return (b) ;
}
```

This function may be translated, intuitively and naively, into elementary instructions, by first placing `a` and `b` into registers, then comparing them:

```
min:
  load       a, reg0
  load       b, reg1
  compare    reg0, reg1
```

Depending on the result of the test – true or false – different continuations are considered. Execution continues using instructions for one or the other of these continuations: we therefore have two possible control paths. In this case, a *conditional jump* instruction must be used to modify the PC value, when required, to select the first instruction of one of the two possible paths.

```
  branchgt  a_gt_b
  load      reg0, reg2
  jump      end
a_gt_b:
  load      reg1, reg2
end:
  return    reg2
```

The `branchgt` instruction loads the location of the instruction at label `a_gt_b` into the PC. If the result of the `compare` instruction is that `reg0` > `reg1`, the next instruction is the one found at this address: `load reg1, reg2`. Otherwise, the next instruction is the one following `branchgt`: `load reg0, reg2`. This is followed by the *unconditional jump* instruction, `jump`, enabling unconditional modification of the PC, loading it with the address of the `end` label. Thus, whatever the result of the comparison, execution finishes with the instruction `return reg2`.

Conditional branching requires the use of a specific memory to determine whether certain conditions have been satisfied by the execution of the previous instruction (overflow, positive result, null result, superiority, etc.). Every CPU contains a dedicated register, the State Register (SR), in which every bit is assigned to signaling one of these conditions. Executing most instructions may modify all or some of the bits in the register. Conditional instructions (both jumps and more "exotic" variants) use the appropriate bit values for execution. Certain ARM® architectures [ARM 10] even permit all instructions to be intrinsically conditional.

Every program is made up of functions that can be called at different points in the program and these calls can be nested. When a function is called, the point where execution should resume once the execution of the function is completed – the *return address* – must be recorded. Consider a program made up of the functions $g() = k() + h()$ and $f() = g() + h()$, featuring several function calls, some of which are nested.

```
g () =
  t11 = k ()
  t12 = h ()
  return t11 + t12

f () =
  v11 = g ()
  v12 = h ()
  return v11 + v12
```

A single register is not sufficient to record the return addresses of the different calls. Calling `k` from `g` must be followed by calling `h` to evaluate `t12`. But this call of `g` was done by `f`, thus its return address in `f` should also be memorized to further evaluation of `v12`. The number of return addresses to record increases with the number of nested calls, and decreases as we leave these calls, suggesting very naturally to save these addresses in a *stack*. Figure 1.3 shows the evolution of a stack structure during successive function calls, demonstrating the need to record multiple return addresses. The state of the stack is shown at every step of the execution, at the moment where the line in the program is being executed.

A dedicated register, the Stack Pointer (SP), always contains the address of the next free slot in the stack (or, alternatively, the address of the last slot used). Thus, in the case of nested calls, the return address is saved at the address indicated by the SP, and the SP is incremented by the size of this address. When the function returns, the PC is loaded with the saved address from the stack, and the SP is decremented accordingly.

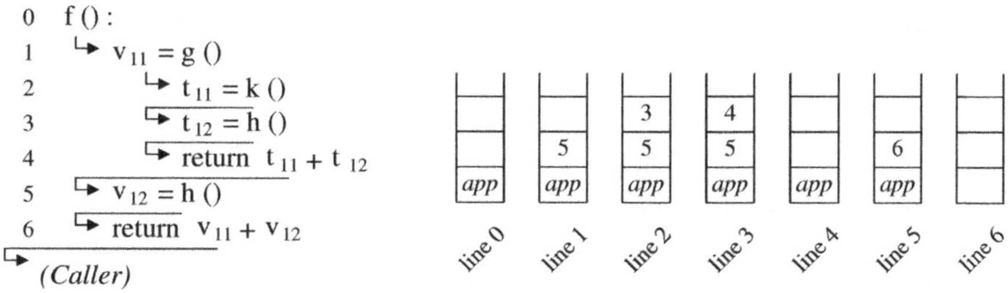

Figure 1.3. *Function calls and return addresses*

In summary, the internal state of a microprocessor is made up of its general registers, the program counter, the state register and the stack pointer. Note, however, that this is a highly simplified vision. There are many different varieties of microprocessors with different internal architectures and/or instruction sets (for example, some do not possess an integer division instruction). Thus, a program written directly using the instruction set of a microprocessor will not be executable using another model of microprocessor, and it will need to be rewritten. The portability of programs written in the assembly language of a given microprocessor is practically null. High-level languages respond to this problem by providing syntactic constructs, which are independent of the target microprocessors. The compiler or the interpreter have to translate these constructs into the language used by the microprocessor.

1.1.4. *Peripheral devices*

As we saw in section 1.1.3, processors execute a constant cycle of fetching, decoding and executing instructions. Computations are carried out using data stored in the memory, either by the program itself or by an input/output mechanism. The results of computations are also stored in the memory, and may be returned to users using this input/output mechanism.

The interest of any programmable system is inherently dependent on input/output capacities through which the system reacts to the external environment and may act on this environment. Even an assembly robot in a car factory, which repeats the same actions again and again, must react to data input from the environment. For example, the pressure of the grip mechanism must stop increasing once it has caught a bolt, and the time it takes to do this will differ depending on the exact position of the bolt.

Input/output systems operate using *peripherals*, ancillary devices that may be electronic, mechanical or a combination of the two. These allow the microprocessor to acquire external information, and to transmit information to the exterior. Computer mice, screens and keyboards are peripherals used with desktop computers, but other elements such as motors, analog/digital acquisition cards, etc. are also peripherals.

If peripherals are present, the microprocessor needs to devote part of its processing time to data acquisition and to the transmission of computed results. This interaction with peripherals may be directly integrated into programs. But in this case, the programs have to integrate regular checking of input peripherals to see if new information is available. It is technically difficult (if not impossible) to include such a monitoring in every program. Furthermore, regular peripheral checks are a waste of time and energy if no new data is available. Finally, there is no guarantee that information would arrive exactly at the moment of checking, as data may be *asynchronously* emitted.

This problem can be avoided by relying on the hardware to indicate the occurrence of new external events, instead of using software to check for these events. The *interrupt* mechanism is used to interrupt the execution of the current code and to launch the interrupt handler associated with the external event. This handler is a section of code, which is not explicitly called by the program being executed; it is located at an address known by the microprocessor. As any program may be interrupted at any point, the processor state, and notably the registers, must be saved before processing the interrupt. The code that is executed to process the interrupt will indeed use the registers and modify the SR, SP and PC. Therefore, previous values of registers must be restored in order to resume execution of the interrupted code. This context saving is carried out partially by the hardware and partially by the software.

1.2. Computers: a high-level view

The low-level vision of a von Neumann machine presented in section 1.1 provides a good overview of the components of a computer and of program execution, without going into detail concerning the operations of electronic components. However, this view is not particularly helpful in the context of everyday programming activity. Programs in binary code, or even assembly code, are difficult to write as they need to take account of every detail of execution; they are, by nature, long and hard to review, understand and debug. The first "high-level" programming languages emerged very shortly after the first computers. These languages assign names to certain values and addresses in the memory, providing a set of instructions that can be split into low-level machine instructions. In other terms, programming languages offer an abstract vision of the computer, enabling users to ignore low-level details while writing a program. The "hello world" program in Figure 1.2 clearly demonstrates the power of abstraction of C compared to the X86 assembly language.

1.2.1. *Modeling computations*

Any program is simply a description, in its own programming language, of a series of computations (including reading and writing), which are the only operations that a computer can carry out. An abstract view of a computer requires an abstract view – we call it a *model* – of the notion of computation. This subject was first addressed well before the emergence of computers, in the late 19th century, by logicians, mathematicians and philosophers, who introduced a range of different approaches to the theory of calculability.

The Turing machine [TUR 95] is a mathematical model of computation introduced in 1936. This machine operates on an infinite memory tape divided into cells and has three instructions: move one cell of the tape right or left, write or read a symbol in the cell or compare the contents of two cells. It has been formally proven that any "imperative" programming language, featuring assignment, a conditional instruction and a `while` loop, has the same power of expression as this Turing machine.

Several other models of the notion of algorithmic computation were introduced over the course of the 20th century, and have been formally proven to be equivalent to the Turing machine. One notable example is Kleene's recursion theory [KLE 52], the basis for the "pure functional" languages, based on the notion of (potentially) recursive functions; hence, these languages also have the same power of expression as the Turing machine. Pure functional and imperative languages have developed in parallel throughout the history of high-level programming, leading to different programming styles.

1.2.2. *High-level languages*

Broadly speaking, the execution of a functional program carries out a series of function calls that lead to the result, with intermediate values stored exclusively in the registers. The execution of an imperative program carries out a sequence of modifications of memory cells named by identifiers, the values in the cells being computed during execution. The most widespread high-level languages include both functional and imperative features, along with various possibilities (modules, object features, etc.) to divide source code into pieces that can be reused.

Whatever the style of programming used, any program written in a high-level language needs to be translated into binary language to be executed. These translations are executed either every time the program is executed – in which case the translation program is known as an *interpreter* or just once, storing the produced binary code – in which case the translator is known as a *compiler*.

As we have seen, high-level languages facilitate the coding of algorithms. They ease reviewing of the source code of a program, as the text is more concise than it would be for the same algorithm in assembly code. This does not, however, imply that users gain a better understanding of the way the program works. To write a program, a precise knowledge of the constructs used – in other terms, their *semantics*, what they do and what they mean – is crucial to understand the source code. Bugs are not always the result of algorithm coding errors, and are often caused by an erroneous interpretation of elements of the language. For example, the incrementation operator ++ in C exists in two forms (`i++` or `++i`), and its understanding is not as simple as it may seem. For example, the program:

```
C
#include <stdio.h>

int main () {
  int i = 0 ;
  printf ("%d\n", i++) ;
  return (0) ;
}
```

will print 0, but if `i++` is replaced with `++i`, the same program will print 1.

There are a number of concepts that are common to all high-level languages: value naming, organization of namespaces, explicit memory management, etc. However, these concepts may be expressed using different syntactic constructs. The field of language semantics covers a set of logico-mathematical theories, which describe these concepts and their properties. Constructing the semantics of a program allows to the formal verification of whether the program possesses all of the required properties.

1.2.3. *From source code to executable programs*

The transition from the program source to its execution is a multistep process. Some of these steps may differ in different languages. In this section, we shall give an overview of the main steps involved in analyzing and transforming source code, applicable to most programming languages.

The source code of a program is made up of one or more text files. Indeed, to ease software architecture, most languages allow source code to be split across several files, known as *compilation units*. Each file is processed separately prior to the final phase, in which the results of processing are combined into one single *executable* file.

1.2.3.1. *Lexical analysis*

Lexical analysis is the first phase of translation: it converts the sequence of characters that is indeed the source file into a sequence of *words*, assigning each to a category. Comments are generally deleted at this stage. Thus, in the following text presumed to be written in C

```
/* This is a comment. */
if [x == 3 int +) cos ($v)
```

lexical analysis will recognize the keyword `if`, the opening bracket, the identifier `x`, the operator `==`, the integer constant `3`, the type identifier `int`, etc. No word in C can contain the character $, so a lexical error will be highlighted when `$v` is encountered.

Lexical analysis may be seen as a form of "spell check", in which each recognized word is assigned to a category (keyword, constant, identifier). These words are referred to as *tokens*.

1.2.3.2. *Syntactic analysis*

Every language follows *grammar*. For example, in English, a sentence is generally considered to be correctly formed if it contains a subject, verb and complement in an understandable order. Programming languages are no exception: *syntactic analysis* verifies that the phrases of a source file conform with the grammar of their language. For example, in C, the keyword `if` must be followed by a bracketed expression, an instruction must end with a semicolon, etc. Clearly, the source text given in the example above in the context of lexical analysis does not respect the syntax of C.

Technically, the syntactic analyzer is in charge of the complete grammatical analysis of the source file. It calls the lexical analyzer every time it requires a token to progress through the analyzed source. Syntactic analysis is thus a form of grammar verification, and it also builds a representation of the source file by a data structure, which is often a tree, called the *abstract syntax tree* (AST). This data structure will be used by all the following phases of compilation, up to the point of execution by an interpreter or the creation of an executable file.

1.2.3.3. *Semantic analyses*

The first two analysis phases of compilation only concern the textual structure of the source. They do not concern the *meaning* of the program, i.e. its *semantics*. Source texts that pass the syntactic analysis phase do not always have meaning. The phrase "the sea eats a derivable rabbit" is grammatically correct, but is evidently nonsense.

The best-known semantic analysis is the typing analysis, which prohibits the combination of elements that are incompatible in nature. Thus, in the previous phase, "derivable" could be applicable to a function, but certainly not to a "rabbit".

Semantic analyses do not reduce to a form of typing analysis but they all interpret the constructs of a program according to the semantics of the chosen language. Semantic analyses may be used to eliminate programs, which leads to execution errors. They may also apply some transformations to program code in order to get an executable file (dependency analysis, closure elimination, etc.). These semantic analyses may be carried out during subsequent passes of source code processing, even after the code generation phase described in the following section.

1.2.3.4. *Code interpretation/generation*

Once the abstract syntax tree (or a derived tree) has been created, there are two options. Either the tree may be executed directly via an *interpreter*, which is a program supplied by the programming language, or the AST is used to generate *object* code files, with the aim of creating an executable file that can be run independently. Let us first focus on the second approach. The interpretation mechanism will be discussed later.

Compilation uses the AST generated from the source file to produce a sequence of instructions to be executed either by the CPU or by a virtual machine (VM). The compilation is correct if the execution of this sequence of instructions gives a result, which conforms to the program's semantics.

Optimization phases may take place during or after object code generation, with the aim of improving its compactness or its execution speed. Modern compilers implement a range of optimizations, which study lies outside the scope of this book. Certain optimizations are "universal", while others may be specific to the CPU for which the code is generated.

The object code produced by the compiler may be either binary code encoding instructions directly or source text in assembly code. In the latter case, a program – known as the *assembler* – must be called to transform this low-level source code into binary code. Generally speaking, assemblers simply produce a mechanical translation of instructions written mnemonically (`mov`, `add`, `jmp`, etc.) into binary representations. However, certain more sophisticated assemblers may also carry out optimization operations at this level.

Assembling mnemonic code into binary code is a very simple operation, which does not alter the structure of the program. The reference manual of the target CPU provides, for each instruction, the meaning of the bits of the corresponding binary word. For example, the reference manual for the MIPS32® architecture [MIP 13] describes the 32-bit binary format of the instruction `ADD rd, rs, rt` (with the effect `rd` ← `rs + rt` on the registers) as:

Bit weight	31 26	25 21	20 16	15 11	10 0
Bit value	000000	num. rs	num. rt	num. rd	000000100000

Figure 1.4. *Coding the ADD instruction in MIPS32®*

Three packets of 6 bits are reserved for encoding the register numbers; the other bits in this word are fixed and encode the instruction. The task of the assembler is to generate such bit patterns according to the instructions encountered in the source code.

1.2.3.5. *Linking*

A single program may be made up of several source files, compiled separately. Once the object code from each source file has been produced, all these codes must be collected into a single executable file. Each object file includes "holes", indicating unknown information at the moment of production of this object code. It is important to know where to find this missing code, when calling functions defined in a different compilation unit, or where to find variables defined in a location outside of the current unit.

The *linker* has to gather all the object files and fill all the holes. Evidently, for a set of object files to lead to an executable file, all holes must be filled; so the code of every function called in the source must be available. The linking process also has to integrate the needed code, if it comes from some libraries, whether from the standard language library or a third-party library. There is one final question to answer, concerning the point at which execution should begin. In certain languages (such as C, C++ and Java), the source code must contain one, and only one, special function, often named `main`, which is called to start the execution. In other languages (such as Python and OCaml), definitions are executed in the order in which they appear, defined by the file ordering during the linking process. Thus, "executing" the definition of a function does not call the function: instead, the "value" of this function is created and stored to be used later when the function is called. This means that programmers have to insert into the source file a call to the function which they consider to be the "starting point" of the execution. This call is usually the final instruction of the last source file processed by the linker.

A simplified illustration of the different transformation passes involved in source code compilation is shown in Figure 1.5.

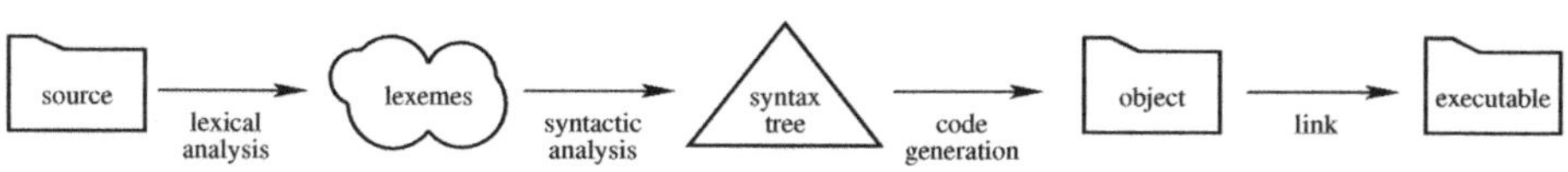

Figure 1.5. *Compilation process*

1.2.3.6. *Interpretation and virtual machines*

As we have seen, informally speaking, an interpreter "executes" a program directly from the AST. Furthermore, it was said that the code generation process may generate code for a virtual machine. In reality, interpreters rarely work directly on the tree; compilation to a virtual machine is often carried out as an intermediate stage. A *virtual machine* (VM) may be seen as a pseudo-microprocessor, with one or more stacks, registers and fairly high-level instructions. The code for a VM is often referred to as *bytecode*. In this case, compilation does not generate a file directly executable by the CPU. Execution is carried out by the *virtual machine interpreter*, a program supplied by the programming language environment. So, the difference between interpretation and compilation is not clear-cut.

There are several advantages of using a VM: the compiler no longer needs to take the specificities of the CPU into account, the code is often more compact and portability is higher. As long as the executable file for the virtual machine interpreter is available on a computer, it will be possible to generate a binary file for the computer in question. The drawback to this approach is that the programs obtained in this way are often slower than programs compiled as "native" machine code.

2

Introduction to Semantics of Programming Languages

This chapter introduces intuitively the notions of name, environment, memory, etc., along with a first formal description of these notions. It allows readers to familiarize themselves with the semantic approach of programming that we share with a number of other authors [ACC 92, DOW 09, DOW 11, FRI 01, WIN 93].

Any high-level programming language uses names to denote the entities handled by programs. These names are generally known as *identifiers*, drawing attention to the fact that they are constructed in accordance with the syntactic rules of the chosen language. They may be used to denote program-specific values or values computed during execution. They may also denote locations (i.e. addresses in the memory), they are then called *mutable variables*. And identifiers can also denote operators, functions, procedures, modules, objects, etc., according to the constructs present in the language. For example, `pi` is often used to denote an approximate value of π; `+` is also an identifier, denoting an addition operator and often placed between the two operands, i.e. in infix position, as in `2 + 3`. The expression `2 * x + 1` uses the identifier `x` and to compute its value, we need to know the value denoted by `x`. Retrieving the value associated with a given identifier is a mechanism at the center of any high-level language. The semantics of a language provides a model of this mechanism, presented – in a simplified form – in section 2.1.

All the formal definitions of languages, instructions, algorithms, etc., given in the following are coded in the programming languages OCaml and Python, trying to paraphrase these definitions and produce very similar versions of code in these two languages, even if developers in these languages may find the programming style used here rather unusual. For readers not introduced to these languages, some very brief explanations are given in the codes' presentation. But almost all features of OCaml and Python will be considered either in this first volume or in the second, where object-oriented programming is considered. We hope that these two encodings of formal notions can help readers who are not truly familiar with mathematical formalism.

2.1. Environment, memory and state

2.1.1. *Evaluation environment*

Let $\mathbf{X}$ be a set of identifiers and $\mathbf{V}$ a set of values. The association of an identifier $x \in \mathbf{X}$ with a value $v \in \mathbf{V}$ is called a *binding* (of the identifier to its value), and a set Env of bindings is called an *execution environment* or *evaluation environment*. $\text{Env}(x)$ denotes the value associated with the identifier x in Env. The set of environments is denoted as $\mathbf{E}$.

In practice, the set of identifiers $\mathbf{X}$ that are actually used is finite: usually, we only consider those identifiers that appear in a program. An environment may thus be represented by a list of bindings, also called Env:

$$[(x_1, \text{Env}(x_1)), (x_2, \text{Env}(x_2)), \cdots, (x_n, \text{Env}(x_n))]$$

where $\{x_1, x_2, \cdots, x_n\}$ denotes a finite subset of $\mathbf{X}$, known as the *domain* of the environment and denoted as $\text{dom}(\text{Env})$. By convention, $\text{Env}(x)$ denotes the value v, which appears in the first (x, v) binding encountered when reading the list Env from the head (here, from left to right).

In this model, a binding can be added to an environment using the operator $\oplus$. By convention, bindings are added at (the left of) the head of the list representing the environment:

$$\begin{aligned}&(x_{new}, v_{new}) \oplus [(x_1, v_1), (x_2, v_2), \cdots, (x_n, v_n)]\\&= [(x_{new}, v_{new}), (x_1, v_1), (x_2, v_2), \cdots, (x_n, v_n)]\end{aligned}$$

Suppose that a certain operation introduces a new binding of an identifier, which is already present in the environment, for example (x_2, v_{new}):

$$\begin{aligned}&(x_2, v_{new}) \oplus [(x_1, v_1), (x_2, v_2), \cdots, (x_n, v_n)]\\&= [(x_2, v_{new}), (x_1, v_1), (x_2, v_2), \cdots, (x_n, v_n)]\end{aligned}$$

The so-obtained environment $(x_2, v_{new}) \oplus \text{Env}$ contains two bindings for x_2. Searching for a binding starts at the head of the environment, and, with our convention, new bindings are added at the head. So the most recent addition, (x_2, v_{new}), will be the first found. The binding (x_2, v_2) is not deleted, but it is said to be *masked* by the new binding (x_2, v_{new}). Several bindings for a single identifier x may therefore exist within the same environment, and the last binding added for x

will be used to determine the associated value of x in the environment. Formally, the environment $(x, v) \oplus \texttt{Env}$ verifies the following property:

$$((x, v) \oplus \texttt{Env})\,(x') = \begin{cases} v & \text{if } x = x' \\ \texttt{Env}(x') & \text{if } x \neq x' \end{cases}$$

By convention, the notation $(x_2, v_2) \oplus (x_1, v_1) \oplus \texttt{Env}$ is used to denote the environment $(x_2, v_2) \oplus ((x_1, v_1) \oplus \texttt{Env})$. For example, $((x, v_2) \oplus (x, v_1) \oplus \texttt{Env})(x) = v_2$

When an environment is represented by a list of bindings, the value $\texttt{Env}(x)$ is found as follows:

Python
```
def valeur_de(env,x):
    for (x1,v1) in env:
        if x==x1: return v1
    return None
```

OCaml
```
let rec valeur_de env x = match env with
   | [] -> None
   | (x1, v1) :: t -> if x = x1 then Some v1 else (valeur_de t x)
val valeur_de: ('a * 'b) list -> 'a -> 'b option
```

If no binding can be found in the environment for a given identifier, this function returns a special value indicating the absence of a binding. In Python, the constant `None` is used to express this absence of value, while in OCaml, the predefined sum type `'a option` is used:

OCaml
```
type 'a option = Some of 'a | None
```

The values of the type `'a option` are either those of the type `'a` or the constant `None`. The transformation of a value v of type `'a` into a value of type `'a option` is done by applying the constructor `Some` to v (see Chapter 5). The value `None` serves to denote the absence of value of type `'a`; more precisely, `None` is a constant that is not a value of type `'a`. This type `'a option` will be used further to denote some kind of meaningless or absent values but that are needed to fully complete some definitions.

The domain of an environment can be computed simply by traversing the list that represents it. A finite set is defined here as the list of all elements with no repetitions.

Python
```
def union_singleton(e,l):
    if e in l: return l
    else: return [e]+l

def dom_env(env):
    r=[]
    for (x,v) in env: r = union_singleton(x,r)
    return r
```

OCaml
```
let rec union_singleton e l = if (List.mem e l) then l else e::l
val union_singleton : 'a -> 'a list -> 'a list
let rec dom_env env = match env with
  | [] -> [] | (x, v) :: t -> (union_singleton x (dom_env t))
val dom_env : ('a * 'b) list -> 'a list
```

Since the value returned by the function `valeur_de` is obtained by traversing the list from its head, adding a new binding (x, v) to an environment is done at the head of the list and the previous bindings of x (if any) are masked, but not deleted.

Python
```
def ajout_liaison_env(env,x,v): return [(x,v)]+env
```

OCaml
```
let ajout_liaison_env env x v = (x, v) :: env
val ajout_liaison_env : ('a * 'b) list -> 'a -> 'b -> ('a * 'b) list
```

2.1.2. *Memory*

The formal model of the memory presented below makes no distinction between the different varieties of physical memory described in Chapter 1. The state of the memory is described by associating a value with the location of the cell in which it is stored. The locations themselves are considered as values, called *references*. As we have seen, high-level languages allow us to name a location c, containing a value v, by an identifier x bound to the reference r of c.

Let $\mathbf{R}$ be a set of references and $\mathbf{V}$ a set of values. The association of a reference $r \in \mathbf{R}$ with a value $v \in \mathbf{V}$ is represented by a pair (r, v), and a set Mem of such pairs is called here a *memory*. $\mathrm{Mem}(r)$ denotes the value stored at the reference r in Mem. Let $\mathbf{M}$ be the set of memories. In practice, the set of references, which is actually used, is finite: once again, only those locations used by a program are generally considered. This means that the memory can be represented by a list, also called Mem:

$$[(r_1, \mathrm{Mem}(r_1)), (r_2, \mathrm{Mem}(r_2)), \cdots, (r_n, \mathrm{Mem}(r_n))]$$

The existence of a pair (r, v) in a memory records that an initialization or a writing operation has been carried out at this location. Every referenced memory cell may be consulted through *reading* and can be assigned a new value by *writing*. In this case, the value previously held in the cell is deleted, it has "crashed".

Writing a value v at an address r transforms a memory $\mathtt{Mem}$ into a memory denoted as $\mathtt{Mem}[r := v]$; if a value was stored at this location r in $\mathtt{Mem}$, then this "old" value is replaced by v; otherwise, a new pair is added to $\mathtt{Mem}$ to take account of the writing operation. There is no masking, contrary to the case of the environments. Writing a new value v at a location r that already contains a value deletes the old value $\mathtt{Mem}(r)$:

$$[(r_1, v_1), \cdots, (r_i, v_i), \cdots, (r_n, v_n)][r_i := v_i'] = [(r_1, v_1), \cdots, (r_i, v_i'), \cdots, (r_n, v_n)]$$

The domain of a memory $\text{dom}(\mathtt{Mem})$ depends on the current environment and represents the space of references, which are accessible (directly or indirectly) from bound and non-masked identifiers in the current execution environment. The addition of a binding (x, r) to an environment $\mathtt{Env}$ has a twofold effect, creating $(x, r) \oplus \mathtt{Env}$ and extending $\mathtt{Mem}$ to $\mathtt{Mem}[r := v]$.

NOTE.– Depending on the language or program in question, the value v may be supplied just when x is introduced, or later, or never. If no value is provided prior to its first use, the result of the program is unpredictable, leading to errors called initialization errors. Indeed, a location always contains a value, which does not need to be suited to the current computation if it has not been explicitly determined by the program.

Note that the addition of a binding (x, r) in an environment $\mathtt{Env}$, of which the domain contains x, may mask a previous binding of x in $\mathtt{Env}$, but will not add a new pair (r, v) to $\mathtt{Mem}$ if r was already present in the domain of $\mathtt{Mem}$. Thus, any list of pairs representing a memory cannot contain two different pairs for the same reference. The memory $\mathtt{Mem}[r := v]$ verifies the following property:

$$\mathtt{Mem}[r := v](r') = \begin{cases} v & \text{if } r = r' \\ \mathtt{Mem}(r') & \text{if } r \neq r' \end{cases}$$

The memory $(\mathtt{Mem}[r_1 := v_1])[r_2 := v_2]$ is denoted as $\mathtt{Mem}[r_1 := v_1][r_2 := v_2]$.

For example, $(\mathtt{Mem}[r := v_1][r := v_2])(r) = v_2$.

The function `valeur_ref` computes the value stored at a given location. If nothing has previously been written at this location, the function returns a special value (`None`), indicating the absence of a known value (i.e. a value resulting from initialization or computation).

Python
```
def valeur_ref(mem,a):
    if len(mem) == 0: return None
    else:
        a1,v1 = mem[0]
        if a == a1: return v1
        else: return valeur_ref(mem[1:],a)
```

OCaml
```
let rec valeur_ref mem a = match mem with
  | [] -> None
  | (a1, v1) :: t -> if a = a1 then Some v1 else (valeur_ref t a)
val valeur_ref : ('a * 'b) list -> 'a -> 'b option
```

The following function writes a value into the memory:

Python
```
def write_mem(mem,a,v):
    if len(mem) == 0: return [(a,v)]
    else:
        a1,v1 = mem[0]
        if a == a1: return [(a1,v)] + mem[1:]
        else: return [(a1,v1)] + write_mem(mem[1:],a,v)
```

OCaml
```
let rec write_mem mem a v = match mem with
  | [] -> [(a, v)]
  | (a1, v1) :: t ->
      if a = a1 then (a1, v) :: t else (a1, v1) :: (write_mem t a v)
val write_mem : ('a * 'b) list -> 'a -> 'b -> ('a * 'b) list
```

2.1.3. *State*

A *state* is defined as a pair $(\text{Env}, \text{Mem}) \in \mathbf{E} \times \mathbf{M}$ such that any reference in the domain of Mem is accessible from a binding in Env. A reference is said to be *accessible* if its value can be read or written from an identifier contained in Env by a series of operations of reading, writing, or reference manipulation.

Given an environment Env, the set of identifiers $\mathbf{X}$ is partitioned into two subsets: $\mathbf{X}^{\text{ref}}_{\text{Env}}$, which contains the identifiers bound to a reference, and $\mathbf{X}^{\text{cst}}_{\text{Env}}$, which contains the others:

$$\mathbf{X}^{\text{ref}}_{\text{Env}} = \{x \in \mathbf{X} \mid \text{Env}(x) \in \mathbf{R}\} \qquad \mathbf{X}^{\text{cst}}_{\text{Env}} = \{x \in \mathbf{X} \mid \text{Env}(x) \in \mathbf{V} \setminus \mathbf{R}\}$$

The value associated with an identifier x in $\mathbf{X}^{\text{ref}}_{\text{Env}}$ is a reference $\text{Env}(x) = r$ where a value $\text{Mem}(r)$ is stored, which can be modified by writing. Identifiers of $\mathbf{X}^{\text{ref}}_{\text{Env}}$ are generally called *mutable variables*.

2.2. Evaluation of expressions

The value of an expression is computed according to an evaluation environment and a memory, i.e. in a given state. This computation is defined by the evaluation semantics of the expression.

2.2.1. *Syntax*

The language of expressions $\mathbf{Exp_1}$ used here will be extended in Chapters 3 and 4. Its syntax is defined in Table 2.1.

$e ::=$ k	Integer constant	$(k \in \mathbb{Z})$
$\mid$ x	Identifier	$(x \in \mathbf{X})$
$\mid$ $e_1 + e_2$	Addition	$(e_1, e_2 \in \mathbf{Exp_1})$
$\mid$ $!x$	Dereferencing	$(x \in \mathbf{X})$

Table 2.1. *Language of expressions* $\mathbf{Exp_1}$

Thus, an expression $e \in \mathbf{Exp_1}$ is either an integer constant $k \in \mathbb{Z}$, an identifier $x \in \mathbf{X}$, an expression obtained by applying an addition operator to two expressions in $\mathbf{Exp_1}$ or an expression of the form $!x$ denoting the value stored in the memory at the location bound to the mutable variable x. Thus, this is an inductive definition of the set $\mathbf{Exp_1}$. Note that $\mathbf{Exp_1}$ does not include an assignment construct. This is a deliberate choice. This point will be discussed in greater detail in section 2.3 by means of an extension of $\mathbf{Exp_1}$.

NOTE.– The symbol $+$ used in defining the syntax of expressions does not denote the integer addition operator. It could be replaced by any other symbol (for example $\boxtimes$). Its meaning will be assigned by the evaluation semantics. The same is true of the constant symbols: for example, the symbol 4 may be interpreted as a natural integer, a relative integer or a character.

EXAMPLE 2.1.– $!x + y$ is an expression of $\mathbf{Exp_1}$ in the same way as $(x + 2) + 3$. Parentheses are used here to structure the expression, they are part of the so-called concrete syntax and will disappear in the AST.

The set $\mathbf{Exp_1}$ of well-formed expressions of the language is defined by induction and expressed directly by a recursive sum type. Types of this kind can be constructed in OCaml, but not in Python; in the latter case, they can be partially simulated by defining a class for each sum-type constructor. Each class must contain a method with arguments corresponding exactly to the arguments of the sum type constructors it implements. An implementation of this type in Python is naïve, and users must ensure that these classes are used correctly. We know that there are possibilities of programming dynamic type verification mechanisms in Python, which simulate strong typing (similar to that used in OCaml) and ensure that the code is used correctly; however, these techniques lie outside of the scope of this book. The objective of all implementations shown in this book is simply to illustrate and intuitively justify the correct handling of concepts. As we have already done, we choose this approach to implement sum types.

Using Python, we define the following classes to represent the constructors of the set $\mathbf{Exp}_1$:

Python
```
class Cste1:
    def __init__(self,cste):
        self.cste = cste
class Var1:
    def __init__(self,symb):
        self.symb = symb
class Plus1:
    def __init__(self,exp1,exp2):
        self.exp1 = exp1
        self.exp2 = exp2
class Bang1:
    def __init__(self,symb):
        self.symb = symb
```

For example, the expression $e_1 = !x + y$ defined in example 2.1 is written as:

Python
```
ex_exp1 = Plus1(Bang1("x"),Var1("y"))
```

Using OCaml, the type of arithmetic expressions is defined directly as:

OCaml
```
type 'a exp1 =
 Cste1 of int | Var1 of 'a | Plus1 of 'a exp1 * 'a exp1 | Bang1 of 'a
```

Values of this type are thus obtained using either the `Cste1` constructor applied to an integer value, in which case they correspond to a constant expression, or using the `Var1` constructor applied to a value of type `'a`, corresponding to the type used to represent identifiers (the type `'a exp1` is thus polymorphic, as it depends on another type), or by applying the `Plus1` constructor to two values of the type `'a exp1`, or by applying the `Bang1` constructor to a value of type `'a`. For example, the expression $e_1 = !x + y$ is written as:

OCaml
```
let ex_exp1 = Plus1 (Bang1 ("x"), Var1 ("y"))
val ex_exp1 : string exp1
```

2.2.2. *Values*

Given a state (Env, Mem), we determine the evaluation semantics of an expression $e \in \mathbf{Exp}_1$ by computing the value of e in this state, i.e. by *evaluating* e in this state. Values may be relative integers or references, hence $\mathbf{V} = \mathbb{Z} \cup \mathbf{R}$. An additional, specific value Err is added to the set $\mathbf{V}$; this result is returned as the value of "meaningless" expressions. The result of the evaluation of an expression in $\mathbf{Exp}_1$ will therefore be a value belonging to the set $\mathbb{V} = \mathbf{V} \cup \{\text{Err}\}$.

Values in $\mathbf{V}$ are either relative integers or references. By defining a sum type, these two collections of values can be grouped into a single type.

Python
```
class CInt1:
    def __init__(self,cst_int):
        self.cst_int = cst_int
```
```
class CRef1:
    def __init__(self,cst_adr):
        self.cst_adr = cst_adr
```

Each class possesses a (object) constructor with the same name as the class: the constant k obtained from integer n (or, respectively, from reference r) is thus written as `CInt1(`n`)` (respectively, `CRef1(`r`)`), and this integer (respectively, reference) can be accessed from (the object) k by writing k`.cst_int` (respectively k`.cst_adr`). With OCaml, the type of elements in $\mathbf{V}$ is defined directly, as follows:

OCaml
```
type 'a const1 = CInt1 of int | CRef1 of 'a
```

A value of this type is obtained either using the constructor `CInt1` applied to an integer value or using the constructor `CRef1` applied to a value of type `'a` corresponding to the type used to represent references.

A type grouping the elements of $\mathbb{V} = \mathbf{V} \cup \{\text{Err}\}$ is defined by applying the same method:

Python
```
class VCste1:
    def __init__(self,cste):
        self.cste = cste
```
```
class Erreur1:
    pass
```

An element v in $\mathbb{V}$ is either a value in $\mathbf{V}$ obtained from a constant k and written as `VCste1(`k`)`, or an object in the class `Erreur1` (`pass` is used here to express the fact that the (object) constructor has no argument). With OCaml, the type of the elements in $\mathbb{V}$ is defined directly as follows:

OCaml
```
type 'a valeurs1 = VCste1 of 'a const1 | Erreur1
```

2.2.3. *Evaluation semantics*

There are several formalisms that may be used to describe the evaluation of an expression. These will be introduced later. Let us construct an evaluation function:

$$[\![\_]\!]_{\_}^{\_} : \mathbf{E} \times \mathbf{M} \times \mathbf{Exp_1} \to \mathbb{V}$$

The evaluation of the expression e in the environment $\mathtt{Env}$ and memory state $\mathtt{Mem}$ is denoted as $[\![e]\!]_{\mathrm{Env}}^{\mathrm{Mem}} = v$ with $v \in \mathbb{V}$. Table 2.2 contains the recursive definition of the function $[\![\_]\!]_{\_}^{\_}$.

$[\![k]\!]_{\mathrm{Env}}^{\mathrm{Mem}} = k$	$(k \in \mathbb{Z})$
$[\![x]\!]_{\mathrm{Env}}^{\mathrm{Mem}} = \mathtt{Env}(x)$	if $x \in \mathbf{X}$ and $x \in \mathrm{dom}(\mathtt{Env})$
$[\![x]\!]_{\mathrm{Env}}^{\mathrm{Mem}} = \mathrm{Err}$	if $x \in \mathbf{X}$ and $x \notin \mathrm{dom}(\mathtt{Env})$
$[\![e_1 + e_2]\!]_{\mathrm{Env}}^{\mathrm{Mem}} = [\![e_1]\!]_{\mathrm{Env}}^{\mathrm{Mem}} + [\![e_2]\!]_{\mathrm{Env}}^{\mathrm{Mem}}$	if $[\![e_1]\!]_{\mathrm{Env}}^{\mathrm{Mem}} \in \mathbb{Z}$ and $[\![e_2]\!]_{\mathrm{Env}}^{\mathrm{Mem}} \in \mathbb{Z}$
$[\![e_1 + e_2]\!]_{\mathrm{Env}}^{\mathrm{Mem}} = \mathrm{Err}$	if $[\![e_1]\!]_{\mathrm{Env}}^{\mathrm{Mem}} \notin \mathbb{Z}$ or $[\![e_2]\!]_{\mathrm{Env}}^{\mathrm{Mem}} \notin \mathbb{Z}$
$[\![!x]\!]_{\mathrm{Env}}^{\mathrm{Mem}} = \mathtt{Mem}(\mathtt{Env}(x))$	if $x \in \mathbf{X}_{\mathrm{Env}}^{\mathrm{ref}}$
$[\![!x]\!]_{\mathrm{Env}}^{\mathrm{Mem}} = \mathrm{Err}$	if $x \notin \mathbf{X}_{\mathrm{Env}}^{\mathrm{ref}}$

Table 2.2. *Evaluation of the expressions of* $\mathbf{Exp_1}$

The value of an integer constant is the integer that it represents. The value of an identifier is that which is bound to it in the environment, or Err. The value of an expression constructed with an addition symbol and two expressions e_1 and e_2 is obtained by adding the relative integers resulting from the evaluations of e_1 and e_2; the result will be Err if e_1 or e_2 is not an integer. The value of $!x$ is the value stored at the reference $\mathtt{Env}(x)$ when x is a mutable variable, and Err otherwise.

Thus, if e is evaluated as a reference, then e can only be an identifier. Furthermore, certain expressions in $\mathbf{Exp_1}$ are syntactically correct, but meaningless: for example, the expression $!x$ when x is not a mutable variable, i.e. when x does not bind a reference in the environment, or $x_1 + x_2$ when x_1 (or x_2) is a mutable variable. On the other hand, $!x + y$ is a meaningful expression that denotes a value when y binds an integer and x binds a reference to an integer.

EXAMPLE 2.2.– Let us evaluate the expression $!x + y$

in the state $\mathtt{Env} = [(x, r_x), (y, 2)]$ and $\mathtt{Mem} = [(r_x, 3)]$:

$$\begin{aligned}[\![!x + y]\!]_{\mathrm{Env}}^{\mathrm{Mem}} &= [\![!x]\!]_{\mathrm{Env}}^{\mathrm{Mem}} + [\![y]\!]_{\mathrm{Env}}^{\mathrm{Mem}} \\ &= \mathtt{Mem}(\mathtt{Env}(x)) + \mathtt{Env}(y) = \mathtt{Mem}(r_x) + 2 = 3 + 2 = 5\end{aligned}$$

The evaluation function $[\![\_]\!]_{\_}^{\_} : \mathbf{E} \times \mathbf{M} \times \mathbf{Exp_1} \rightarrow \mathbb{V}$ is obtained directly as follows:

Python
```
def eval_exp1(env,mem,e):
    if isinstance(e,Cste1): return VCste1(CInt1(e.cste))
    if isinstance(e,Var1):
        x = valeur_de(env,e.symb)
        if isinstance(x,CInt1) or isinstance(x,CRef1): return VCste1(x)
        return Erreur1()
    if isinstance(e,Plus1):
        ev1 = eval_exp1(env,mem,e.exp1)
        if isinstance(ev1,Erreur1): return Erreur1()
        v1 = ev1.cste
        ev2 = eval_exp1(env,mem,e.exp2)
        if isinstance(ev2,Erreur1): return Erreur1()
        v2 = ev2.cste
        if isinstance(v1,CInt1) and isinstance(v2,CInt1):
            return VCste1(CInt1(v1.cst_int + v2.cst_int))
        return Erreur1()
    if isinstance(e,Bang1):
        x = valeur_de(env,e.symb)
        if isinstance(x,CRef1):
            y = valeur_ref(mem,x.cst_adr)
            if y is None: return Erreur1()
            return VCste1(y)
        return Erreur1()
    raise ValueError
```

OCaml
```
let rec eval_exp1 env mem e = match e with
  | Cste1 n -> VCste1 (CInt1 n)
  | Var1 x ->
      (match valeur_de env x with Some v -> VCste1 v | _ -> Erreur1)
  | Plus1 (e1, e2) -> (
      match ((eval_exp1 env mem e1), (eval_exp1 env mem e2)) with
      | (VCste1 (CInt1 n1), VCste1 (CInt1 n2)) -> VCste1 (CInt1 (n1 + n2))
  | _ -> Erreur1)
  | Bang1 x -> (match valeur_de env x with
      | Some (CRef1 a) ->
          (match valeur_ref mem a with  Some v -> VCste1 v | _ -> Erreur1)
      | _ -> Erreur1)
val eval_exp1 : ('a * 'b const1) list -> ('b * 'b const1) list -> 'a exp1
                -> 'b valeurs1
```

Considering example 2.2, we obtain:

Python
```
ex_env1 = [("x",CRef1("rx")),("y",CInt1(2))]
ex_mem1 = [("rx",CInt1(3))]
>>> (eval_exp1(ex_env1,ex_mem1,ex_exp1)).cste.cst_int
5
```

```
OCaml
let ex_env1 = [ ("x", CRef1 ("rx")); ("y", CInt1 (2)) ]
val ex_env1 : (string * string const1) list
let ex_mem1 = [ ("rx", CInt1 (3)) ]
val ex_mem1 : (string * 'a const1) list
# (eval_exp1 ex_env1 ex_mem1 ex_exp1) ;;
- : string valeurs1 = VCste1 (CInt1 5)
```

2.3. Definition and assignment

2.3.1. *Defining an identifier*

The language $\mathbf{Def}_1$ extends $\mathbf{Exp}_1$ by adding definitions of identifiers. There are two constructs that make it possible to introduce an identifier naming a mutable or non-mutable variable (as defined in section 2.1.3). Note that, in both cases, the initial value must be provided. This value corresponds to a constant or to the result of a computation specified by an expression $e \in \mathbf{Exp}_1$. These constructs modify the current state of the system; after computing $[\![e]\!]^{\mathrm{Mem}}_{\mathrm{Env}}$, the next step in evaluating $\mathsf{let}\ x = e;$ is to add the binding $(x, [\![e]\!]^{\mathrm{Mem}}_{\mathrm{Env}})$ to the environment, while the evaluation of $\mathsf{var}\ x = e;$ adds a binding (x, r_x) to the environment and writes the value $[\![e]\!]^{\mathrm{Mem}}_{\mathrm{Env}}$ to the reference r_x. In this case, we assume that the location denoted by the reference r_x is computed by an external mechanism responsible for memory allocation.

$d ::= \mathsf{let}\ x = e;$	Definition of a non-mutable variable	$(x \in \mathbf{X},\ e \in \mathbf{Exp}_1)$
$\mid\ \mathsf{var}\ x = e;$	Definition of a mutable variable	$(x \in \mathbf{X},\ e \in \mathbf{Exp}_1)$

Table 2.3. *Language* $\mathbf{Def}_1$ *of definitions*

The evaluation of a definition is expressed as follows:

$$
\begin{array}{l}
(\mathrm{Env}, \mathrm{Mem}) \xrightarrow{\mathsf{let}\ x = e;}_{\mathbf{Def}_1} \left((x, [\![e]\!]^{\mathrm{Mem}}_{\mathrm{Env}}) \oplus \mathrm{Env}, \mathrm{Mem}\right) \\
(\mathrm{Env}, \mathrm{Mem}) \xrightarrow{\mathsf{var}\ x = e;}_{\mathbf{Def}_1} \left((x, r_x) \oplus \mathrm{Env}, \mathrm{Mem}[r_x := [\![e]\!]^{\mathrm{Mem}}_{\mathrm{Env}}]\right)
\end{array}
\qquad [2.1]
$$

This evaluation $\rightarrow_{\mathbf{Def}_1}$ defines a relation between a state, a definition and a resulting state, or, in formal terms:

$$\rightarrow_{\mathbf{Def}_1} \subseteq (\mathbf{E} \times \mathbf{M}) \times \mathbf{Def}_1 \times (\mathbf{E} \times \mathbf{M})$$

Starting with a finite sequence of definitions $\mathbf{d} = [d_1; \cdots; d_n]$ and an initial state $(\mathrm{Env}_0, \mathrm{Mem}_0)$, this relation produces the state $(\mathrm{Env}_n, \mathrm{Mem}_n)$:

$$(\mathrm{Env}_0, \mathrm{Mem}_0) \xrightarrow{d_1}_{\mathbf{Def}_1} (\mathrm{Env}_1, \mathrm{Mem}_1) \xrightarrow{d_2}_{\mathbf{Def}_1} \cdots \xrightarrow{d_n}_{\mathbf{Def}_1} (\mathrm{Env}_n, \mathrm{Mem}_n)$$

This sequence of transitions may, more simply, be noted $(\text{Env}_0, \text{Mem}_0) \overset{\mathbf{d}}{\rightarrow}^{\star}_{\mathbf{Def}_1} (\text{Env}_n, \text{Mem}_n)$.

EXAMPLE 2.3.– Starting with a memory with no accessible references and an "empty" environment, the sequence [var $y = 2$; let $x = !y + 3$;] builds the following state:

$$([\,], [\,]) \xrightarrow{\text{var } y = 2;}_{\mathbf{Def}_1} ([(y, r_y)], [(r_y, 2)])$$

$$([(y, r_y)], [(r_y, 2)]) \xrightarrow{\text{let } x = !y + 3;}_{\mathbf{Def}_1} ([(x, 5), (y, r_y)], [(r_y, 2)])$$

In the environment $\text{Env} = [(x, 5), (y, r_y)]$, we obtain $\mathbf{X}^{\text{ref}}_{\text{Env}} = \{y\}$ and $\mathbf{X}^{\text{cst}}_{\text{Env}} = \{x\}$.

NOTE.– In the definition of the two transitions in [2.1], we presume that the result of the evaluation of the expression e, denoted as $[\![e]\!]^{\text{Mem}}_{\text{Env}}$, is not an error result. In the case of an error, no state will be produced and the evaluation stops.

The abstract syntax of language $\mathbf{Def}_1$ may be defined as follows:

Python
```
class Let_def1:
    def __init__(self,var,exp):
        self.var = var
        self.exp = exp
```

```
class Var_def1:
    def __init__(self,var,exp):
        self.var = var
        self.exp = exp
```

OCaml
```
type 'a def1 = Let_def1 of 'a * 'a exp1 | Var_def1 of 'a * 'a exp1
```

We choose to construct a value corresponding to a reference using a constructor applied to an identifier.

Python
```
class Ref_Var1:
    def __init__(self,idvar):
        self.idvar = idvar
```

OCaml
```
type 'a refer = Ref_Var1 of 'a
```

Hence, r_x will be represented by `Ref_Var1(''x'')`. As the relation $\rightarrow_{\mathbf{Def}_1}$ defines a function, it can be implemented directly as follows:

Python
```
def trans_def1(st,d):
    (env,mem) = st
    if isinstance(d,Let_def1):
        v = eval_exp1(env,mem,d.exp)
        if isinstance(v,VCste1):
            return (ajout_liaison_env(env,d.var,v.cste),mem)
        raise ValueError
    if isinstance(d,Var_def1):
        v = eval_exp1(env,mem,d.exp)
        if isinstance(v,VCste1):
            r = Ref_Var1(d.var)
            return (ajout_liaison_env(env,d.var,CRef1(r)),
                    write_mem(mem,r,v.cste))
        raise ValueError
    raise ValueError
```

OCaml
```
let trans_def1 (env, mem) d = match d with
  | Let_def1 (x, e) -> (match eval_exp1 env mem e with
      | VCste1 v -> ((ajout_liaison_env env x v), mem)
      | Erreur1 -> failwith "Erreur")
  | Var_def1 (x, e) -> (match eval_exp1 env mem e with
      | VCste1 v -> ((ajout_liaison_env env x (CRef1 (Ref_Var1 x))),
                     (write_mem mem (Ref_Var1 x) v))
      | Erreur1 -> failwith "Erreur")
val trans_def1 :
 ('a * 'a refer const1) list * ('a refer * 'a refer const1) list
 -> 'a def1
 -> ('a * 'a refer const1) list * ('a refer * 'a refer const1) list
```

By iterating this function, we obtain an implementation of $\rightarrow^{\star}_{\mathbf{Def}_1}$.

Python
```
def trans_def1_exec(st,ld):
    (env,mem) = st
    if len(ld) == 0: return (env,mem)
    else: return trans_def1_exec(trans_def1((env,mem),ld[0]),ld[1:])
```

OCaml
```
let trans_def1_exec (env, mem) ld = (List.fold_left trans_def1 (env, mem) ld)
val trans_def1_exec :
    ('a * 'a refer const1) list * ('a refer * 'a refer const1) list
-> 'a def1 list
-> ('a * 'a refer const1) list * ('a refer * 'a refer const1) list
```

Now, considering example 2.3, we obtain:

```
Python
ex_ld0 = [Var_def1("y",Cste1(2)), Let_def1("x",Plus1(Bang1("y"),Cste1(3)))]
(ex_e0,ex_m0) = trans_def1_exec(([],[]),ex_ld0)
>>> eval_exp1(ex_e0,ex_m0,Var1("x")).cste.cst_int
5
>>> eval_exp1(ex_e0,ex_m0,Bang1("y")).cste.cst_int
2
```

```
OCaml
let ex_ld0 = [ Var_def1 ("y", Cste1 2);
               Let_def1 ("x", Plus1 (Bang1 "y", Cste1 3)) ]
val ex_ld0 : string def1 list
# (trans_def1_exec ([], []) ex_ld0) ;;
- : (string * string refer const1) list *
    (string refer * string refer const1) list
= ([("x", CInt1 5); ("y", CRef1 (Ref_Var1 "y"))], [(Ref_Var1 "y", CInt1 2)])
```

2.3.2. *Assignment*

The language $\mathbf{Lang}_1$ extends $\mathbf{Def}_1$ by adding *assignment*. The syntax of an assignment instruction is:

$$x := e$$

where $x \in \mathbf{X}$ and $e \in \mathbf{Exp}_1$. When the mutable variable x is already bound in the current environment, this instruction enables us to modify the value of $!x$. Formally, execution of the instruction $x := e$ modifies the memory of the current state, and it is described by the following transition:

$$(\text{Env}, \text{Mem}) \xrightarrow{x:=e}_{\mathbf{Lang}_1} (\text{Env}, \text{Mem}[\text{Env}(x) := [\![e]\!]^{\text{Mem}}_{\text{Env}}])$$

NOTE.– Once again, if the identifier x is not bound in the environment or if the evaluation of e results in an error, no state is generated and evaluation stops.

EXAMPLE 2.4.– Based on the state obtained in example 2.3, the following two assignments can be executed:

$$([(x,5),(y,r_y)],[(r_y,2)])$$
$$\xrightarrow{y:=!y+x}_{\mathbf{Lang}_1} ([(x,5),(y,r_y)],[(r_y,7)]) \xrightarrow{y:=8}_{\mathbf{Lang}_1} ([(x,5),(y,r_y)],[(r_y,8)])$$

Representing the abstract syntax of the assignment $x := e$ by the pair (x, e), the relation $\rightarrow_{\mathbf{Lang}_1}$ and the iteration of this relation from a sequence of assignments are implemented as follows:

Python
```
def trans_lang1(st,a):
    (env,mem) = st
    (x,e) = a
    v = valeur_de(env,x)
    if isinstance(v,CRef1):
        ve = eval_exp1(env,mem,e)
        if isinstance(ve,VCste1):
            return (env,write_mem(mem,v.cst_adr,ve.cste))
    raise ValueError

def trans_lang1_exec(st,la):
    (env,mem) = st
    if len(la) == 0:return (env,mem)
    else: return trans_lang1_exec(trans_lang1((env,mem),la[0]),la[1:])
```

OCaml
```
let trans_lang1 (env,mem) (x, e) = match valeur_de env x with
  | Some (CRef1 (Ref_Var1 y)) -> (match eval_exp1 env mem e with
      | VCste1 v -> (env, (write_mem mem (Ref_Var1 y) v))
      | Erreur1 -> failwith "Eval error")
  | _ -> failwith "Undefined var"
val trans_lang1 :
    ('a * 'b refer const1) list * ('b refer * 'b refer const1) list
-> 'a * 'a exp1
-> ('a * 'b refer const1) list * ('b refer * 'b refer const1) list

let trans_lang1_exec (env, mem) la =
    (List.fold_left trans_lang1 (env, mem) la)
val trans_lang1_exec :
    ('a * 'b refer const1) list * ('b refer * 'b refer const1) list
-> ('a * 'a exp1) list
-> ('a * 'b refer const1) list * ('b refer * 'b refer const1) list
```

Considering example 2.4, we obtain:

Python
```
ex_la0 = [("y",Plus1(Bang1("y"),Var1("x"))), ("y",Cste1(8))]
(ex_e1,ex_m1) = trans_lang1_exec((ex_e0,ex_m0),ex_la0)
>>> eval_exp1(ex_e1,ex_m1,Bang1("y")).cste.cst_int
8
```

OCaml
```
let ex_la0 = [ ("y", Plus1 (Bang1 "y",Var1 "x")); ("y", Cste1 8) ]
val ex_la0 : (string * string exp1) list
# (trans_lang1_exec (trans_def1_exec ([],[]) ex_ld0) ex_la0);;
- : (string * string refer const1) list *
    (string refer * string refer const1) list
= ([("x", CInt1 5); ("y", CRef1 (Ref_Var1 "y"))], [(Ref_Var1 "y", CInt1 8)])
```

NOTE.– In the presentation above, assignment concerns only mutable variables. Similarly, the dereferencing operator ! can only be applied to a reference to obtain the stored value at this location. Thus, the assignment of a value to a mutable variable from a given reference requires here the use of the dedicated syntactic structures: `x := ! x + 1`. However, in many languages, assignment does not explicitly mention the dereferencing operator, and the syntactic use of variables is identical on both sides of the assignment: `x = x + 1`. Hence, an identifier x denotes two different notions: the value $\text{Env}(x)$ on the left of the assignment symbol, and the value $\text{Mem}(\text{Env}(x))$ on the right of the assignment symbol.

These languages mask the different roles of a variable according to its position in the assignment. When a variable x is positioned to the left of the assignment, it is known as an *l-value* and it denotes a location where the value to be assigned should be stored. In this way, it acts as a pointer. The variable x on the right of the assignment is known as the *r-value*, and implicitly acts as a dereferenced pointer: the value to fetch is found at the location denoted by the variable. Thus, in the expression `x = x + 1`, even if `x` is used in the same way from a syntactic perspective, the instance on the right implicitly denotes ! `x`. In some languages, variable declaration implicitly involves the creation of a reference, unless otherwise stated; the name of the variable represents the location where the compiler will store the values assigned to it.

2.4. Exercises

Exercise 2.1

Consider the state etat_0 defined by:

$$\text{Env}_0 = [(x, 3), (y, r_y), (x, 8), (z, r_z)] \text{ and } \text{Mem}_0 = [(r_z, 2), (r_y, 5)]$$

1) Compute $\mathbf{X}^{\text{ref}}_{\text{Env}_0}$ and $\mathbf{X}^{\text{cst}}_{\text{Env}_0}$.

2) Give a sequence of definitions to obtain the state etat_0 from the empty state $([\,], [\,])$. Are there other sequences that can be used to obtain this state?

3) Compute $[\![(!y + !z) + x]\!]^{\text{Mem}_0}_{\text{Env}_0}$.

4) Give three examples of expressions that generate three different error types when evaluated in the state etat_0.

5) Consider the sequence

$$\text{etat}_0 \xrightarrow{y:=x}_{\mathbf{Lang}_1} \text{etat}_1 \xrightarrow{\text{let } y = x;}_{\mathbf{Def}_1} \text{etat}_2.$$

Determine the states $\text{etat}_1 = (\text{Env}_1, \text{Mem}_2)$ and $\text{etat}_2 = (\text{Env}_2, \text{Mem}_2)$ and the sets $\mathbf{X}^{\text{ref}}_{\text{Env}_1}$, $\mathbf{X}^{\text{cst}}_{\text{Env}_1}$, $\mathbf{X}^{\text{ref}}_{\text{Env}_1}$ and $\mathbf{X}^{\text{cst}}_{\text{Env}_1}$.

6) Consider the sequence

$$\text{etat}_0 \xrightarrow{\text{let } x = y;}_{\mathbf{Def}_1} \text{etat}_3 \xrightarrow{x:=4}_{\mathbf{Lang}_1} \text{etat}_4.$$

Determine the states $\text{etat}_3 = (\text{Env}_3, \text{Mem}_3)$ and $\text{etat}_4 = (\text{Env}_4, \text{Mem}_4)$. Does $[\![!y]\!]^{\text{Mem}_0}_{\text{Env}_0} = [\![!y]\!]^{\text{Mem}_4}_{\text{Env}_4}$?

Exercise 2.2

The two constructs $x++$ and $++x$:

$$e = k \mid x \mid e_1 + e_2 \mid !x \mid x++ \mid ++x$$

are added to the language $\mathbf{Exp}_1$. In these two new constructs, the identifier x must be a mutable variable. Intuitively, we see that the evaluation of the expression $x++$ produces the value stored in the location denoted by x and increments the value in the memory by 1. The expression $++x$ is evaluated differently: the value stored at the location denoted by x is incremented by 1, and this new value is the result of the evaluation.

1) Define an evaluation function:

$$[\![\_]\!]^{\_}_{\_} : \mathbf{E} \times \mathbf{M} \times \mathbf{Exp}_1 \to \mathbb{V} \times (\mathbf{E} \times \mathbf{M})$$

for expressions in the language $\mathbf{Exp}_1$, extended so that $[\![e]\!]^{\text{Mem}}_{\text{Env}} = (v, (\text{Env}', \text{Mem}'))$ expresses the fact that evaluation of the expression e in the state (Env, Mem) transforms this state into a state $(\text{Env}', \text{Mem}')$ and produces the value v.

2) Let $\text{Env} = [(x, r_x)]$ and $\text{Mem} = [(r_x, 6)]$. Compute:

$$[\![!x + (x++)]\!]^{\text{Mem}}_{\text{Env}} \quad \text{et} \quad [\![(x++) + !x]\!]^{\text{Mem}}_{\text{Env}}$$

Does $[\![!x + (x++)]\!]^{\text{Mem}}_{\text{Env}} = [\![(x++) + !x]\!]^{\text{Mem}}_{\text{Env}}$?

3) Show that, for any state (Env, Mem):

$$[\![++x]\!]^{\text{Mem}}_{\text{Env}} = [\![(x++) + 1]\!]^{\text{Mem}}_{\text{Env}}$$

Now, we extend the language $\mathbf{Lang}_1$ by considering the expressions of the extended $\mathbf{Def}_1$ language and adding the construction $x+ := e$. In informal terms, the execution of this instruction in a state (Env, Mem) consists of first finding the value v_x stored at the reference v in Mem, then evaluating the expression e in this state to obtain its value v_e and a new state $(\text{Env}', \text{Mem}')$, and finally, assigning to v the result of the addition of v_e and v_x. If at least one of the two values v_e and v_x is not an integer, the execution fails.

4) Redefine the relation $\rightarrow_{\mathbf{Lang}_1}$ for the extended $\mathbf{Lang}_1$ language.

5) Let $\text{Env}_0 = [(x, r_x)]$ and $\text{Mem}_0 = [(r_x, 2)]$. Determine the state etat_1 such that $(\text{Env}_0, \text{Mem}_0) \xrightarrow{x+:=++x}_{\mathbf{Lang}_1} \text{etat}_1$.

6) Do the assignments $x := x++$, $x := ++x$ and $x+ := 1$ produce the same states when executed in the same state?

Now, we extend the language [illegible] by considering the expressions of the [illegible]

3

Semantics of Functional Features

In this chapter, we examine the semantics of the constructs that form the kernel of functional languages. We shall present several methods for expressing the meaning of these constructs, along with different ways of describing their semantics. The functional language studied here corresponds to the functional kernel of MLstyle languages. Many other published presentations of the semantics of functional features of languages exist, for example [DOW 11, FRI 01].

3.1. Syntactic aspects

3.1.1. *Syntax of a functional kernel*

The language studied here, denoted as $\mathbf{Exp_2}$, includes only one syntactic category: expressions. Its abstract syntax is defined from a set $\mathbf{X}$ of (symbols of) variables, the set $\mathbf{K} = \mathbb{Z} \cup \mathbb{B}$ of constant values (where $\mathbb{Z}$ is the set of relative integers and $\mathbb{B}$ the set of booleans) and a set of primitive operators.

Table 3.1 presents the nine construction rules that enable the inductive definition of the set $\mathbf{Exp_2}$. An expression in this language can therefore be:

1) a constant denoting a value of $\mathbf{K}$, which will be either a relative integer or $true$ or $false$;

2) an identifier, here called a variable to maintain consistency with the usual terminology of expressions;

3) an application of a unary operator to an expression;

4) an application of a binary operator to two expressions with infix notation;

5) a conditional expression;

6) an expression let $x = e_1$ in e_2 which declares the variable x, called the *local variable* in the expression e_2, and defines it by the expression e_1 (in other words, the value of e_1 is named x in the expression e_2);

7) a function of a variable x, the body of which is an expression e, obtained by the *abstraction* of x in e. The variable x is generally called the parameter (or the *formal parameter*) of the function;

8) a recursive function, designated by an identifier f, of which the parameter is x and the body is an expression e, which may contain recursive calls to f;

9) an application of an expression e_1 to an expression e_2 (the expression e_2 is called the *argument* or the *actual parameter* of e_1).

$e ::= v$	Constant $(v \in \mathbf{K})$	(1)
$\mid x$	Variable $(x \in \mathbf{X})$	(2)
$\mid \text{op } e_1$	Application of a primitive unary operator $(\text{op} \in \{\text{not}\})$	(3)
$\mid e_1 \text{ op } e_2$	Application of a primitive binary operator $(\text{op} \in \{+, -, *, /, \text{and}, \text{or}, =, \leq\})$	(4)
$\mid$ if e_1 then e_2 else e_3	Conditional expression	(5)
$\mid$ let $x = e_1$ in e_2	Local binding $(x \in \mathbf{X})$	(6)
$\mid$ fun $x \mapsto e$	Function definition $(x \in \mathbf{X})$	(7)
$\mid$ rec $f\ x = e$	Recursive function definition $(x, f \in \mathbf{X})$	(8)
$\mid (e_1\ e_2)$	Application of a function	(9)

Table 3.1. *Syntax of the language of expressions* $\mathbf{Exp}_2$

3.1.2. *Abstract syntax tree*

Each expression in this language can be represented by a tree, the *abstract syntax tree* (AST), every node of which is annotated with the label of the applied construction rule. For example, the expression:

$$\text{fun } x_1 \mapsto x_1 + x_2 \qquad [3.1]$$

is obtained by abstraction of x_1 in the expression $x_1 + x_2$, which becomes the body of the function constructed in this way. Similarly, the expression e_{fact}:

$$\text{rec } fact\ n = \text{if } n \leq 1 \text{ then } 1 \text{ else } (fact\,(n - 1)) * n \qquad [3.2]$$

recursively defines the function computing $n!$ by abstraction of n in the function body.

Its AST is:

$$
(8)\dfrac{(5)\dfrac{(4)\dfrac{(2)\dfrac{}{n}\quad(1)\dfrac{}{1}}{n \le 1}\quad (1)\dfrac{}{1}\quad (4)\dfrac{(9)\dfrac{(2)\dfrac{}{fact}\quad (4)\dfrac{(2)\dfrac{}{n}\quad(1)\dfrac{}{1}}{n - 1}}{(fact\,(n - 1))}\quad (2)\dfrac{}{n}}{(fact\,(n - 1)) * n}}{\text{if } n \le 1 \text{ then } 1 \text{ else } (fact\,(n - 1)) * n}}{\text{rec } fact\ n = \text{if } n \le 1 \text{ then } 1 \text{ else } (fact\,(n - 1)) * n}
$$

All operators introduced in the programs, written in Python and OCaml, end with 2 in reference to the name $\mathbf{Exp}_2$. The set of constants $\mathbf{K}$ contains the relative integers and booleans, and can be represented as follows:

Python
```
class CInt2:
    def __init__(self,cst_int):
        self.cst_int = cst_int

class CBool2:
    def __init__(self,cst_bool):
        self.cst_bool = cst_bool
```

OCaml
```
type const2 = CInt2 of int | CBool2 of bool
```

Let us define the operators used in expressions of $\mathbf{Exp}_2$:

Python
```
class Not2:
    pass
class Plus2:
    pass
class Minus2:
    pass
class Mult2:
    pass
class Div2:
    pass

class And2:
    pass
class Or2:
    pass
class Eq2:
    pass
class Leq2:
    pass
```

OCaml
```
type unary_op2 = Not2
type bin_op2 = Plus2 | Minus2 | Mult2 | Div2 | And2 | Or2 | Eq2 | Leq2
```

Expressions of $\mathbf{Exp}_2$ can now be defined as follows:

Python
```
class Cste2:
    def __init__(self,cste):
        self.cste = cste
class Var2:
    def __init__(self,symb):
        self.symb = symb
class U_Op2:
    def __init__(self,op,exp1):
        self.op = op
        self.exp1 = exp1

class Letin2:
    def __init__(self,symb,exp1,exp2):
        self.symb = symb
        self.exp1 = exp1
        self.exp2 = exp2
class Fun2:
    def __init__(self,param,corps):
        self.param = param
        self.corps = corps
```

```
class B_Op2:
    def __init__(self,op,exp1,exp2):
        self.op = op
        self.exp1 = exp1
        self.exp2 = exp2
class If2:
    def __init__(self,exp1,exp2,exp3):
        self.exp1 = exp1
        self.exp2 = exp2
        self.exp3 = exp3
class Rec2:
    def __init__(self,nom,param,corps):
        self.nom = nom
        self.param = param
        self.corps = corps
class App2:
    def __init__(self,exp1,exp2):
        self.exp1 = exp1
        self.exp2 = exp2
```

OCaml
```
type 'a exp2 =
  | Cste2 of const2                        | Var2 of 'a
  | U_Op2 of unary_op2 * 'a exp2           | B_Op2 of bin_op2 * 'a exp2 * 'a exp2
  | If2 of 'a exp2 * 'a exp2 * 'a exp2     | Letin2 of 'a * 'a exp2 * 'a exp2
  | Fun2 of 'a * 'a exp2                   | Rec2 of 'a * 'a * 'a exp2
  | App2 of 'a exp2 * 'a exp2
```

For example, the expressions fun $x_1 \mapsto x_1 + x_2$, let $x_1 = x_1 + x_2$ in $x_1 + x_3$ and the expression e_{fact} defined in [3.2] are written as:

Python
```
ex_exp1 = Fun2("x1",B_Op2(Plus2(),Var2("x1"),Var2("x2")))
ex_exp2 = Letin2("x1",B_Op2(Plus2(),Var2("x1"),Var2("x2")),
                 B_Op2(Plus2(),Var2("x1"),Var2("x3")))
a_fact = \
       Rec2("fact","n",
            If2(B_Op2(Leq2(),Var2("n"),Cste2(CInt2(1))),
                Cste2(CInt2(1)),
                B_Op2(Mult2(),App2(Var2("fact"),
                                   B_Op2(Minus2(),Var2("n"),Cste2(CInt2(1)))),
                      Var2("n"))))
```

OCaml
```
let ex_exp1 = Fun2 ("x1", B_Op2 (Plus2, Var2 "x1", Var2 "x2"))
val ex_exp1 : string exp2
let ex_exp2 =
  Letin2 ("x1", B_Op2 (Plus2, Var2 "x1", Var2 "x2"),
      B_Op2 (Plus2, Var2 "x1", Var2 "x3"))
val ex_exp2 : string exp2
let a_fact =
  Rec2
    ("fact", "n",
     If2 (B_Op2 (Leq2, Var2 "n", Cste2 (CInt2 1)),
        Cste2 (CInt2 1),
        B_Op2 (Mult2,
                 App2 (Var2 "fact", B_Op2 (Minus2, Var2 "n", Cste2 (CInt2 1))),
            Var2 "n")))
val a_fact : string exp2
```

3.1.3. *Reasoning by induction over expressions*

The set $\mathbf{Exp}_2$ is defined inductively: each expression $e \in \mathbf{Exp}_2$ is obtained by applying the construction rules of $\mathbf{Exp}_2$ a finite number of times. This process means that a property P can be proved over the elements of $\mathbf{Exp}_2$ by applying the *structural induction* scheme shown in Box 3.1: we simply prove the property P for each construct assuming that P is satisfied by each component of the construct. This reasoning by induction will be used several times in the remainder of the book.

if $\forall v \in \mathbf{K}\ P(v)$
and $\forall x \in \mathbf{X}\ P(x)$
and $\forall e_1 \in \mathbf{Exp}_2\ P(e_1) \Rightarrow P(\text{op } e_1)$
and $\forall e_1, e_2 \in \mathbf{Exp}_2\ (P(e_1) \text{ and } P(e_2)) \Rightarrow P(e_1 \text{ op } e_2)$
and $\forall e_1, e_2, e_3 \in \mathbf{Exp}_2\ (P(e_1) \text{ and } P(e_2) \text{ and } P(e_3)) \Rightarrow P(\textsf{if } e_1 \textsf{ then } e_2 \textsf{ else } e_3)$
and $\forall x \in \mathbf{X}\ \forall e_1, e_2 \in \mathbf{Exp}_2\ (P(e_1) \text{ and } P(e_2)) \Rightarrow P(\textsf{let } x = e_1 \textsf{ in } e_2)$
and $\forall x \in \mathbf{X}\ \forall e \in \mathbf{Exp}_2\ P(e) \Rightarrow P(\textsf{fun } x \mapsto e)$
and $\forall x, f \in \mathbf{X}\ \forall e \in \mathbf{Exp}_2\ P(e) \Rightarrow P(\textsf{rec } f\ x = e)$
and $\forall e_1, e_2 \in \mathbf{Exp}_2\ (P(e_1) \text{ and } P(e_2)) \Rightarrow P((e_1\ e_2))$
then $\forall e \in \mathbf{Exp}_2\ P(e)$

Box 3.1. *Structural induction scheme over* $e \in \mathbf{Exp}_2$

3.1.4. *Declaration of variables, bound and free variables*

An identifier can be used several times within the same expression, and each of its uses is called an *occurrence* of the identifier. The occurrences of the same identifier do not all play the same role. In construct (6) $\textsf{let } x = e_1 \textsf{ in } e_2$, the identifier x is declared and bound to the value of e_1 in the environment, which will be used to evaluate e_2. Construct (7) abstracts the variable x in the expression e. This parameter x is said to be *bound* (or *abstracted*) in the expression $\textsf{fun } x \mapsto e$. It can be renamed without modifying the definition of the function; thus, it is sometimes called a *dummy variable*. For example, replacing x with z in $\textsf{fun } x \mapsto x + y$ gives $\textsf{fun } z \mapsto z + y$, which is equivalent. Evidently, we cannot replace x with y since y already occurs in the function body $x + y$. In construct (8), the identifier f is declared and the variable x is bound in the expression e. For example, the expression:

$$\textsf{let } x_{(1)} = 25 \textsf{ in let } f = \textsf{fun } x_{(2)} \mapsto x_{(3)} + 3 \textsf{ in } (f\ 7) + x_{(4)} \qquad [3.3]$$

contains four occurrences of x, indexed from 1 to 4. Occurrence 2 abstracts occurrence 3 in the body of f. These occurrences of the dummy variable x may be renamed. Occurrence 1 declares the identifier x, which is used to give a value to occurrence 4.

When an occurrence of a variable in an expression is neither declared nor bound, it is said to be *free* in this expression. For $e \in \mathbf{Exp}_2$, the set $\mathcal{F}(e)$ of variables, which has at least one free occurrence in e, is defined inductively by:

$$
\begin{array}{lr}
\mathcal{F} : \mathbf{Exp}_2 \to \wp(\mathbf{X}) & \\
\mathcal{F}(v) = \emptyset & (v \in \mathbf{K}) \\
\mathcal{F}(x) = \{x\} & (x \in \mathbf{X}) \\
\mathcal{F}(\mathsf{op}\ e_1) = \mathcal{F}(e_1) & \\
\mathcal{F}(e_1\ \mathsf{op}\ e_2) = \mathcal{F}(e_1) \cup \mathcal{F}(e_2) & \\
\mathcal{F}(\mathsf{if}\ e_1\ \mathsf{then}\ e_2\ \mathsf{else}\ e_3) = \mathcal{F}(e_1) \cup \mathcal{F}(e_2) \cup \mathcal{F}(e_3) & \\
\mathcal{F}(\mathsf{let}\ x = e_1\ \mathsf{in}\ e_2) = \mathcal{F}(e_1) \cup (\mathcal{F}(e_2) \setminus \{x\}) & \\
\mathcal{F}(\mathsf{fun}\ x \mapsto e) = \mathcal{F}(e) \setminus \{x\} & \\
\mathcal{F}(\mathsf{rec}\ f\ x = e) = \mathcal{F}(e) \setminus \{x, f\} & \\
\mathcal{F}((e_1\ e_2)) = \mathcal{F}(e_1) \cup \mathcal{F}(e_2) &
\end{array}
$$

Expressions that do not contain a free variable are said to be *closed*. This is the case for a function $\mathsf{fun}\ x \mapsto e$, which is said to be closed if it is a constant function ($\mathcal{F}(e) = \emptyset$ and x does not occur in e) or if $\mathcal{F}(e) = \{x\}$ (i.e. the body e has no free variable except x). For example, the function $\mathsf{fun}\ x \mapsto x + 1$ is closed. This is also the case for $\mathsf{fun}\ x_1 \mapsto \mathsf{fun}\ x_2 \mapsto x_1 + x_2$, but not for $\mathsf{fun}\ x_2 \mapsto x_1 + x_2$ since x_1 is free in this instance: the notion is not stable when switching to a subexpression (the fact that an expression is closed does not necessarily imply that its subexpressions are also closed).

EXAMPLE 3.1.– For the expressions e_{fact} defined in [3.2] and e_{25} defined in [3.3], we have $\mathcal{F}(e_{fact}) = \mathcal{F}(e_{25}) = \emptyset$. In expression e_{25}, the occurrence 3 of x is free in the subexpression $x + 3$ but becomes bound in the expression $\mathsf{fun}\ x \mapsto x + 3$. The following two examples are not closed expressions:

$$\mathcal{F}(\mathsf{fun}\ x_1 \mapsto x_1 + x_2) = \{x_2\}$$

$$\mathcal{F}(\mathsf{let}\ x_1 = x_1 + x_2\ \mathsf{in}\ x_1 + x_3) = \{x_1, x_2, x_3\}$$

Intuitively, to evaluate an expression, we need to know the values of its variables. The value of a free variable in an expression is searched for in the evaluation environment. For example, to evaluate the expression $x + 1$, we need to know the value of x. The expression $\mathsf{let}\ x = 3\ \mathsf{in}\ 2 * x$, on the other hand, is closed and can be evaluated without "external" information concerning the (bound) variables it contains.

The set of free variables of an expression in $\mathbf{Exp}_2$ can be defined recursively as follows:

Python
```
def free_var2(e):
    if isinstance(e,Cste2): return emptyset
    if isinstance(e,Var2): return [e.symb]
    if isinstance(e,U_Op2): return free_var2(e.exp1)
    if isinstance(e,B_Op2): return union(free_var2(e.exp1),free_var2(e.exp2))
    if isinstance(e,If2):
        return union(union(free_var2(e.exp1),free_var2(e.exp2)),
                     free_var2(e.exp3))
    if isinstance(e,Letin2):
        return union(free_var2(e.exp1),
                     diff_set(free_var2(e.exp2),[e.symb]))
    if isinstance(e,Fun2): return diff_set(free_var2(e.corps),[e.param])
    if isinstance(e,Rec2): return diff_set(free_var2(e.corps),[e.param,e.nom])
    if isinstance(e,App2): return union(free_var2(e.exp1),free_var2(e.exp2))
    raise ValueError
```

OCaml
```
let rec free_var2 e = match e with
  | Cste2 _ -> emptyset
  | Var2 x -> [x]
  | U_Op2 (op, e1) -> free_var2 e1
  | B_Op2 (op, e1, e2) -> union (free_var2 e1) (free_var2 e2)
  | If2 (e1, e2, e3) ->
      union (union (free_var2 e1) (free_var2 e2)) (free_var2 e3)
  | Letin2 (x, e1, e2) -> union (free_var2 e1) (diff_set (free_var2 e2) [x])
  | Fun2 (x, e1) -> diff_set (free_var2 e1) [x]
  | Rec2 (f, x, e1) -> diff_set (free_var2 e1) [x;f]
  | App2 (e1, e2) -> union (free_var2 e1) (free_var2 e2)
val free_var2 : 'a exp2 -> 'a list
```

Considering the expressions in example 3.1, we obtain:

Python
```
>>> free_var2(ex_exp1)
['x2']
>>> free_var2(ex_exp2)
['x2', 'x1', 'x3']
>>> free_var2(a_fact)
[]
```

OCaml
```
# free_var2 ex_exp1 ;;
- : string list = ["x2"]
# free_var2 ex_exp2 ;;
- : string list = ["x1"; "x2"; "x3"]
# free_var2 a_fact ;;
- : string list = []
```

3.2. Execution semantics: evaluation functions

The execution semantics (or *dynamic semantics*) of a programming language describes the execution of all syntactically correct programs in this language. In this section, we study the "pure" functional language $\mathbf{Exp_2}$. As operations acting on memory are not provided by this language, only the notion of an execution environment is required. Note that, unlike memory (with modifiable contents), an execution environment (also called evaluation environment) can only be extended by a new binding of an identifier to its value; any existing bindings for this identifier are simply masked by the new binding.

A program in $\mathbf{Exp_2}$ is an expression. Its execution consists of evaluating the expression, i.e. of computing its value. To do this, we begin by defining an evaluation function, as in Chapter 2:

$$[\![\_]\!]_{\_} : \mathbf{E} \times \mathbf{Exp_2} \to \mathbb{V}$$

The value $[\![e]\!]_{\text{Env}}$ is the result of the evaluation of expression e in environment Env. Here, an environment $\text{Env} \in \mathbf{E}$ is a list of bindings $(x, v) \in \mathbf{X} \times \mathbf{V}$. The set $\mathbf{V}$ (defined in section 3.2.2) is the set of values associated with the variables, and $\mathbb{V}$ is the union of $\mathbf{V}$ and of a set of special values used to indicate evaluation failures.

3.2.1. *Evaluation errors*

We have chosen to distinguish between three different kinds of errors produced during the evaluation of an expression $e \in \mathbf{Exp_2}$.

1) The value of a free occurrence of a variable in e must be found in the current execution environment. If this variable does not belong to the environment domain, evaluation of e returns an error denoted as Err_U (with U signifying unbound).

2) Arithmetic errors, such as overflow or mathematically undefined values, are denoted as Err_E (E for exception, the mechanism used by the processor to indicate an "exceptional" situation which makes it impossible to continue normal execution). For our purposes, we shall limit our considerations to division by zero.

3) The expressions of $\mathbf{Exp_2}$ denote either integers, booleans or functions. Thus, attempts may be made to add a boolean and an integer ($3 + true$), or to apply an integer to an expression $(8\,(x + 1))$. Such expressions are syntactically correct, but cannot denote a value in $\mathbf{V}$. In our model, the evaluation of such expressions leads to an error called typing error Err_T. Detection of typing errors prior to execution will be presented in Chapter 5. Note that certain languages assign semantics to expressions, which we consider to be invalid in $\mathbf{Exp_2}$. In Python (v3.6), for example, the expression `3 + True` returns the value 4.

The set of errors considered here is thus $\mathbb{E} = \{\mathrm{Err_U}, \mathrm{Err_E}, \mathrm{Err_T}\}$. This choice is somewhat arbitrary: it would be possible to use only one sort of error, or, on the contrary, to make a finer distinction between different causes of errors. This can be done by modifying the error handling rules.

In $\mathbf{Exp}_2$, the evaluation of some expressions e may not terminate, and in this case, the value $[\![e]\!]_{\mathrm{Env}}$ is undefined: no value – not even an error value – will be returned. The evaluation function $[\![\_]\!]_{\_}$ is thus a partial function. The formal processing of non-termination lies outside the scope of this book.

3.2.2. *Values*

The set $\mathbf{V}$ contains integers, booleans and the values of functions called *closures*, presented in section 3.2.4. For a non-recursive function (fun $x \mapsto e$), the closure is a triple $\langle x, e, \mathrm{Env}\rangle$ made up of the name of the function parameter (x), the function body (e) and the environment in which the definition of the function is evaluated (Env). For a recursive function (rec $f\ x = e$), the closure is a quadruple $\langle f, x, e, \mathrm{Env}\rangle$ made up of the name of the function (f), the name of the function parameter (x), the function body (e) and the environment in which the definition of the function is evaluated (Env). Hence, the set of closures $\mathbb{F}$ is defined by:

$$\mathbb{F} = (\mathbf{X} \times \mathbf{Exp}_2 \times \mathbf{E}) \cup (\mathbf{X} \times \mathbf{X} \times \mathbf{Exp}_2 \times \mathbf{E})$$

Every value bound to an identifier in an environment is an element in the set $\mathbf{V} = \mathbb{Z} \cup \mathbb{B} \cup \mathbb{F}$. The set $\mathbb{V} = \mathbf{V} \cup \mathbb{E}$ comprises all possible results of the evaluation of an expression.

Sets $\mathbb{E}$, $\mathbf{V}$ and $\mathbb{V}$ can be defined as follows:

Python
```
class E_Typ2:
    pass
class E_Exec2:
    pass
class E_Undef2:
    pass
class VInt2:
    def __init__(self,val_int):
        self.val_int = val_int
class VBool2:
    def __init__(self,val_bool):
        self.val_bool = val_bool
class Clos2:
    def __init__(self,param,corps,envdef):
```

```
        self.param = param
        self.corps = corps
        self.envdef = envdef
class ClosRec2:
    def __init__(self,nomf,param,corps,envdef):
        self.nomf = nomf
        self.param = param
        self.corps = corps
        self.envdef = envdef
class Val2:
    def __init__(self,val):
        self.val = val
class Err2:
    def __init__(self,err):
        self.err = err
```

OCaml
```
type erreur2 = E_Typ2 | E_Exec2 | E_Undef2
type 'a valeur2 =
  | VInt2 of int | VBool2 of bool
  | Clos2 of 'a * 'a exp2 * ('a * 'a valeur2) list
  | ClosRec2 of 'a * 'a * 'a exp2 * ('a * 'a valeur2) list
type 'a val_err2 = Val2 of 'a valeur<2 | Err2 of erreur2
```

b	$true$	$false$	$b \in \mathbf{V} \setminus \mathbb{B}$
$\text{not}_{\mathbf{V}}\ b$	$false$	$true$	Err_{T}

$\text{and}_{\mathbf{V}}$	$true$	$false$	$b_2 \in \mathbf{V} \setminus \mathbb{B}$
$true$	$true$	$false$	Err_{T}
$false$	$false$	$false$	Err_{T}
$b_1 \in \mathbf{V} \setminus \mathbb{B}$	Err_{T}	Err_{T}	Err_{T}

$\text{or}_{\mathbf{V}}$	$true$	$false$	$b_2 \in \mathbf{V} \setminus \mathbb{B}$
	$true$	$true$	Err_{T}
	$true$	$false$	Err_{T}
	Err_{T}	Err_{T}	Err_{T}

$\begin{pmatrix} +_{\mathbf{V}} \\ -_{\mathbf{V}} \\ *_{\mathbf{V}} \end{pmatrix}$	$v_2 \in \mathbb{Z}$	$v_2 \in \mathbf{V} \setminus \mathbb{Z}$
$v_1 \in \mathbb{Z}$	$v_1 \begin{pmatrix} + \\ - \\ * \end{pmatrix} v_2$	Err_{T}
$v_1 \in \mathbf{V} \setminus \mathbb{Z}$	Err_{T}	Err_{T}

$/_{\mathbf{V}}$	$v_2 \in \mathbb{Z}$	$v_2 \in \mathbf{V} \setminus \mathbb{Z}$
	v_1/v_2 if $v_2 \neq 0$ Err_{E} if $v_2 = 0$	Err_{T}
	Err_{T}	Err_{T}

$=_{\mathbf{V}}$	$v_2 \in \mathbb{Z}$	$v_2 \in \mathbb{B}$	$v_2 \in \mathbb{F}$
$v_1 \in \mathbb{Z}$	$true$ if $v_1 = v_2$ $false$ if $v_1 \neq v_2$	Err_{T}	Err_{T}
$v_1 \in \mathbb{B}$	Err_{T}	$true$ if $v_1 = v_2$ $false$ if $v_1 \neq v_2$	Err_{T}
$v_1 \in \mathbb{F}$	Err_{T}	Err_{T}	Err_{T}

$\leq_{\mathbf{V}}$	$v_2 \in \mathbb{Z}$	$v_2 \notin \mathbb{Z}$
$v_1 \in \mathbb{Z}$	$true$ if $v_1 \leq v_2$ $false$ if $v_1 > v_2$	Err_{T}
$v_1 \notin \mathbb{Z}$	Err_{T}	Err_{T}

Table 3.2. *Interpretation of operators*

3.2.3. *Interpretation of operators*

The primitive operators in $\mathbf{Exp}_2$ are symbols that are used to construct new expressions from existing expressions. To evaluate an expression, we must decide which operation $\mathrm{op}_{\mathbf{V}}$ on the values of $\mathbf{V}$ is represented by each symbol op; in other terms, we must interpret these symbols. The interpretation of the operators used in $\mathbf{Exp}_2$ is defined in Table 3.2.

The interpretation of primitive operators is defined as follows:

Python
```
def apply2_unary_op(op,v):
    if isinstance(op,Not2):
        if isinstance(v,VBool2): return Val2(VBool2(not v.val_bool))
        return Err2(E_Typ2)
    raise ValueError

def apply2_binary_op(op,v1,v2):
    if isinstance(op,Plus2):
        if isinstance(v1,VInt2) and isinstance(v2,VInt2):
            return Val2(VInt2(v1.val_int + v2.val_int))
        return Err2(E_Typ2)
    if isinstance(op,Minus2):
        if isinstance(v1,VInt2) and isinstance(v2,VInt2):
            return Val2(VInt2(v1.val_int - v2.val_int))
        return Err2(E_Typ2)
    if isinstance(op,Mult2):
        if isinstance(v1,VInt2) and isinstance(v2,VInt2):
            return Val2(VInt2(v1.val_int * v2.val_int))
        return Err2(E_Typ2)
    if isinstance(op,Div2):
        if isinstance(v1,VInt2) and isinstance(v2,VInt2):
            if v2.val_int == 0: return Err2(E_Exec2)
            return Val2(VInt2(v1.val_int / v2.val_int))
        return Err2(E_Typ2)
    if isinstance(op,And2):
        if isinstance(v1,VBool2) and isinstance(v2,VBool2):
            return Val2(VBool2(v1.val_bool and v2.val_bool))
        return Err2(E_Typ2)
    if isinstance(op,Or2):
        if isinstance(v1,VBool2) and isinstance(v2,VBool2):
            return Val2(VBool2(v1.val_bool or v2.val_bool))
        return Err2(E_Typ2)
    if isinstance(op,Eq2):
        if isinstance(v1,VBool2) and isinstance(v2,VBool2):
            return Val2(VBool2(v1.val_bool == v2.val_bool))
        if isinstance(v1,VInt2) and isinstance(v2,VInt2):
            return Val2(VBool2(v1.val_int == v2.val_int))
        return Err2(E_Typ2)
```

```
    if isinstance(op,Leq2):
        if isinstance(v1,VInt2) and isinstance(v2,VInt2):
            return Val2(VBool2(v1.val_int <= v2.val_int))
        return Err2(E_Typ2)
    raise ValueError
```

OCaml

```
let apply2_unary_op op v = match op with
  | Not2 -> (match v with
      | VBool2 b -> Val2 (VBool2 (not b))
      | _ -> Err2 E_Typ2)
val apply2_unary_op : unary_op2 -> 'a valeur2 -> 'b val_err2

let apply2_binary_op op v1 v2 = match op with
  | Plus2 -> (match (v1, v2) with
      | (VInt2 n1, VInt2 n2) -> Val2 (VInt2 (n1 + n2))
      | _ -> Err2 E_Typ2)
  | Minus2 -> (match (v1, v2) with
      | (VInt2 n1, VInt2 n2) -> Val2 (VInt2 (n1 - n2))
      | _ -> Err2 E_Typ2)
  | Mult2 -> (match (v1, v2) with
      | (VInt2 n1, VInt2 n2) -> Val2 (VInt2 (n1 * n2))
      | _ -> Err2 E_Typ2)
  | Div2 -> (match (v1, v2) with
      | (VInt2 n1, VInt2 n2) ->
          if n2 = 0 then Err2 E_Exec2 else Val2 (VInt2 (n1 / n2))
      | _ -> Err2 E_Typ2)
  | And2 -> (match (v1, v2) with
      | (VBool2 b1, VBool2 b2) -> Val2 (VBool2 (b1 && b2))
      | _ -> Err2 E_Typ2)
  | Or2 -> (match (v1, v2) with
      | (VBool2 b1, VBool2 b2) -> Val2 (VBool2(b1 || b2))
      | _ -> Err2 E_Typ2)
  | Eq2 -> (match (v1, v2) with
      | (VInt2 n1, VInt2 n2) -> Val2 (VBool2 (n1 = n2))
      | (VBool2 b1, VBool2 b2) -> Val2 (VBool2 (b1 = b2))
      | _ -> Err2 E_Typ2)
  | Leq2 -> (match (v1, v2) with
      | (VInt2 n1, VInt2 n2) -> Val2 (VBool2 (n1 <= n2))
      | _ -> Err2 E_Typ2)
val apply2_binary_op : bin_op2 -> 'a valeur2 -> 'b valeur2 -> 'c val_err2
```

3.2.4. *Closures*

The language $\mathbf{Exp}_2$ allows us to define functions and apply them to actual parameters, also called *arguments*. Let f be a function with formal parameter x and body e_f. To evaluate $(f\ e)$ in a current environment Env_a, we must evaluate e in this environment to obtain a value v. The body e_f must then be evaluated in an environment Env in which x is bound to v: this binding allows us to replace the formal parameter with the actual parameter. So what is this environment Env? Let f be the expression $\mathsf{fun}\ x_1 \mapsto x_1 + x_2$: evaluation of $(f\,2)$ creates the binding $(x_1, 2)$, and it remains to find the value of x_2 – but in which environment? Two choices are possible: this value will be supplied either by the environment Env_a in which application $(f\,2)$ is evaluated, or by the environment in which f was defined. In the first case, we speak of *dynamic scope*, whereas in the second case, we speak of *static* or *lexical scope*. For $\mathbf{Exp}_2$, we adopt lexical scope: closures therefore record the environment in which functions are defined. In this way, the values of free variables of the function body (except the formal parameter) are those at the function definition stage, i.e. those bound to these variables in the *definition environment* of the function.

Compilers generally work with lexical scope, whereas interpreters often choose dynamic scope. Compilers supply optimized implementations of the closure notion, avoiding code duplication. In a language using lexical scope, the function body is only evaluated when the function in question is applied to an actual parameter. This does not, however, imply that the source code of this body is retained; the function body is translated into object code, which contains the instructions needed to record the definition environment (at least the bindings of the free variables of the function in this environment), along with instructions that bind the formal parameter to the value of the actual parameter, provided by a call. Execution of application of a function to an actual parameter leads to execution of the object code of its body, in the right environment.

3.2.5. *Evaluation of expressions*

Table 3.3 defines the evaluation function for the expressions in $\mathbf{Exp}_2$ by induction, following the rules of abstract syntax. Let us examine the rows in this table. The value of a constant is the constant itself. Binding a variable in an environment provides the value of that variable. To evaluate an expression of the form op e or $e_1 \,\text{op}\, e_2$, we must evaluate the subexpressions e or e_1 and e_2. The evaluation of a conditional expression $\textsf{if}\ e_1\ \textsf{then}\ e_2\ \textsf{else}\ e_3$ depends on the result returned by the evaluation of e_1. To evaluate $\textsf{let}\ x = e_1\ \textsf{in}\ e_2$, we evaluate e_1 getting v, then extend the current environment with the binding (x, v) to enable evaluation of e_2. The value of a function is a closure. If this function is recursive then the closure will, in addition, contain the name of the function f. In order to evaluate an application $(e_1\ e_2)$, we must first evaluate e_1 and e_2 in the current environment Env. The function body is then evaluated in the environment Env_f contained in the closure, extended by the binding (x, v). In the case of a recursive function f, adding the name f to the closure means that we can evaluate its body e_f in an environment where f is bound to the closure. Another way of constructing the closure of a recursive function f (used in section 3.3.3) consists of extending the definition environment of f to include the binding of f with its own closure, making it possible to evaluate recursive calls:

$$[\![\textsf{rec}\ f\ x = e]\!]_{\text{Env}} = F \quad \text{with} \quad F = \langle x, e, (f, F) \oplus \text{Env} \rangle$$

These two approaches are equivalent.

e	$[\![e]\!]_{\text{Env}}$	
v	v	$(v \in \mathbf{K})$
x	$\text{Env}(x)$	if $x \in \text{dom}(\text{Env})$ $(x \in \mathbf{X})$
	Err_{U}	if $x \notin \text{dom}(\text{Env})$
op e_1	$\text{op}_{\mathbf{V}}\ [\![e_1]\!]_{\text{Env}}$	if $[\![e_1]\!]_{\text{Env}} \in \mathbf{V}$ $(\text{op} \in \{\text{not}\})$
	$[\![e_1]\!]_{\text{Env}}$	if $[\![e_1]\!]_{\text{Env}} \in \mathbb{E}$
e_1 op e_2	$[\![e_1]\!]_{\text{Env}}\ \text{op}_{\mathbf{V}}\ [\![e_2]\!]_{\text{Env}}$	$(\text{op} \in \{+, -, *, /, \text{and}, \text{or}, =, \leq\})$ if $[\![e_1]\!]_{\text{Env}} \in \mathbf{V}$ and $[\![e_2]\!]_{\text{Env}} \in \mathbf{V}$
	$[\![e_1]\!]_{\text{Env}}$	if $[\![e_1]\!]_{\text{Env}} \in \mathbb{E}$
	$[\![e_2]\!]_{\text{Env}}$	if $[\![e_1]\!]_{\text{Env}} \in \mathbf{V}$ and $[\![e_2]\!]_{\text{Env}} \in \mathbb{E}$
if e_1 then e_2 else e_3	$[\![e_2]\!]_{\text{Env}}$	if $[\![e_1]\!]_{\text{Env}} = \mathit{true}$
	$[\![e_3]\!]_{\text{Env}}$	if $[\![e_1]\!]_{\text{Env}} = \mathit{false}$
	Err_{T}	if $[\![e_1]\!]_{\text{Env}} \in \mathbf{V} \setminus \mathbb{B}$
	$[\![e_1]\!]_{\text{Env}}$	if $[\![e_1]\!]_{\text{Env}} \in \mathbb{E}$
let $x = e_1$ in e_2	$[\![e_2]\!]_{(x,v)\oplus\text{Env}}$	if $[\![e_1]\!]_{\text{Env}} = v \in \mathbf{V}$
	$[\![e_1]\!]_{\text{Env}}$	if $[\![e_1]\!]_{\text{Env}} \in \mathbb{E}$
fun $x \mapsto e$	$\langle x, e, \text{Env}\rangle$	
rec $f\ x = e$	$\langle f, x, e, \text{Env}\rangle$	
$(e_1\ e_2)$	$[\![e_f]\!]_{(x,v)\oplus\text{Env}_f}$	if $[\![e_1]\!]_{\text{Env}} = \langle x, e_f, \text{Env}_f\rangle$ and $[\![e_2]\!]_{\text{Env}} = v \in \mathbf{V}$
	$[\![e_f]\!]_{(f,\langle f,x,e_f,\text{Env}_f\rangle)\oplus(x,v)\oplus\text{Env}_f}$	if $[\![e_1]\!]_{\text{Env}} = \langle f, x, e_f, \text{Env}_f\rangle$ and $[\![e_2]\!]_{\text{Env}} = v \in \mathbf{V}$
	$[\![e_2]\!]_{\text{Env}}$	if $[\![e_1]\!]_{\text{Env}} \in \mathbb{F}$ and $[\![e_2]\!]_{\text{Env}} \in \mathbb{E}$
	Err_{T}	if $[\![e_1]\!]_{\text{Env}} \notin \mathbb{F}$

Table 3.3. *Evaluation of expressions in* $\mathbf{Exp}_2$

REMARK.– Unlike the language $\mathbf{Def}_1$ presented in Chapter 2, $\mathbf{Exp}_2$ does not supply the construct let $x = e$; required to add a binding to the current environment. However, the construct let $x = e_1$ in e_2 makes it possible to mimic it. To evaluate an expression e in an environment $[(x_1, v_1), (x_2, v_2), \cdots, (x_n, v_n)]$, we simply consider the following expression:

let $x_1 = v_1$ in
let $x_2 = v_2$ in
...
let $x_n = v_n$ in e

EXAMPLE 3.2.– The expression defined in [3.3] is evaluated as follows:

$$\begin{aligned}
&[\![\text{let } x = 25 \text{ in let } f = \text{fun } x \mapsto x + 3 \text{ in } (f\,7) + x]\!]_{\text{Env}_0} \\
&= [\![\text{let } f = \text{fun } x \mapsto x + 3 \text{ in } (f\,7) + x]\!]_{\text{Env}_1} \\
&= [\![(f\,7) + x]\!]_{\text{Env}_2} = [\![(f\,7)]\!]_{\text{Env}_2} + [\![x]\!]_{\text{Env}_2} = [\![x + 3]\!]_{(x,7)\oplus\text{Env}_1} + 25 \\
&= 7 + 3 + 25 = 35
\end{aligned}$$

where:

$$\begin{aligned}
&\text{Env}_1 = (x, 25) \oplus \text{Env}_0 \text{ since } [\![25]\!]_{\text{Env}_0} = 25 \\
&\text{Env}_2 = (f, \langle x, x + 3, \text{Env}_1\rangle) \oplus \text{Env}_1 \text{ since } [\![\text{fun } x \mapsto x + 3]\!]_{\text{Env}_1} = \langle x, x + 3, \text{Env}_1\rangle
\end{aligned}$$

Note that each occurrence of the variable x is processed in a way adapted to its meaning.

EXAMPLE 3.3.– The expression $(e_{fact}\,3)$ where e_{fact} is the expression defined in [3.2] is evaluated as follows:

$$\begin{aligned}
&[\![(\underbrace{(\text{rec } fact\; n = \text{if } n \leq 1 \text{ then } 1 \text{ else } (fact\,(n - 1)) * n)}_{e_{fact}}\; 3)]\!]_{\text{Env}_0} \\
&= [\![\text{if } n \leq 1 \text{ then } 1 \text{ else } (fact\,(n - 1)) * n]\!]_{\text{Env}_1} \\
&(\text{since } [\![e_{fact}]\!]_{\text{Env}_0} = F_{fact} \text{ and } [\![3]\!]_{\text{Env}_0} = 3) \\
&= [\![(fact\,(n - 1)) * n]\!]_{\text{Env}_1} \\
&(\text{since } [\![n \leq 1]\!]_{\text{Env}_1} = false \text{ since } [\![n]\!]_{\text{Env}_1} = 3) \\
&= [\![(fact\,(n - 1))]\!]_{\text{Env}_1} * [\![n]\!]_{\text{Env}_1} \\
&= [\![\text{if } n \leq 1 \text{ then } 1 \text{ else } (fact\,(n - 1)) * n]\!]_{\text{Env}_2} * [\![n]\!]_{\text{Env}_1} \\
&(\text{since } [\![fact]\!]_{\text{Env}_1} = F_{fact} \text{ and } [\![n - 1]\!]_{\text{Env}_1} = [\![n]\!]_{\text{Env}_1} - 1 = 3 - 1 = 2) \\
&= [\![(fact\,(n - 1)) * n]\!]_{\text{Env}_2} * [\![n]\!]_{\text{Env}_1} \\
&(\text{since } [\![n \leq 1]\!]_{\text{Env}_2} = false \text{ since } [\![n]\!]_{\text{Env}_2} = 2) \\
&= [\![(fact\,(n - 1))]\!]_{\text{Env}_2} * [\![n]\!]_{\text{Env}_2} * [\![n]\!]_{\text{Env}_1} \\
&= [\![\text{if } n \leq 1 \text{ then } 1 \text{ else } (fact\,(n - 1)) * n]\!]_{\text{Env}_3} * [\![n]\!]_{\text{Env}_2} * [\![n]\!]_{\text{Env}_1} \\
&(\text{since } [\![fact]\!]_{\text{Env}_2} = F_{fact} \text{ and } [\![n - 1]\!]_{\text{Env}_2} = [\![n]\!]_{\text{Env}_2} - 1 = 2 - 1 = 1) \\
&= [\![1]\!]_{\text{Env}_3} * [\![n]\!]_{\text{Env}_2} * [\![n]\!]_{\text{Env}_1} \\
&(\text{since } [\![n \leq 1]\!]_{\text{Env}_3} = true \text{ since } [\![n]\!]_{\text{Env}_3} = 1) \\
&= 1 * 2 * 3 = 6
\end{aligned}$$

where:

$$\begin{aligned}
&F_{fact} = \langle fact, n, \text{if } n \leq 1 \text{ then } 1 \text{ else } (fact\,(n - 1)) * n, \text{Env}_0\rangle \\
&\text{Env}_1 = (fact, F_{fact}) \oplus (n, 3) \oplus \text{Env}_0 \\
&\text{Env}_2 = (fact, F_{fact}) \oplus (n, 2) \oplus \text{Env}_0 \\
&\text{Env}_3 = (fact, F_{fact}) \oplus (n, 1) \oplus \text{Env}_0
\end{aligned}$$

Note that each recursive call is evaluated in the environment Env_0 contained in the closure F_{fact}, extended with the binding of n to its current value.

The evaluation function for the expressions of $\mathbf{Exp}_2$ shown in Table 3.3 can be directly defined as follows:

```
Python
def eval2(env,e):
    if isinstance(e,Cste2):
        if isinstance(e.cste,CInt2): return Val2(VInt2(e.cste.cst_int))
        if isinstance(e.cste,VBool2): return Val2(VBool2(e.cste.cst_bool))
    if isinstance(e,Var2):
        x = valeur_de(env,e.symb)
        if x is None: return Err2(E_Undef2)
        return Val2(x)
    if isinstance(e,U_Op2):
        a = eval2(env,e.exp1)
        if isinstance(a,Val2): return apply2_unary_op(e.op,a.val)
        return a
    if isinstance(e,B_Op2):
        a1 = eval2(env,e.exp1)
        if isinstance(a1,Val2):
            a2 = eval2(env,e.exp2)
            if isinstance(a2,Val2):
                return apply2_binary_op(e.op,a1.val,a2.val)
            return a2
        return a1
    if isinstance(e,If2):
        a1 = eval2(env,e.exp1)
        if isinstance(a1,Val2):
            if isinstance(a1.val,VBool2):
                if a1.val.val_bool == True: return eval2(env,e.exp2)
                return eval2(env,e.exp3)
            return Err2(E_Typ2)
        return a1
    if isinstance(e,Letin2):
        a1 = eval2(env,e.exp1)
        if isinstance(a1,Val2):
            ne = ajout_liaison_env(env,e.symb,a1.val)
            return eval2(ne,e.exp2)
        return a1
    if isinstance(e,Fun2): return Val2(Clos2(e.param,e.corps,env))
    if isinstance(e,Rec2):
        return Val2(ClosRec2(e.nom,e.param,e.corps,env))
    if isinstance(e,App2):
        a1 = eval2(env,e.exp1)
        if isinstance(a1,Val2):
            if isinstance(a1.val,Clos2):
                a2 = eval2(env,e.exp2)
                if isinstance(a2,Val2):
                    ne = ajout_liaison_env(a1.val.envdef,a1.val.param,a2.val)
                    return eval2(ne,a1.val.corps)
                return a2
    if isinstance(e,Fun2): return Val2(Clos2(e.param,e.corps,env))
    if isinstance(e,Rec2):
        return Val2(ClosRec2(e.nom,e.param,e.corps,env))
```

```
    if isinstance(e,App2):
        a1 = eval2(env,e.exp1)
        if isinstance(a1,Val2):
            if isinstance(a1.val,Clos2):
                a2 = eval2(env,e.exp2)
                if isinstance(a2,Val2):
                    ne = ajout_liaison_env(a1.val.envdef,a1.val.param,a2.val)
                    return eval2(ne,a1.val.corps)
                return a2
            if isinstance(a1.val,ClosRec2):
                a2 = eval2(env,e.exp2)
                if isinstance(a2,Val2):
                    ne = ajout_liaison_env(ajout_liaison_env(a1.val.envdef,
                                                             a1.val.nomf,
                                                             a1.val),
                                           a1.val.param,a2.val)
                    return eval2(ne,a1.val.corps)
                return a2
            return Err2(E_Typ2)
        return a1
    raise ValueError
```

OCaml
```
let rec eval2 env e = match e with
  | Cste2 (CInt2 n) -> Val2 (VInt2 n)
  | Cste2 (CBool2 n) -> Val2 (VBool2 n)
  | Var2 x -> (match valeur_de env x with
      | Some v -> Val2 v | None -> Err2 E_Undef2)
  | U_Op2 (op, e1)-> (match eval2 env e1 with
      | Val2 v -> apply2_unary_op op v | z -> z)
  | B_Op2 (op, e1, e2) -> (match eval2 env e1 with
      | Val2 v1 -> (match eval2 env e2 with
          | Val2 v2 -> apply2_binary_op op v1 v2 | z -> z)
      | z -> z)
  | If2 (e1, e2, e3) -> (match eval2 env e1 with
      | Val2 (VBool2 true) -> eval2 env e2
      | Val2 (VBool2 false) -> eval2 env e3
      | Val2 x -> Err2 E_Typ2
      | z -> z)
  | Letin2 (x, e1, e2) -> (match eval2 env e1 with
      | Val2 v -> eval2 (ajout_liaison_env env x v) e2 | z -> z)
  | Fun2 (x, e1) -> Val2 (Clos2 (x, e1, env))
  | Rec2 (f, x, e1) -> Val2 (ClosRec2 (f, x, e1, env))
  | App2 (e1, e2) -> (match eval2 env e1 with
      | Val2 (Clos2 (pf, cf, ef)) -> (match eval2 env e2 with
          | Val2 v -> eval2 (ajout_liaison_env ef pf v) cf | z -> z)
      | Val2 (ClosRec2 (nf, pf, cf, ef)) -> (match eval2 env e2 with
          | Val2 v ->
              eval2 (ajout_liaison_env
                       (ajout_liaison_env
                          ef nf (ClosRec2(nf,pf,cf,ef))) pf v) cf
          | z -> z)
      | _ -> Err2 E_Typ2)
val eval2 : ('a * 'a valeur2) list -> 'a exp2 -> 'a val_err2
```

The evaluation of the expressions in examples 3.2 and 3.3 is obtained in the following way:

```
Python
e25 = Letin2("x",Cste2(CInt2(25)),
              Letin2("f",Fun2("x",B_Op2(Plus2(),Var2("x"),Cste2(CInt2(3)))),
                     B_Op2(Plus2(),App2(Var2("f"),Cste2(CInt2(7))),Var2("x"))))
>>> (eval2([],e25)).val.val_int
35
>>> (eval2([],App2(a_fact,Cste2(CInt2(3))))).val.val_int
6
```

```
OCaml
let e25 =
  Letin2 ("x", Cste2 (CInt2 25),
          Letin2("f", Fun2 ("x", B_Op2 (Plus2, Var2 "x", Cste2 (CInt2 3))),
                 B_Op2 (Plus2, App2 (Var2 "f", Cste2 (CInt2 7)), Var2 "x")))
val e25 : string exp2
# (eval2 [] e25);;
- : string val_err2 = Val2 (VInt2 35)
# (eval2 [] (App2(a_fact,Cste2(CInt2(3)))));;
- : string val_err2 = Val2 (VInt2 6)
```

3.2.5.1. *Order of evaluation of subexpressions*

The order in which the subexpressions of an expression are evaluated has an effect on the result of the final evaluation. The evaluation function defined in Table 3.3 arbitrarily establishes the order of evaluation of the subexpressions of $e_1 \,\mathrm{op}\, e_2$: we have chosen to evaluate e_1 and then e_2. If evaluation of e_1 triggers an error, then e_2 will not be evaluated. The implementation of a mathematically commutative binary operator thus loses this property. For example, if a variable x does not belong to the domain of environment $\textsc{Env}$, then $[\![(3\,/\,0) + x]\!]_{\mathrm{Env}} \neq [\![x + (3\,/\,0)]\!]_{\mathrm{Env}}$. Indeed, we have:

$$[\![(3\,/\,0) + x]\!]_{\mathrm{Env}} = [\![3\,/\,0]\!]_{\mathrm{Env}} = \mathrm{Err_E} \quad [\![x + (3\,/\,0)]\!]_{\mathrm{Env}} = [\![x]\!]_{\mathrm{Env}} = \mathrm{Err_U}$$

The same goes for some so-called "lazy" boolean operators. For example, when evaluating $e_1 \,\mathrm{and}\, e_2$, if e_1 is evaluated as $false$, the result $false$ may be returned without e_2 being evaluated (as the conjunction value will be $false$ in any case); hence, execution errors in e_2 will not be detected.

On the other hand, there is no specified evaluation order for e_1 and e_2 in the evaluation of application $(e_1\, e_2)$. Nevertheless, when the semantics is implemented in a compiler or an interpreter, a choice must be made. For example, in the interpreter `eval2` presented earlier, e_1 is evaluated before e_2. In some "lazy" languages, argument e_2 of a function is evaluated only when necessary (if e_1 is a constant function, there is no need to evaluate the argument; this means that execution errors in the argument will not be detected).

For other expressions, evaluation order is dictated by the construct semantics. For example, evaluation of $\textsf{if}\ e_1\ \textsf{then}\ e_2\ \textsf{else}\ e_3$ begins with the evaluation of e_1 and continues with evaluation of either e_2 or e_3 depending on the returned boolean value. Evidently, if the evaluation of expression e_1 results in an error, then no subexpression will be evaluated.

3.2.5.2. *Lexical and dynamic scope*

The same program may produce different results depending on the chosen scope mechanism. Consider the following expression:

$$\begin{array}{l}\text{let } x_2 = 3 \text{ in} \\ \text{let } f = \textsf{fun}\ x_1 \mapsto x_1 + x_2 \text{ in} \\ \text{let } x_2 = 8 \text{ in } (f\,2)\end{array} \qquad [3.4]$$

Using lexical scope, this expression is evaluated as follows:

$$\begin{array}{l}[\![\textsf{let}\ x_2 = 3\ \textsf{in let}\ f = \textsf{fun}\ x_1 \mapsto x_1 + x_2\ \textsf{in let}\ x_2 = 8\ \textsf{in}\ (f\,2)]\!]_{\text{Env}} \\ = [\![\textsf{let}\ f = \textsf{fun}\ x_1 \mapsto x_1 + x_2\ \textsf{in let}\ x_2 = 8\ \textsf{in}\ (f\,2)]\!]_{(x_2,3)\oplus\text{Env}} \\ = [\![\textsf{let}\ x_2 = 8\ \textsf{in}\ (f\,2)]\!]_{(f,\langle x_1, x_1+x_2, (x_2,3)\oplus\text{Env}\rangle)\oplus(x_2,3)\oplus\text{Env}} \\ = [\![(f\,2)]\!]_{(x_2,8)\oplus(f,\langle x_1, x_1+x_2, (x_2,3)\oplus\text{Env}\rangle)\oplus(x_2,3)\oplus\text{Env}} \\ = [\![x_1 + x_2]\!]_{(x_1,2)\oplus(x_2,3)\oplus\text{Env}} = 2 + 3 = 5\end{array}$$

Using dynamic scope, the evaluation rules shown in Table 3.3 are used, except those concerning definition and application of functions: in this case, the definition environment of the function is not recorded in the closure. If an application is evaluated in an environment Env_a, the function body is evaluated in the extension of Env_a with the binding of the formal parameter to the value of the argument. In the previous example, this evaluation is done as follows:

$$\begin{array}{l}[\![\textsf{let}\ x_2 = 3\ \textsf{in let}\ f = \textsf{fun}\ x_1 \mapsto x_1 + x_2\ \textsf{in let}\ x_2 = 8\ \textsf{in}\ (f\,2)]\!]_{\text{Env}} \\ = [\![\textsf{let}\ f = \textsf{fun}\ x_1 \mapsto x_1 + x_2\ \textsf{in let}\ x_2 = 8\ \textsf{in}\ (f\,2)]\!]_{(x_2,3)\oplus\text{Env}} \\ = [\![\textsf{let}\ x_2 = 8\ \textsf{in}\ (f\,2)]\!]_{(f,\langle x_1, x_1+x_2\rangle)\oplus(x_2,3)\oplus\text{Env}} \\ = [\![(f\,2)]\!]_{(x_2,8)\oplus(f,\langle x_1, x_1+x_2\rangle)\oplus(x_2,3)\oplus\text{Env}} \\ = [\![x_1 + x_2]\!]_{(x_1,2)\oplus(x_2,8)\oplus(f,\langle x_1, x_1+x_2\rangle)\oplus(x_2,3)\oplus\text{Env}} = 2 + 8 = 10\end{array}$$

The dynamic scope mechanism presents two drawbacks. First, if the body of a function contains free variables, its meaning depends on the current value (in Env_a) of the free variables that may change from one call to another. This is particularly problematic when attempting to review source code: the meaning of the function being modified along the execution. Furthermore, a free variable may be bound to a value of a different type in the course of execution, and this may lead to a typing error when evaluating the function application. For example, if x_2 in expression [3.4] is associated with a boolean value instead of the integer value 8, then evaluating $(f\,2)$ will involve adding 2 to a boolean, resulting in an error. However, during the definition of function f, the value associated with x_2 was an integer, 3, and the application of the operator $+$ in the body of f does not pose any problem in this case. Function f was thus "correctly typed" during definition.

Using the lexical scope mechanism, the values associated with the free occurrences of variables in a function body are determined by the environment in which the function is defined. Thus, in the body of f in the previous example, the value of x_2 will always be 3 and expression [3.4] will always be evaluated as 5. The lexical scope mechanism will be used throughout the remainder of the chapter.

3.3. Execution semantics: operational semantics

Operational semantics describes the specification of a language's compiler through a set of *inference rules* as:

$$(R)\frac{h_1 \cdots h_n}{c}(I)$$

This rule, named R, means that, if hypotheses $h_1, \cdots, h_n$ are true, we can deduce c, or, in an equivalent form, that c can be deduced by establishing $h_1, \cdots, h_n$. I corresponds to additional information (condition for application of the rule, or intermediate computation). For example, consider a rule specifying the evaluation of the addition of two expressions:

$$(R)\frac{\text{Env} \vdash e_1 \rightsquigarrow v_1 \qquad \text{Env} \vdash e_2 \rightsquigarrow v_2}{\text{Env} \vdash e_1 + e_2 \rightsquigarrow v}(v = v_1 + v_2)$$

In this case, $\text{Env} \vdash e \rightsquigarrow v$ expresses the fact that, in environment Env, the evaluation of expression e returns a value v. The order of evaluation of e_1 and e_2 is not specified, and the two rules below leave the choice of the error propagation to the (developer of the) compiler:

$$(R_1)\frac{\text{Env} \vdash e_1 \rightsquigarrow v_1}{\text{Env} \vdash e_1 + e_2 \rightsquigarrow v_1}(v_1 \in \mathbb{E}) \quad (R_2)\frac{\text{Env} \vdash e_2 \rightsquigarrow v_2}{\text{Env} \vdash e_1 + e_2 \rightsquigarrow v_2}(v_2 \in \mathbb{E})$$

This is an evaluation relation, not a function. Function $[\![\_]\!]_{\_}$, defined in section 3.2, implements rule R_1.

Function `eval2` presented in section 3.2 defines a very precise evaluation strategy, establishing an order for evaluating subexpressions. This function therefore provides a fully deterministic definition of the evaluation.

Function $[\![\_]\!]_{\_}$, on the other hand, leaves certain choices open (for example the order in which the subexpressions of an application are to be evaluated) while fixing others (such as the order of evaluation of the subexpressions of an operator).

NOTE.– When the result of the evaluation of an expression depends on the choice of implementation used by the compiler to evaluate subexpressions, the expression itself is considered to be *semantically incorrect*. The value associated with the expression depends on the compiler. The reference handbooks of certain programming languages detail these situations.

Operational semantics allows us to safely ignore certain constraints, such as the order of evaluation of arguments in an operator. For example, a compiler designer may prefer not to follow an imposed order for argument computation, often for reasons relating to optimization. The obligations of a compiler may be specified by using a semantics that establishes the meaning of constructs in a language, while maintaining a degree of freedom regarding certain implementation choices. The evaluation order for arguments of a commutative binary operator may, for example, be left unspecified in the semantics. This means that several different values may be associated with the same expression, for example if the semantics contains the two rules R_1 and R_2 defined above.

3.3.1. *Simple expressions*

The evaluation relation for simple expressions in language $\mathbf{Exp_2}$, denoted as $\rightsquigarrow$, is defined using the inference system presented in Box 3.2, We have chosen to define an operational semantics, which respects the choices used in the function $[\![\_]\!]_{\_}$. This results in two equivalent semantics.

$$(F_1)\frac{}{\text{Env} \vdash v \rightsquigarrow v}(v \in \mathbf{K}) \qquad (F_2)\frac{\text{Env} \vdash e_1 \rightsquigarrow v_1}{\text{Env} \vdash \text{op } e_1 \rightsquigarrow \text{op}_{\mathbf{V}}\ v_1}(v_1 \in \mathbf{V})$$

$$(F_3)\frac{\text{Env} \vdash e_1 \rightsquigarrow v_1}{\text{Env} \vdash \text{op } e_1 \rightsquigarrow v_1}(v_1 \in \mathbb{E}) \quad (F_4)\frac{\text{Env} \vdash e_1 \rightsquigarrow v_1 \quad \text{Env} \vdash e_2 \rightsquigarrow v_2}{\text{Env} \vdash e_1 \text{ op } e_2 \rightsquigarrow v_1 \text{ op}_{\mathbf{V}}\ v_2}\begin{pmatrix} v_1 \in \mathbf{V} \\ v_2 \in \mathbf{V} \end{pmatrix}$$

$$(F_5)\frac{\text{Env} \vdash e_1 \rightsquigarrow v_1}{\text{Env} \vdash e_1 \text{ op } e_2 \rightsquigarrow v_1}(v_1 \in \mathbb{E}) \quad (F_6)\frac{\text{Env} \vdash e_1 \rightsquigarrow v_1 \quad \text{Env} \vdash e_2 \rightsquigarrow v_2}{\text{Env} \vdash e_1 \text{ op } e_2 \rightsquigarrow v_2}\begin{pmatrix} v_1 \in \mathbf{V} \\ v_2 \in \mathbb{E} \end{pmatrix}$$

$$(F_7)\frac{\text{Env} \vdash e_1 \rightsquigarrow true \quad \text{Env} \vdash e_2 \rightsquigarrow v}{\text{Env} \vdash \textsf{if } e_1 \textsf{ then } e_2 \textsf{ else } e_3 \rightsquigarrow v} \qquad (F_8)\frac{\text{Env} \vdash e_1 \rightsquigarrow false \quad \text{Env} \vdash e_3 \rightsquigarrow v}{\text{Env} \vdash \textsf{if } e_1 \textsf{ then } e_2 \textsf{ else } e_3 \rightsquigarrow v}$$

$$(F_9)\frac{\text{Env} \vdash e_1 \rightsquigarrow v}{\text{Env} \vdash \textsf{if } e_1 \textsf{ then } e_2 \textsf{ else } e_3 \rightsquigarrow \text{Err}_{\text{T}}}(v \in \mathbf{V} \setminus \mathbb{B})$$

$$(F_{10})\frac{\text{Env} \vdash e_1 \rightsquigarrow v}{\text{Env} \vdash \textsf{if } e_1 \textsf{ then } e_2 \textsf{ else } e_3 \rightsquigarrow v}(v \in \mathbb{E})$$

Box 3.2. *Operational semantics of simple expressions in* $\mathbf{Exp_2}$

Rule F_1 expresses the fact that a constant in the language is evaluated as itself. Rules F_2, F_3, F_4, F_5 and F_6 indicate that the evaluation of the application of a unary or binary operator to one or two expressions is obtained by applying this operator to the result of the evaluation of the subexpressions involved. Of course, if an error is encountered when evaluating one of these expressions, any application of the operator will return the same error. Rules F_7, F_8, F_9 and F_{10} deal with the evaluation of conditional expressions of the form if e_1 then e_2 else e_3. Depending on the result of the evaluation of e_1, either expression e_2 or e_3 will be evaluated. Once again, if an error is encountered when evaluating e_1, evaluation of the conditional expression will result in the same error. Furthermore, if the evaluation of e_1 returns a non-boolean value, then a type error ($\mathrm{Err_T}$) will occur when evaluating the conditional expression.

3.3.2. *Call-by-value*

DEFINITION 3.1.– *Call-by-value: the* call-by-value *mechanism defines a strategy for evaluating function applications, indicating that the argument (i.e. the actual parameter) must always be evaluated. This strategy is extended to the expressions* let $x = e_1$ in e_2, *where* e_1 *is always evaluated.*

The rules set out in Box 3.3 define the operational semantics of $\mathbf{Exp_2}$. These rules apply a call-by-value mechanism.

$$(F_{11})\frac{}{\mathrm{Env} \vdash x \rightsquigarrow \mathrm{Env}(x)}(x \in \mathrm{dom}(\mathrm{Env})) \quad (F_{12})\frac{}{\mathrm{Env} \vdash x \rightsquigarrow \mathrm{Err_U}}(x \notin \mathrm{dom}(\mathrm{Env}))$$

$$(F_{13})\frac{\mathrm{Env} \vdash e_1 \rightsquigarrow v_1 \quad (x, v_1) \oplus \mathrm{Env} \vdash e_2 \rightsquigarrow v_2}{\mathrm{Env} \vdash \mathsf{let}\ x = e_1\ \mathsf{in}\ e_2 \rightsquigarrow v_2}(v_1 \in \mathbf{V})$$

$$(F_{14})\frac{\mathrm{Env} \vdash e_1 \rightsquigarrow v_1}{\mathrm{Env} \vdash \mathsf{let}\ x = e_1\ \mathsf{in}\ e_2 \rightsquigarrow v_1}(v_1 \in \mathbb{E})$$

$$(F_{15})\frac{}{\mathrm{Env} \vdash \mathsf{fun}\ x \mapsto e \rightsquigarrow \langle x, e, \mathrm{Env}\rangle} \quad (F_{16})\frac{}{\mathrm{Env} \vdash \mathsf{rec}\ f\ x = e \rightsquigarrow \langle f, x, e, \mathrm{Env}\rangle}$$

$$(F_{17})\frac{\begin{array}{c}\mathrm{Env} \vdash e_1 \rightsquigarrow \langle x, e_f, \mathrm{Env}_f\rangle \quad \mathrm{Env} \vdash e_2 \rightsquigarrow v_2 \\ (x, v_2) \oplus \mathrm{Env}_f \vdash e_f \rightsquigarrow v\end{array}}{\mathrm{Env} \vdash (e_1\, e_2) \rightsquigarrow v}(v_2 \notin \mathbb{E})$$

$$(F_{18})\frac{\begin{array}{c}\mathrm{Env} \vdash e_1 \rightsquigarrow \langle f, x, e_f, \mathrm{Env}_f\rangle \quad \mathrm{Env} \vdash e_2 \rightsquigarrow v_2 \\ (f, \langle f, x, e_f, \mathrm{Env}_f\rangle) \oplus (x, v_2) \oplus \mathrm{Env}_f \vdash e_f \rightsquigarrow v\end{array}}{\mathrm{Env} \vdash (e_1\, e_2) \rightsquigarrow v}(v_2 \notin \mathbb{E})$$

$$(F_{19})\frac{\mathrm{Env} \vdash e_1 \rightsquigarrow v_1 \quad \mathrm{Env} \vdash e_2 \rightsquigarrow v_2}{\mathrm{Env} \vdash (e_1\, e_2) \rightsquigarrow v_2}(v_1 \in \mathbb{F}, v_2 \in \mathbb{E})$$

$$(F_{20})\frac{\mathrm{Env} \vdash e_1 \rightsquigarrow v}{\mathrm{Env} \vdash (e_1\, e_2) \rightsquigarrow \mathrm{Err_T}}(v \notin \mathbb{F})$$

Box 3.3. *Operational semantics of* $\mathbf{Exp_2}$*: call-by-value*

Rules F_{11} and F_{12} are used to obtain the value associated with a variable x in an environment Env. If this variable does not belong to the domain of Env, the value Err_{U} is returned. Rules F_{13} and F_{14} describe the evaluation of expressions of the form $\mathsf{let}\ x = e_1\ \mathsf{in}\ e_2$ in an environment Env. First, expression e_1 in Env is evaluated, leading to a value noted v_1. Next, e_2 is evaluated in Env, extended with the binding (x, v_1), to obtain the final result. If an error occurs when evaluating expression e_1, then evaluating expression $\mathsf{let}\ x = e_1\ \mathsf{in}\ e_2$ will result in the same error. Rules F_{15} and F_{16} define the value of a function, i.e. a closure containing the current environment (the one in which the definition of the function takes place).

Rules F_{17} and F_{18} describe the evaluation of an application $(e_1\, e_2)$. This evaluation will not produce an error if e_1 is evaluated as a closure and if the evaluation of e_2 does not result in an error. Rule F_{17} applies to the case of an expression e_1, which is evaluated as a non-recursive function, i.e. as a closure of the form $\langle x, e_f, \text{Env}_f\rangle$. Argument e_2 is evaluated in the current environment (in which the call takes place), Env, giving a value v_2. The body of function e_1, i.e. e_f, is evaluated in Env_f, the definition environment of e_1 extended by the binding of the formal parameter to the value of the actual parameter, (x, v_2).

Rule F_{18} applies to cases where e_1 is evaluated as a recursive function, i.e. a closure of the form $\langle f, x, e_f, \text{Env}_f\rangle$. The evaluation of this application is similar to that defined by rule F_{17}, but Env_f contains an additional binding $(f, \langle f, x, e_f, \text{Env}_f\rangle)$, used to evaluate applications corresponding to recursive calls to the function in e_f. Note the use of the lexical binding mechanism: the free variables in a function body keep the values they had when the function was defined, no matter what happens later.

EXAMPLE 3.4.– Again, let us consider the expression defined in [3.3]:

$$\mathsf{let}\ x = 25\ \mathsf{in}\ \underbrace{\mathsf{let}\ f = \underbrace{\mathsf{fun}\ x \mapsto x + 3}_{e_2}\ \mathsf{in}\ (f\ 7) + x}_{e_1}$$

The evaluation tree for this expression is given as:

$$
(F_{13})\dfrac{(F_1)\dfrac{}{\text{Env}_0 \vdash 25 \leadsto 25} \qquad (F_{13})\dfrac{(F_{15})\dfrac{}{\text{Env}_1 \vdash e_2 \leadsto \underbrace{\langle x, x+3, \text{Env}_1\rangle}_{F_f}} \quad \underset{1}{\nabla}}{\underbrace{(x, 25) \oplus \text{Env}_0}_{\text{Env}_1} \vdash e_1 \leadsto 35}}{\text{Env}_0 \vdash \mathsf{let}\ x = 25\ \mathsf{in}\ \mathsf{let}\ f = \mathsf{fun}\ x \mapsto x + 3\ \mathsf{in}\ (f\ 7) + x \leadsto 35}
$$

where ∇_1 is the evaluation tree:

$$(F_4)\dfrac{(F_{17})\dfrac{(F_{11})\dfrac{}{\mathrm{Env}_2 \vdash f \rightsquigarrow F_f} \quad (F_1)\dfrac{}{\mathrm{Env}_2 \vdash 7 \rightsquigarrow 7} \quad \underset{2}{\nabla}}{\mathrm{Env}_2 \vdash (f\,7) \rightsquigarrow 10} \quad (F_{11})\dfrac{}{\mathrm{Env}_2 \vdash x \rightsquigarrow 25}}{\underbrace{(f, F_f) \oplus \mathrm{Env}_1}_{\mathrm{Env}_2} \vdash (f\,7) + x \rightsquigarrow 35}$$

where ∇_2 is the evaluation tree:

$$(F_4)\dfrac{(F_{11})\dfrac{}{(x,7) \oplus \mathrm{Env}_1 \vdash x \rightsquigarrow 7} \quad (F_{11})\dfrac{}{(x,7) \oplus \mathrm{Env}_1 \vdash 3 \rightsquigarrow 3}}{(x,7) \oplus \mathrm{Env}_1 \vdash x + 3 \rightsquigarrow 10}$$

EXAMPLE 3.5.– The evaluation of $(e_{fact}\,3)$ where e_{fact} is the expression defined in [3.2]:

$$\underbrace{(\text{rec } fact\ n = \underbrace{\text{if } n \leq 1 \text{ then } 1 \text{ else } \underbrace{(fact\,(n - 1)) * n}_{e_2}}_{e_1})}_{e_{fact}}$$

is done as follows:

$$(F_{18})\dfrac{(F_{16})\dfrac{}{\mathrm{Env}_0 \vdash e_{fact} \rightsquigarrow \underbrace{\langle fact, n, e_1, \mathrm{Env}_0 \rangle}_{F_{fact}}} \quad (F_1)\dfrac{}{\mathrm{Env}_0 \vdash 3 \rightsquigarrow 3} \quad \underset{1}{\nabla}}{\mathrm{Env}_0 \vdash (e_{fact}\,3) \rightsquigarrow 6}$$

where ∇_1 is the evaluation tree:

$$(F_8)\dfrac{(F_4)\dfrac{(F_4)\dfrac{}{\mathrm{Env}_1 \vdash n \rightsquigarrow 3} \quad (F_1)\dfrac{}{\mathrm{Env}_1 \vdash 1 \rightsquigarrow 1}}{\mathrm{Env}_1 \vdash n \leq 1 \rightsquigarrow false} \quad (F_4)\dfrac{\underset{2}{\nabla} \quad (F_{11})\dfrac{}{\mathrm{Env}_1 \vdash n \rightsquigarrow 3}}{\mathrm{Env}_1 \vdash e_2 \rightsquigarrow 6}}{\underbrace{(fact, F_{fact}) \oplus (n,3) \oplus \mathrm{Env}_0}_{\mathrm{Env}_1} \vdash e_1 \rightsquigarrow 6}$$

where ∇_2 is the evaluation tree:

$$(F_{18})\dfrac{(F_{11})\dfrac{}{\mathrm{Env}_1 \vdash fact \rightsquigarrow F_{fact}} \quad (F_4)\dfrac{(F_{11})\dfrac{}{\mathrm{Env}_1 \vdash n \rightsquigarrow 3} \quad (F_1)\dfrac{}{\mathrm{Env}_1 \vdash 1 \rightsquigarrow 1}}{\mathrm{Env}_1 \vdash n - 1 \rightsquigarrow 2} \quad \underset{3}{\nabla}}{\mathrm{Env}_1 \vdash (fact\,(n - 1)) \rightsquigarrow 2}$$

where ∇_3 is the evaluation tree:

$$
(F_8)\dfrac{(F_4)\dfrac{(F_4)\dfrac{}{\text{Env}_2 \vdash n \leadsto 2} \quad (F_1)\dfrac{}{\text{Env}_2 \vdash 1 \leadsto 1}}{\text{Env}_2 \vdash n \leq 1 \leadsto false} \quad (F_4)\dfrac{\nabla_4 \quad (F_{11})\dfrac{}{\text{Env}_2 \vdash n \leadsto 2}}{\text{Env}_2 \vdash e_2 \leadsto 2}}{\underbrace{(fact, F_{fact}) \oplus (n, 2) \oplus \text{Env}_0}_{\text{Env}_2} \vdash e_1 \leadsto 2}
$$

where ∇_4 is the evaluation tree:

$$
(F_{18})\dfrac{(F_{11})\dfrac{}{\text{Env}_2 \vdash fact \leadsto F_{fact}} \quad (F_4)\dfrac{(F_{11})\dfrac{}{\text{Env}_2 \vdash n \leadsto 2} \quad (F_1)\dfrac{}{\text{Env}_2 \vdash 1 \leadsto 1}}{\text{Env}_2 \vdash n - 1 \leadsto 1} \quad \nabla_5}{\text{Env}_2 \vdash (fact\,(n \;-\; 1)) \leadsto 2}
$$

where ∇_5 is the evaluation tree:

$$
(F_8)\dfrac{(F_4)\dfrac{(F_4)\dfrac{}{\text{Env}_3 \vdash n \leadsto 1} \quad (F_1)\dfrac{}{\text{Env}_3 \vdash 1 \leadsto 1}}{\text{Env}_3 \vdash n \leq 1 \leadsto true} \quad (F_1)\dfrac{}{\text{Env}_3 \vdash 1 \leadsto 1}}{\underbrace{(fact, F_{fact}) \oplus (n, 1) \oplus \text{Env}_0}_{\text{Env}_3} \vdash e_1 \leadsto 1}
$$

REMARK.– Termination of the evaluation mechanism: using the syntactic construct $\mathsf{rec}\ f\ x = e$, expressions may be defined for which evaluation does not terminate. For example, from expression e_{foo} defined by

$$\mathsf{rec}\ f\ x = (f\ x) \qquad [3.5]$$

we can construct the expression:

$$\mathsf{let}\ foo = (\mathsf{rec}\ f\ x = (f\ x))\ \mathsf{in}\ (foo\,2) \qquad [3.6]$$

The evaluation of this expression does not terminate. The construction of the evaluation tree for this expression is never-ending:

$$
(F_{13})\dfrac{(F_{16})\dfrac{}{\text{Env} \vdash \mathsf{rec}\ f\ x = (f\ x) \leadsto \underbrace{\langle f, x, (f\ x), \text{Env}\rangle}_{F_{foo}}} \quad \nabla_1}{\text{Env} \vdash \mathsf{let}\ foo = \mathsf{rec}\ f\ x = (f\ x)\ \mathsf{in}\ (foo\,2) \leadsto}
$$

where the construction of ∇_1 is given as:

$$
(F_{18})\dfrac{(F_{11})\dfrac{}{\text{Env}_1 \vdash foo \leadsto F_{foo}} \quad (F_1)\dfrac{}{\text{Env}_1 \vdash 2 \leadsto 2} \quad \nabla_2}{\underbrace{(foo, F_{foo}) \oplus \text{Env}}_{\text{Env}_1} \vdash (foo\,2) \leadsto}
$$

where the construction of ∇_2 is given as:

$$(F_{18})\dfrac{(F_{11})\dfrac{}{\text{Env}_2 \vdash f \rightsquigarrow F_{foo}} \quad (F_{11})\dfrac{}{\text{Env}_2 \vdash x \rightsquigarrow 2} \quad (F_{18})\dfrac{\vdots}{\text{Env}_2 \vdash (f\ x) \rightsquigarrow}}{\underbrace{(f, F_{foo}) \oplus (x, 2) \oplus \text{Env}}_{\text{Env}_2} \vdash (f\ x) \rightsquigarrow}$$

This semantics can only be used to describe terminating executions. In other terms, $\text{Env} \vdash e \rightsquigarrow v$ expresses the fact that v is the value of e obtained by a finite evaluation tree, constructed using the inference rules presented in Box 3.2 and Box 3.3. There is no such evaluation tree for expression [3.6]. This semantics is called *big-step operational semantics*: $\text{Env} \vdash e \rightsquigarrow v$ does not describe the computation, only the result. To describe and reason using computations, including non-terminating computations, the semantics may be formalized 'as *small-step semantics* (or reduction semantics), a subject discussed at length in Chapter 4.

3.3.3. *Recursive and mutually recursive functions*

Two occurrences of e_f occur on either side of the symbol $\vdash$ in rule F_{18}, which concerns the application of a recursive function. At first glance, this may seem paradoxical: evaluating e_f on the right leads us to evaluate e_f on the left. As we saw in section 3.2.5, this question may be expressed directly by using rule F_{16} and binding f to the closure:

$$F = \langle x, e, (f, F) \oplus \text{Env} \rangle$$

Once again, F appears on both sides of the equals sign. The existence of value F can be proven using the mathematical theory of the least fixpoint (LFP), briefly presented in section 4.7.1; a more detailed description can be found in [JAU 16]. Taking this approach, rules F_{16}, F_{17} and F_{18} can be condensed into two rules:

$$(N_{16})\dfrac{}{\text{Env} \vdash \textsf{rec}\ f\ x = e \rightsquigarrow \underbrace{\langle x, e, \underbrace{(f, F) \oplus \text{Env}}_{\text{Env}_f} \rangle}_{=F}}$$

$$(N_{17\_18})\dfrac{\begin{array}{ll}\text{Env} \vdash e_1 \rightsquigarrow \langle x, e, \text{Env}_f \rangle & \text{Env} \vdash e_2 \rightsquigarrow v_2 \\ (x, v_2) \oplus \text{Env}_f \vdash e \rightsquigarrow v & \end{array}}{\text{Env} \vdash (e_1\ e_2) \rightsquigarrow v}$$

Applications of recursive and non-recursive functions are based on the same mechanism. The LFP theory allows us to be sure that the evaluation of any expression with a finite number of recursive calls will terminate. Furthermore, the same theory can be used to prove that the two approaches define the same semantics for $\textbf{Exp}_2$.

Now, let us consider the case of mutually recursive functions. For example, the following functions determine the parity of an integer:

$$\textsf{rec}\ \text{even}\ n = \textsf{if}\ n = 0\ \textsf{then}\ true\ \textsf{else}\ \text{odd}(n-1)$$
$$\textsf{rec}\ \text{odd}\ n = \textsf{if}\ n = 0\ \textsf{then}\ false\ \textsf{else}\ \text{even}(n-1)$$

The body of function even contains a call to function odd, of which the body also contains a call to function even. These two functions are said to be *mutually recursive*, and some programming languages use a syntactic construct to express this form of dependency between two functions. When two functions are mutually recursive, it is impossible to give them two independent values: for example, the definition environment contained in the closure associated with function even must contain a binding for function odd and vice versa. One simple way of solving this problem is to handle it syntactically. For example, the construct

$$\textsf{let\_rec}\ f\ x = e_f\ \textsf{and}\ g\ y = e_g;$$

can be used to define two mutually recursive functions f and g. Evaluation of this construct simultaneously adds the bindings of f and g to the current environment.

Considering the approach presented at the start of this section for closures, the evaluation of these definitions extends the current environment Env as follows:

$$(g, G) \oplus (f, F) \oplus \text{Env} \ \text{ with } \begin{array}{l} F = \langle x, e_f, (g, G) \oplus (f, F) \oplus \text{Env} \rangle \\ G = \langle y, e_g, (g, G) \oplus (f, F) \oplus \text{Env} \rangle \end{array}$$

3.3.4. *Call-by-name*

DEFINITION 3.2.– *Call-by-name: the* call-by-name *mechanism defines a strategy for evaluating function applications and other expressions. The subexpressions of an expression* e *will only be evaluated if their values are needed for the evaluation of* e*.*

To evaluate $(f\ e)$, we evaluate f in the current environment Env. The expression e is only evaluated if the evaluation of the body of f in the definition environment Env_f of f requires a value for e in Env. The pair $\lfloor e, \text{Env} \rfloor$, called a *frozen expression*, must therefore be retained. An environment thus becomes a list of associations, where the elements are pairs $(x, \lfloor e, \text{Env}_e \rfloor)$, where x is a "frozen" variable, e is the expression defining x and Env_e is the environment in which e is to be evaluated:

$$\text{Env} = [(x_1, \lfloor e_1, \text{Env}_{e_1} \rfloor), \cdots, (x_n, \lfloor e_n, \text{Env}_{e_n} \rfloor)]$$

The notion of closure is thus generalized to all expressions, not only functions.

Big-step operational semantics with the call-by-name strategy is defined in Box 3.2 and Box 3.4. The evaluation rules for simple expressions, defined in Box 3.2, are valid for both call strategies. For example, the evaluation of a conditional expression is always dependent on the evaluation of the condition, and the evaluation of an arithmetic expression requires us to evaluate at least one operand. The rules shown in Box 3.4 use the new notion of environment. To evaluate a variable x in an environment Env, which associates $\lfloor e_x, \text{Env}_x \rfloor$ with x, we simply evaluate e_x in the environment Env_x, i.e. apply the rules F'_{11} and F'_{12}. To evaluate an expression of the form $\mathsf{let}\ x = e_1\ \mathsf{in}\ e_2$ in an environment Env, we evaluate e_2 in the environment $(x, \lfloor e_1, \text{Env} \rfloor) \oplus \text{Env}$; this is an application of rule F'_{13}. The evaluation of a function results in an additional closure. The formulation presented in section 3.3.3 is again used here for rules F'_{14} and F'_{15}. The application of a function f to an argument e in an environment Env first evaluates f in Env as a closure $\langle x, e_f, \text{Env}_f \rangle$, and then evaluates e_f in environment $(x, \lfloor e, \text{Env} \rfloor) \oplus \text{Env}_f$, following rule F'_{16}.

$$(F'_{11})\frac{\text{Env}_x \vdash e_x \rightsquigarrow v}{\text{Env} \vdash x \rightsquigarrow v}((x, \lfloor e_x, \text{Env}_x \rfloor) \in \text{Env}) \quad (F'_{12})\frac{}{\text{Env} \vdash x \rightsquigarrow \text{Err}_{\text{U}}}(x \notin \text{dom}(\text{Env}))$$

$$(F'_{13})\frac{(x, \lfloor e_1, \text{Env} \rfloor) \oplus \text{Env} \vdash e_2 \rightsquigarrow v}{\text{Env} \vdash \mathsf{let}\ x = e_1\ \mathsf{in}\ e_2 \rightsquigarrow v}$$

$$(F'_{14})\frac{}{\text{Env} \vdash \mathsf{fun}\ x \mapsto e \rightsquigarrow \langle x, e, \text{Env} \rangle}$$

$$(F'_{15})\frac{}{\text{Env} \vdash \mathsf{rec}\ f\ x = e \rightsquigarrow \langle x, e, (f, \lfloor \mathsf{rec}\ f\ x = e, \text{Env} \rfloor) \oplus \text{Env} \rangle}$$

$$(F'_{16})\frac{\text{Env} \vdash e_1 \rightsquigarrow \langle x, e_f, \text{Env}_f \rangle \quad (x, \lfloor e_2, \text{Env} \rfloor) \oplus \text{Env}_f \vdash e_f \rightsquigarrow v}{\text{Env} \vdash (e_1\ e_2) \rightsquigarrow v}$$

$$(F'_{17})\frac{\text{Env} \vdash e_1 \rightsquigarrow v}{\text{Env} \vdash (e_1\ e_2) \rightsquigarrow \text{Err}_{\text{T}}}(v \notin \mathbb{F})$$

Box 3.4. *Operational semantics of* $\mathbf{Exp}_2$*: call-by-name*

3.3.5. *Call-by-value versus call-by-name*

The two call strategies do not necessarily produce the same results. For example, the evaluation of expression $((\mathsf{fun}\ x \mapsto 3)\,(8\,/\,0))$ using a call-by-value strategy operates as follows:

$$(F_{19})\frac{(F_{15})\dfrac{}{\text{Env} \vdash \mathsf{fun}\ x \mapsto 3 \rightsquigarrow \langle x, 3, \text{Env} \rangle} \quad (F_4)\dfrac{(F_1)\dfrac{}{\text{Env} \vdash 8 \rightsquigarrow 8} \quad (F_1)\dfrac{}{\text{Env} \vdash 0 \rightsquigarrow 0}}{\text{Env} \vdash 8\,/\,0 \rightsquigarrow \text{Err}_{\text{E}}}}{\text{Env} \vdash ((\mathsf{fun}\ x \mapsto 3)\,(8\,/\,0)) \rightsquigarrow \text{Err}_{\text{E}}}$$

If a call-by-value strategy produces a finite evaluation tree, then a call-by-name strategy will also produce a finite tree. On the other hand, obtaining a finite tree using a call-by-name strategy does not imply that a finite tree will necessarily be produced using call-by-value. For example, the evaluation of the expression $((\mathsf{fun}\ x \mapsto 3)\,(e_{foo}\,2))$ does not terminate using a call-by-value strategy (see section 3.3.2), whereas the value 3 is obtained using call-by-name:

$$(F'_{16})\dfrac{(F'_{14})\dfrac{}{\text{Env} \vdash \mathsf{fun}\ x \mapsto 3 \leadsto \langle x, 3, \text{Env}\rangle} \quad (F_1)\dfrac{}{(x, \lfloor (e_{foo}\,2), \text{Env}\rfloor) \oplus \text{Env} \vdash 3 \leadsto 3}}{\text{Env} \vdash ((\mathsf{fun}\ x \mapsto 3)\,(e_{foo}\,2)) \leadsto 3}$$

The two strategies also differ in their capacity to share evaluations of subexpressions. For example, using call-by-value, $((\mathsf{fun}\ x \mapsto x + x)\,e)$ is evaluated as follows:

$$(F_{17})\dfrac{(F_{15})\dfrac{}{\text{Env}\vdash e_f \leadsto \langle x, x+x, \text{Env}\rangle} \quad (R)\dfrac{\vdots}{\text{Env}\vdash e \leadsto v_e} \quad \nabla}{\text{Env}\vdash ((\underbrace{\mathsf{fun}\ x\mapsto x+x}_{e_f})\,e)\leadsto v}$$

where ∇ is the tree:

$$(F_4)\dfrac{(F_{11})\dfrac{}{\text{Env}_1\vdash x\leadsto v_e}\quad (F_{11})\dfrac{}{\text{Env}_1\vdash x\leadsto v_e}}{\underbrace{(x,v_e)\oplus\text{Env}}_{\text{Env}_1}\vdash x+x\leadsto v}$$

Here, the value v is obtained by computing the value v_e of e only once. Using a call-by-name mechanism, on the other hand, we obtain:

$$(F'_{16})\dfrac{(F'_{14})\dfrac{}{\text{Env}\vdash e_f\leadsto\langle x,x+x,\text{Env}\rangle}\quad (F_4)\dfrac{(F_{11})\dfrac{\overset{\nabla}{\text{Env}\vdash e\leadsto v_e}}{\text{Env}_1\vdash x\leadsto v_e}\quad (F_{11})\dfrac{\overset{\nabla}{\text{Env}\vdash e\leadsto v_e}}{\text{Env}_1\vdash x\leadsto v_e}}{\underbrace{(x,\lfloor e,\text{Env}\rfloor)\oplus\text{Env}}_{\text{Env}_1}\vdash x+x\leadsto v}}{\text{Env}\vdash((\underbrace{\mathsf{fun}\ x\mapsto x+x}_{e_f})\,e)\leadsto v}$$

The value v_e of e is computed twice to obtain the value v of the expression $x + x$. The *call-by-need* strategy is an implementation of the call-by-name strategy which attempts to maximize sharing of subexpressions, in order to evaluate the arguments of functions once only.

REMARK.– In languages that allow side effects (reading/writing/assignment), call-by-value is often used because this strategy makes it possible to predict when side effects will occur; this is not the case with a call-by-name strategy. Compilers for languages using a call-by-name strategy research and evaluate subexpressions producing side effects using a call-by-value strategy.

3.4. Evaluation functions versus evaluation relations

In this section, we examine the determinism of the evaluation function with the call-by-value strategy. This leads us to study a new form of induction.

3.4.1. *Status of the evaluation function*

The question examined here is whether, for every expression $e \in \mathbf{Exp}_2$, there exists a single value $v \in \mathbb{V}$ such that $[\![e]\!]_{\text{Env}} = v$. The answer is no. If e is an expression that requires evaluation of the application of a non-terminating recursive function (such as expression [3.6], called "foo", in section 3.3.2), then we cannot obtain a value $v \in \mathbb{V}$ such that $[\![e]\!]_{\text{Env}} = v$. So $[\![\_]\!]_{\_}$ is not a total function. However, if such a value v exists, it is provably unique (thus $[\![\_]\!]_{\_}$ is a partial function):

$$\forall e \in \mathbf{Exp}_2 \ \underbrace{\forall \text{Env} \in \mathbf{E} \ \forall v, v' \in \mathbb{V} \ \left([\![e]\!]_{\text{Env}} = v \text{ and } [\![e]\!]_{\text{Env}} = v'\right) \Rightarrow v = v'}_{P(e)}$$

Let us call this property P. We may try to prove that P holds for all expressions $e \in \mathbf{Exp}_2$ by induction over e, using the induction scheme defined in Box 3.1. This is possible for most of the syntactic constructs, for example for expressions of the form $\text{let } x = e_1 \text{ in } e_2$. Let Env be an environment and v and v' two values. If $[\![\text{let } x = e_1 \text{ in } e_2]\!]_{\text{Env}} = v$ and $[\![\text{let } x = e_1 \text{ in } e_2]\!]_{\text{Env}} = v'$, then we can prove that $v = v'$. First, note that by induction hypothesis, $P(e_1)$ is verified, and thus if $[\![e_1]\!]_{\text{Env}} = v_1$ and $[\![e_1]\!]_{\text{Env}} = v_1'$, then $v_1 = v_1'$. Two cases are now possible. If $v_1 \in \mathbf{V}$, then, by definition, $v = [\![e_2]\!]_{(x,v_1)\oplus\text{Env}}$ and $v' = [\![e_2]\!]_{(x,v_1')\oplus\text{Env}}$ and, since $v_1 = v_1'$, by induction hypothesis [1] over e_2, we obtain $v = v'$. Otherwise, if $v_1 \in \mathbb{E}$, then by definition, $v = v_1$ and $v' = v_1'$, and since $v_1 = v_1'$, we directly obtain $v = v'$.

Similar reasoning can be used to prove P for all constructs with the exception of function applications, i.e. expressions of the form $(e_1\, e_2)$. In this case, the only available induction hypotheses are related to the subexpressions e_1 and e_2 in $(e_1\, e_2)$. However, when e_1 is evaluated as a closure containing the expression e_f, the evaluation of the application $(e_1\, e_2)$ leads to an evaluation of the body e_f, which is

1 Note that the order of quantifications in the statement P impacts the form of the induction hypotheses. For example, quantifying first over the environment ($\forall \text{Env} \in \mathbf{E} \ \forall e \in \mathbf{Exp}_2 \forall v, v' \in \mathbb{V}$) allows us to use the induction hypothesis over e_1 since e_1 is evaluated in Env but forbids us to use the induction hypothesis over e_2 as e_2 is evaluated in an environment that is different to Env.

not a subexpression of $(e_1\, e_2)$ (for example, e_1 may be only the name of the function). We cannot suppose that P holds for e_f and we cannot conclude. Thus, it is not possible to prove $P(e)$ through reasoning by structural induction over the expression e. This property cannot be proved by induction over the structure of expressions and requires an induction over the "size" of the computations involved in evaluation. Considering the evaluation function $[\![\_]\!]_{\_}$, this can be done using the least fixpoint theory for continuous functions, as described in section 4.7.1. The property will be proved here using big-step operational semantics, which allows us to reason by induction over the evaluation trees; this is a way to consider the "size" of the computations involved in evaluation.

3.4.2. *Induction over evaluation trees*

Our aim here is to prove a property P which states that if an expression is evaluated as a value, then this value is unique. Formally, within the framework of big-step operational semantics, we wish to prove the following proposition:

PROPOSITION 3.1.– Determinism:

$$\forall e \in \mathbf{Exp}_2\ \forall \text{Env} \in \mathbf{E}\ \forall v, v' \in \mathbb{V}\ \ (\text{Env} \vdash e \rightsquigarrow v \text{ and } \text{Env} \vdash e \rightsquigarrow v') \Rightarrow v = v'$$

The set of evaluation (or inference) trees for the expressions in $\mathbf{Exp}_2$ can be defined inductively, which means that we can reason by induction over these trees, i.e. over the "size" of the computations carried out during evaluations. More generally, given an inference system $\mathbb{S}$, i.e. a set of inference rules of the form

$$(R)\frac{e_1 \cdots e_n}{e}$$

the set $\mathbb{A}_\mathbb{S}$ of elements e that can be inferred (sometimes called the set of theorems of $\mathbb{S}$) is defined inductively as follows:

– for every axiom in $\mathbb{S}$, i.e. for every rule of the form $(R)\frac{-}{e}$, the conclusion e belongs to $\mathbb{A}_\mathbb{S}$;

– if $e_1, \cdots, e_n$ are elements in $\mathbb{A}_\mathbb{S}$, and if

$$(R)\frac{e_1 \cdots e_n}{e}$$

is a rule in $\mathbb{S}$, then $e \in \mathbb{A}_{\mathbb{S}}$. Indeed if $e_1, \cdots, e_n$ are elements in $\mathbb{A}_{\mathbb{S}}$, then there exist trees $\nabla_1, \cdots, \nabla_n$ for each of these elements, and we can therefore construct the following tree:

$$(R)\frac{\overset{\nabla_1}{e_1} \quad \cdots \quad \overset{\nabla_n}{e_n}}{e}$$

Proving a property P over the elements of $\mathbb{A}_{\mathbb{S}}$ comes down to proving that this property holds for all elements concluding an axiom in $\mathbb{S}$ and proving, for each rule

$$(R)\frac{e_1 \cdots e_n}{e}$$

in $\mathbb{S}$ that if all $e_1, \cdots, e_n$ satisfy the property P, then e also satisfies the property P. This results in an inductive reasoning scheme built on the inference tree of e. In the case of evaluation trees for the expressions in $\mathbf{Exp}_2$, this enables us to reason by induction over the computations carried out during the evaluation.

Using this technique, and taking $\mathbb{S}_{\mathbf{Exp_2}}$ to be the inference system defining the big-step operational semantics of the expressions of $\mathbf{Exp}_2$, proposition 3.1 can be proved by considering the following equivalent formulation:

$$\forall \text{Env} \vdash e \rightsquigarrow v \in \mathbb{A}_{\mathbb{S}_{\mathbf{Exp_2}}} \quad \underbrace{(\forall v' \in \mathbb{V}\ \text{Env} \vdash e \rightsquigarrow v' \in \mathbb{A}_{\mathbb{S}_{\mathbf{Exp_2}}} \Rightarrow v = v')}_{P(\text{Env}\vdash e \rightsquigarrow v)} \qquad [3.7]$$

PROOF.– Let P be the property to prove and let us show that this property is satisfied by all evaluation trees $\text{Env} \vdash e \rightsquigarrow v \in \mathbb{A}_{\mathbb{S}_{\mathbf{Exp_2}}}$.

If $\text{Env} \vdash e \rightsquigarrow v$ has been obtained using rule F_1, then $e = v$ is a constant in $\mathbf{K}$, and since F_1 is the only rule that can be used to evaluate a constant, $v = v'$.

If $\text{Env} \vdash e \rightsquigarrow v$ has been obtained using rule F_2, then $e = \text{op}\ e_1$, $\text{Env} \vdash e_1 \rightsquigarrow v_1$, $v_1 \in \mathbf{V}$ and $v = \text{op}_{\mathbf{V}}\ v_1$. By induction hypothesis, if $\text{Env} \vdash e_1 \rightsquigarrow v_1'$, then $v_1' = v_1 \in \mathbf{V}$ and since F_2 is the only applicable rule for evaluating e, we obtain $v' = \text{op}_{\mathbf{V}}\ v_1' = \text{op}_{\mathbf{V}}\ v_1 = v$.

If $\text{Env} \vdash e \rightsquigarrow v$ has been obtained using rule F_3, then $e = \text{op}\ e_1$, $\text{Env} \vdash e_1 \rightsquigarrow v_1$, $v_1 \in \mathbb{E}$ and $v = v_1$. By induction hypothesis, if $\text{Env} \vdash e_1 \rightsquigarrow v_1'$, then $v_1' = v_1 \in \mathbb{E}$. F_3 is the only applicable rule for evaluating e, and we obtain $v' = v_1' = v_1 = v$.

If $\text{Env} \vdash e \rightsquigarrow v$ has been obtained using rule F_4, then $e = e_1\, \text{op}\, e_2$, $\text{Env} \vdash e_1 \rightsquigarrow v_1$, $\text{Env} \vdash e_2 \rightsquigarrow v_2$, $v_1, v_2 \in \mathbf{V}$ and $v = v_1\, \text{op}_{\mathbf{V}}\, v_2$. By induction hypothesis, if $\text{Env} \vdash e_1 \rightsquigarrow v_1'$, then $v_1' = v_1 \in \mathbf{V}$ and if $\text{Env} \vdash e_2 \rightsquigarrow v_2'$, then $v_2' = v_2 \in \mathbf{V}$. Rule F_4 is the only applicable rule for evaluating e and we obtain $v' = v_1'\, \text{op}_{\mathbf{V}}\, v_2' = v_1\, \text{op}_{\mathbf{V}}\, v_2 = v$.

If $\text{Env} \vdash e \rightsquigarrow v$ has been obtained using rule F_5, then $e = e_1 \,\mathrm{op}\, e_2$, $\text{Env} \vdash e_1 \rightsquigarrow v_1$, $v_1 \in \mathbb{E}$ and $v = v_1$. By induction hypothesis, if $\text{Env} \vdash e_1 \rightsquigarrow v_1'$, then $v_1' = v_1 \in \mathbb{E}$. Rule F_5 is the only applicable rule for evaluating e and we obtain $v' = v_1' = v_1 = v$.

If $\text{Env} \vdash e \rightsquigarrow v$ has been obtained using rule F_6, then $e = e_1 \,\mathrm{op}\, e_2$, $\text{Env} \vdash e_1 \rightsquigarrow v_1$, $\text{Env} \vdash e_2 \rightsquigarrow v_2$, $v_1 \in \mathbf{V}$, $v_2 \in \mathbb{E}$ and $v = v_2$. By induction hypothesis, if $\text{Env} \vdash e_1 \rightsquigarrow v_1'$, then $v_1' = v_1 \in \mathbf{V}$ and if $\text{Env} \vdash e_2 \rightsquigarrow v_2'$, then $v_2' = v_2 \in \mathbb{E}$. Rule F_6 is the only applicable rule for evaluating e and we obtain $v' = v_2' = v_2 = v$.

If $\text{Env} \vdash e \rightsquigarrow v$ has been obtained using rule F_7 (respectively, F_8), then e is the conditional expression $\mathsf{if}\ e_1\ \mathsf{then}\ e_2\ \mathsf{else}\ e_3$, $\text{Env} \vdash e_1 \rightsquigarrow \mathit{true}$ (respectively, $\text{Env} \vdash e_1 \rightsquigarrow \mathit{false}$) and $\text{Env} \vdash e_2 \rightsquigarrow v$ (respectively, $\text{Env} \vdash e_3 \rightsquigarrow v$). By induction hypothesis, if:

$$\text{Env} \vdash e_1 \rightsquigarrow v_1' \text{ then } v_1' = \mathit{true} \text{ (respectively } v_1' = \mathit{false})$$

and if:

$$\text{Env} \vdash e_2 \rightsquigarrow v' \text{ (respectively, } \text{Env} \vdash e_3 \rightsquigarrow v') \text{ then } v' = v$$

Rule F_7 (respectively, F_8) is the only applicable rule for evaluating e and we obtain $v' = v$.

If $\text{Env} \vdash e \rightsquigarrow v$ has been obtained using rule F_9 (respectively, F_{10}), then $e = \mathsf{if}\ e_1\ \mathsf{then}\ e_2\ \mathsf{else}\ e_3$, $\text{Env} \vdash e_1 \rightsquigarrow v_1$, $v_1 \in \mathbf{V} \setminus \mathbb{B}$ (respectively, $v_1 \in \mathbb{E}$) and $v = \text{Err}_\text{T}$ (respectively, $v = v_1$). By induction hypothesis, if $\text{Env} \vdash e_1 \rightsquigarrow v_1'$, then $v_1' = v_1$. Rule F_9 (respectively, F_{10}) is the only applicable rule for evaluating e and we obtain $v' = v$.

If $\text{Env} \vdash e \rightsquigarrow v$ has been obtained using rule F_{11} (respectively, F_{12}), then $e = x$ is a variable in $\mathbf{X}$ such that $x \in \text{dom}(\text{Env})$ (respectively, $x \notin \text{dom}(\text{Env})$) and $v = \text{Env}(x)$ (respectively, $v = \text{Err}_\text{U}$). Rule F_{11} (respectively, F_{12}) is the only applicable rule for evaluating e and we obtain $v' = v$.

If $\text{Env} \vdash e \rightsquigarrow v$ has been obtained using rule F_{13}, then $e = \mathsf{let}\ x = e_1\ \mathsf{in}\ e_2$, $\text{Env} \vdash e_1 \rightsquigarrow v_1$, $(x, v_1) \oplus \text{Env} \vdash e_2 \rightsquigarrow v_2$, $v_1 \in \mathbf{V}$ and $v = v_2$. By induction hypothesis, if $\text{Env} \vdash e_1 \rightsquigarrow v_1'$, then $v_1' = v_1 \in \mathbf{V}$ and if $(x, v_1) \oplus \text{Env} \vdash e_2 \rightsquigarrow v_2'$, then $v_2' = v_2$. Rule F_{13} is the only applicable rule for evaluating e, and since $v_1' = v_1$, then $(x, v_1') \oplus \text{Env} = (x, v_1) \oplus \text{Env}$ and we obtain $v' = v_2' = v_2 = v$.

If $\text{Env} \vdash e \rightsquigarrow v$ has been obtained using rule F_{14}, then $e = \mathsf{let}\ x = e_1\ \mathsf{in}\ e_2$, $\text{Env} \vdash e_1 \rightsquigarrow v_1$, $v_1 \in \mathbb{E}$ and $v = v_1$. By induction hypothesis, if $\text{Env} \vdash e_1 \rightsquigarrow v_1'$, then $v_1' = v_1 \in \mathbb{E}$ and F_{14} is the only applicable rule for evaluating e and we obtain $v' = v_1' = v_1 = v$.

If $\text{Env} \vdash e \rightsquigarrow v$ has been obtained using rule F_{15} (respectively, F_{16}), then $e = \textsf{fun}\ x \mapsto e_f$ and $v = \langle x, e_f, \text{Env}\rangle$ (respectively, $e = \textsf{rec}\ f\ x = e_f$ and $v = \langle f, x, e_f, \text{Env}\rangle$) and since F_{15} (respectively, F_{16}) is the only applicable rule for evaluating a function (respectively, a recursive function), then $v = v'$.

If $\text{Env} \vdash e \rightsquigarrow v$ has been obtained using rule F_{17} (respectively, F_{18}), then $e = (e_1\, e_2)$ and:

$$\begin{array}{l}\text{Env} \vdash e_1 \rightsquigarrow \langle x, e_f, \text{Env}_f\rangle \text{ (respectively, } \text{Env} \vdash e_1 \rightsquigarrow \langle f, x, e_f, \text{Env}_f\rangle)\\ \text{Env} \vdash e_2 \rightsquigarrow v_2\\ (x, v_2) \oplus \text{Env}_f \vdash e_f \rightsquigarrow v \text{ (respectively, } (f, \langle f, x, e_f, \text{Env}_f\rangle) \oplus (x, v_2) \oplus \text{Env}_f \vdash e_f \rightsquigarrow v)\end{array}$$

By induction hypothesis, if:

$$\text{Env} \vdash e_1 \rightsquigarrow v_1' \text{ then } v_1' = \langle x, e_f, \text{Env}_f\rangle \text{ (respectively} v_1' = \langle f, x, e_f, \text{Env}_f\rangle)$$

and if:

$$\text{Env} \vdash e_2 \rightsquigarrow v_2' \text{ then } v_2' = v_2$$

Hence:

$$\begin{array}{l}(x, v_2') \oplus \text{Env}_f = (x, v_2) \oplus \text{Env}_f\\ \text{(respectively } (f, v_1') \oplus (x, v_2') \oplus \text{Env}_f = (f, \langle f, x, e_f, \text{Env}_f\rangle) \oplus (x, v_2) \oplus \text{Env}_f)\end{array}$$

The only applicable rule is F_{17} (respectively, F_{18}) and, by induction hypothesis, since:

$$\begin{array}{l}(x, v_2) \oplus \text{Env}_f \vdash e_f \rightsquigarrow v\\ \text{(respectively, } (f, \langle f, x, e_f, \text{Env}_f\rangle) \oplus (x, v_2) \oplus \text{Env}_f \vdash e_f \rightsquigarrow v)\end{array}$$

we obtain $v = v'$.

If $\text{Env} \vdash e \rightsquigarrow v$ has been obtained using rule F_{19}, then $e = (e_1\, e_2)$, $\text{Env} \vdash e_1 \rightsquigarrow v_1$, $\text{Env} \vdash e_2 \rightsquigarrow v_2$, $v_1 \in \mathbb{F}$, $v_2 \in \mathbb{E}$ and $v = v_2$. By induction hypothesis, if $\text{Env} \vdash e_1 \rightsquigarrow v_1'$, then $v_1' = v_1 \in \mathbb{F}$ and if $\text{Env} \vdash e_2 \rightsquigarrow v_2'$, then $v_2' = v_2 \in \mathbb{E}$. Rule F_{19} is the only applicable rule for evaluating e and we obtain $v' = v_2' = v_2 = v$.

If $\text{Env} \vdash e \rightsquigarrow v$ has been obtained using rule F_{20}, then $e = (e_1\, e_2)$, $\text{Env} \vdash e_1 \rightsquigarrow v_1$, $v_1 \notin \mathbb{F}$ and $v = \text{Err}_{\mathbb{T}}$. By induction hypothesis, if $\text{Env} \vdash e_1 \rightsquigarrow v_1'$, then $v_1' = v_1 \notin \mathbb{F}$. Rule F_{20} is the only applicable rule for evaluating e and we obtain $v' = v$. □

The proposition below establishes an equivalence between the semantics defined by the evaluation function $[\![\_]\!]_{\_}$ and the operational semantics with call-by-value defined in Box 3.2 and Box 3.3.

PROPOSITION 3.2.– Equivalence:

$$\forall e \in \mathbf{Exp_2}\ \forall \text{Env} \in \mathbf{E}\ \forall v \in \mathbb{V} \quad [\![e]\!]_{\text{Env}} = v \Leftrightarrow \text{Env} \vdash e \rightsquigarrow v$$

The implication $\Rightarrow$ can be proved using the LFP theory but it is out of the scope of this book. The proof of the implication $\Leftarrow$ is obtained using the same technique applied for proposition 3.1, by induction over the evaluation tree for $\text{Env} \vdash e \rightsquigarrow v$. For each rule:

$$(F_i)\frac{\text{Env}_1 \vdash e_1 \rightsquigarrow v_1 \quad \cdots \quad \text{Env}_n \vdash e_n \rightsquigarrow v_n}{\text{Env} \vdash e \rightsquigarrow v}$$

of the inference system $\mathbb{S}_{\mathbf{Exp_2}}$, which defines the big-step operational semantics of expressions in $\mathbf{Exp_2}$, we must prove that $[\![e]\!]_{\text{Env}} = v$, under the induction hypothesis that $[\![e_1]\!]_{\text{Env}_1} = v_1$, etc., $[\![e_n]\!]_{\text{Env}_n} = v_n$. This proof is immediate, as the rules for evaluating the operational semantics were defined following exactly the same choices used to define the evaluation function.

3.5. Semantic properties

The usual techniques used to prove semantic properties over expressions in a programming language are presented as follows.

3.5.1. *Equivalent expressions*

DEFINITION 3.3.– *Equivalent expressions: two expressions are equivalent if and only if, in each environment, their evaluations return the same value:*

$$\forall e_1, e_2 \in \mathbf{Exp_2} \qquad e_1 \equiv e_2 \Leftrightarrow \left(\forall \text{Env} \in \mathbf{E}\ [\![e_1]\!]_{\text{Env}} = [\![e_2]\!]_{\text{Env}}\right) \qquad [3.8]$$

In this definition, we used the evaluation function $[\![\_]\!]_{\_}$ for the evaluation of the expressions. The relation $\rightsquigarrow$ for evaluating big-step operational semantics can also be used to define the notion of equivalent expressions:

$$\begin{array}{l}\forall e_1, e_2 \in \mathbf{Exp_2} \\ e_1 \equiv e_2 \Leftrightarrow (\forall \text{Env} \in \mathbf{E}\ \forall v \in \mathbb{V}\ \text{Env} \vdash e_1 \rightsquigarrow v \Leftrightarrow \text{Env} \vdash e_2 \rightsquigarrow v)\end{array} \qquad [3.9]$$

Using this second definition, proving that $e_1 \equiv e_2$ amounts to proving how the evaluation tree for $\text{Env} \vdash e_2 \rightsquigarrow v$ can be obtained from the tree for $\text{Env} \vdash e_1 \rightsquigarrow v$ and vice versa. Following proposition 3.2, these two approaches are equivalent. We can easily prove that $\equiv$ defines a congruence, i.e. an equivalence relation compatible with the internal composition laws of $\mathbf{Exp_2}$. For example, if $e_1 \equiv e_2$ and $e'_1 \equiv e'_2$, then $(e_1 + e'_1) \equiv (e_2 + e'_2)$.

EXAMPLE 3.6.– We can prove that $\mathsf{let}\ x = e_1\ \mathsf{in}\ e_2 \equiv ((\mathsf{fun}\ x \mapsto e_2)\, e_1)$. Using definition [3.8], this equivalence is proved via proof by cases, using the evaluation of e_1.

If $[\![e_1]\!]_{\text{Env}} = v \in \mathbf{V}$, then $[\![\mathsf{let}\ x = e_1\ \mathsf{in}\ e_2]\!]_{\text{Env}} = [\![e_2]\!]_{(x,v)\oplus\text{Env}}$ and since $[\![\mathsf{fun}\ x \mapsto e_2]\!]_{\text{Env}} = \langle x, e_2, \text{Env}\rangle$, we obtain $[\![((\mathsf{fun}\ x \mapsto e_2)\, e_1)]\!]_{\text{Env}} = [\![e_2]\!]_{(x,v)\oplus\text{Env}}$, which concludes the proof.

If $[\![e_1]\!]_{\text{Env}} = v \in \mathbb{E}$, it is also possible to complete the proof. In this case, we have $[\![\mathsf{let}\ x = e_1\ \mathsf{in}\ e_2]\!]_{\text{Env}} = v = [\![((\mathsf{fun}\ x \mapsto e_2)\, e_1)]\!]_{\text{Env}}$.

The equivalence can be established by considering definition [3.9]. First, take a tree $\text{Env} \vdash \mathsf{let}\ x = e_1\ \mathsf{in}\ e_2 \rightsquigarrow v$. There are two possibilities. If this tree has been obtained by applying rule F_{13}, it will take the form:

$$(F_{13})\dfrac{\overset{\nabla_1}{\text{Env} \vdash e_1 \rightsquigarrow v_1} \quad \overset{\nabla_2}{(x, v_1) \oplus \text{Env} \vdash e_2 \rightsquigarrow v}}{\text{Env} \vdash \mathsf{let}\ x = e_1\ \mathsf{in}\ e_2 \rightsquigarrow v}(v_1 \in \mathbf{V}) \qquad [3.10]$$

and a tree for $\text{Env} \vdash ((\mathsf{fun}\ x \mapsto e_2)\, e_1) \rightsquigarrow v$ can be constructed easily from trees ∇_1 and ∇_2:

$$(F_{17})\dfrac{(F_{15})\dfrac{}{\text{Env} \vdash f \rightsquigarrow \langle x, e_2, \text{Env}\rangle} \quad \overset{\nabla_1}{\text{Env} \vdash e_1 \rightsquigarrow v_1} \quad \overset{\nabla_2}{(x, v_1) \oplus \text{Env} \vdash e_2 \rightsquigarrow v}}{\text{Env} \vdash (\underbrace{(\mathsf{fun}\ x \mapsto e_2)}_{f}\, e_1) \rightsquigarrow v}(v_1 \notin \mathbb{E}) \qquad [3.11]$$

Second, tree $\text{Env} \vdash \mathsf{let}\ x = e_1\ \mathsf{in}\ e_2 \rightsquigarrow v$ has been obtained using rule F_{14}:

$$(F_{14})\dfrac{\overset{\nabla}{\text{Env} \vdash e_1 \rightsquigarrow v}}{\text{Env} \vdash \mathsf{let}\ x = e_1\ \mathsf{in}\ e_2 \rightsquigarrow v}(v \in \mathbb{E}) \qquad [3.12]$$

Once again, a tree for $\text{Env} \vdash ((\mathsf{fun}\ x \mapsto e_2)\, e_1) \rightsquigarrow v$ can be constructed easily from ∇:

$$(F_{19})\dfrac{(F_{15})\dfrac{}{\text{Env} \vdash f \rightsquigarrow \langle x, e_2, \text{Env}\rangle} \quad \overset{\nabla}{\text{Env} \vdash e_1 \rightsquigarrow v}}{\text{Env} \vdash (\underbrace{(\mathsf{fun}\ x \mapsto e_2)}_{f}\, e_1) \rightsquigarrow v}(v \in \mathbb{E}) \qquad [3.13]$$

Reciprocally, a tree for $\text{Env} \vdash ((\mathsf{fun}\ x \mapsto e_2)\, e_1) \rightsquigarrow v$ can be used to construct a tree for $\text{Env} \vdash \mathsf{let}\ x = e_1\ \mathsf{in}\ e_2 \rightsquigarrow v$ since trees [3.10] and [3.12] can be obtained from trees [3.11] and [3.13], respectively.

3.5.2. *Equivalent environments*

As we said informally in section 3.1.4, the result of evaluating an expression e in an environment Env depends only on the values associated with the free variables of e by Env. In other terms, the evaluation result of e is the same in all environments, which associate the same values with the free variables of e. Giving a formal definition of this property requires the introduction of the notion of equivalent environments over a set of variables as follows:

DEFINITION 3.4.– *Equivalent environments: let $F \subseteq \mathbf{X}$ be a set of variables and Env_1 and Env_2 two environments:*

$$\text{Env}_1 \equiv_F \text{Env}_2 \Leftrightarrow \forall x \in F\ \text{Env}_1(x) = \text{Env}_2(x)$$

Obviously, given two subsets of $\mathbf{X}$, F_1 and F_2, if $F_1 \subseteq F_2$ and $\text{Env}_1 \equiv_{F_2} \text{Env}_2$, then $\text{Env}_1 \equiv_{F_1} \text{Env}_2$. It is now possible to express, formally, that for any expression e, the value of $[\![e]\!]_{\text{Env}}$ depends only on the values bound in Env to the free variables of e ($\mathcal{F}(e)$).

PROPOSITION 3.3.– For any expression $e \in \mathbf{Exp}_2$:

$$\begin{array}{l} \forall \text{Env}_1, \text{Env}_2 \in \mathbf{E} \\ \quad \text{Env}_1 \equiv_{\mathcal{F}(e)} \text{Env}_2 \\ \quad \Rightarrow \left(\begin{array}{l} [\![e]\!]_{\text{Env}_1} = [\![e]\!]_{\text{Env}_2} \in \mathbb{V} \setminus \mathbb{F} \\ \text{or } [\![e]\!]_{\text{Env}_1} = \langle x, e', \text{Env}'_1 \rangle \text{ and } [\![e]\!]_{\text{Env}_2} = \langle x, e', \text{Env}'_2 \rangle \\ \quad \text{and } \text{Env}'_1 \equiv_{\mathcal{F}(e') \setminus \{x\}} \text{Env}'_2 \\ \text{or } [\![e]\!]_{\text{Env}_1} = \langle f, x, e', \text{Env}'_1 \rangle \text{ and } [\![e]\!]_{\text{Env}_2} = \langle f, x, e', \text{Env}'_2 \rangle \\ \quad \text{and } \text{Env}'_1 \equiv_{\mathcal{F}(e') \setminus \{x\}} \text{Env}'_2 \end{array} \right) \end{array}$$

The proof of this proposition is long and complex. Exercise 3.5 establishes this property for a language containing no functional expressions. In this case, proof is obtained by induction over the syntactic structure of the expression. In cases where functional values are present, this induction is no longer sufficient, and the proof requires the use of theorems of the fixpoint theory.

3.6. Exercises

Exercise 3.1

Given an environment Env, compute:

$$[\![\text{fun } x \mapsto \text{let } y = 3 \text{ in } x + y]\!]_{\text{Env}} \text{ and } [\![\text{let } y = 3 \text{ in fun } x \mapsto x + y]\!]_{\text{Env}}$$

Exercise 3.2

Let Env be the following environment defined from an expression $e \in \mathbf{Exp_2}$ and an initial environment Env_0:

$$\text{Env} = (x, 4) \oplus (f, \langle y, e, (x, 3) \oplus \text{Env}_0 \rangle) \oplus (x, 3) \oplus \text{Env}_0$$

1) Give a condition for the expression e such that:

$$[\![\text{let } f = \text{fun } y \mapsto e \text{ in } (f\,x)]\!]_{\text{Env}} = [\![(f\,x)]\!]_{\text{Env}}$$

2) Compute $[\![\text{let } f = \text{fun } x \mapsto x + (f\,x) \text{ in } (f\,x)]\!]_{\text{Env}}$ when $e = y + x$.

Exercise 3.3

Consider the following expression e:

$$\text{let } x = 3 \text{ in } (\text{let } f = e_1 \text{ in } (\text{let } g = e_2 \text{ in } (\text{let } h = e_3 \text{ in } e_4)))$$

where:

e_1: fun $x \mapsto (\text{let } y = x \text{ in } y)$
e_2: let $y = x$ in fun $x \mapsto y$
e_3: fun $k \mapsto$ fun $x \mapsto (k\,x)$

In this exercise, the call-by-value strategy is applied to evaluate expressions.

1) Build a "partial" evaluation tree for the expression e in an environment Env_0 to determine the environment Env_4, which will be used to evaluate expression e_4.

2) Complete the tree for $e_4 = ((h\,f)\,2)$.

3) Complete the tree for $e_4 = ((h\,g)\,2)$.

Exercise 3.4

Consider the following expression e_0:

$$\text{let } x = 2 \text{ in } (\text{let } f = e_1 \text{ in } e)$$

where e_1 is the expression fun $y \mapsto y + x$. In this exercise, the call-by-name strategy is applied to evaluate expressions.

1) Build a "partial" evaluation tree for the expression e_0 in an environment Env_0 to determine the environment Env_2, which will be used to evaluate expression e.

2) Complete the tree for the following expression e: $\mathsf{let}\ x = 3\ \mathsf{in}\ (f\ x)$.

3) Complete the tree for the following expression e:

$$\mathsf{let}\ x = 3\ \mathsf{in}\ ((\mathsf{fun}\ y \mapsto y\ +\ x)\ x)$$

4) Complete the tree for the following expression e: $\mathsf{let}\ g = e_2\ \mathsf{in}\ ((g\ (x\ -\ 1))\ f)$, where e_2 is the expression:

$$\mathsf{rec}\ g\ z = \mathsf{fun}\ k \mapsto \mathsf{if}\ z\ =\ 0\ \mathsf{then}\ (k\ x)\ \mathsf{else}\ ((g\ (z\ -\ 1))\ k)$$

Exercise 3.5

In this exercise, we consider the subset of $\mathbf{Exp}_2$ containing only simple arithmetic expressions, retaining only constructs (1), (2), (4) and (6) from Box 3.1 with $\mathbf{K} = \mathbb{Z}$ and op $\in \{+, -, *\}$. The syntax of this language, denoted as E_A, is thus:

$$e ::= k \mid x \mid e_1 + e_2 \mid e_1 - e_2 \mid e_1 * e_2 \mid \mathsf{let}\ x = e_1\ \mathsf{in}\ e_2$$

For the sake of simplicity, we suppose here that the environments used to evaluate any expression e always contain a binding for each free variable occurring in e; thus, we do not need to account for cases of error. Hence, the operational semantics for evaluating the expressions in E_A can be defined from the set of values $\mathbb{V} = \mathbf{V} = \mathbb{Z}$ (since $\mathbb{E} = \emptyset$), and by only considering rules F_1, F_4, F_{11} and F_{13} from Box 3.2 and Box 3.3. Prove that the result of the evaluation of an expression $a \in E_A$ in an environment Env only depends on the value associated with the free variables of a in environment Env.

$$\begin{array}{l}\forall a \in E_A\ \forall \mathrm{Env}_1, \mathrm{Env}_2 \in \mathbf{E}\ \ \mathrm{Env}_1 =_{\mathcal{F}(a)} \mathrm{Env}_2 \\ \Rightarrow (\mathrm{Env}_1 \vdash a \rightsquigarrow n \Leftrightarrow \mathrm{Env}_2 \vdash a \rightsquigarrow n)\end{array}$$

Exercise 3.6

In this exercise, we consider the language E_A defined in exercise 3.5 without the *let-in* construct. Replacing a variable $x \in \mathbf{X}$ by an expression $e' \in E_A$ in an expression $e \in E_A$ leads to an expression inductively defined by:

$$e[x \leftarrow e'] = \begin{cases} k & \text{if } e = k \in \mathbb{Z} \\ e' & \text{if } e = x \in \mathbf{X} \\ y & \text{if } e = y \in \mathbf{X} \text{ et } y \neq x \\ e_1[x \leftarrow e']\,\mathrm{op}\,e_2[x \leftarrow e'] & \text{if } e = e_1\,\mathrm{op}\,e_2 \end{cases}$$

Prove that:

$$\forall e, e' \in E_A \ \forall x \in \mathbf{X} \ \forall \text{Env} \in \mathbf{E} \ [\![e[x \leftarrow e']]\!]_{\text{Env}} = [\![e]\!]_{(x,[\![e']\!]_{\text{Env}})\oplus\text{Env}}$$

Exercise 3.7

Small-step operational semantics. In this exercise, we consider the language E_A defined in exercise 3.5 without the *let-in* construct (the operational semantics is defined using only rules F_1, F_4 and F_{11} from Box 3.2 and Box 3.3). A small-step operational semantics can be defined to evaluate the expressions in E_A. The small-step semantics allows us to take into account the steps leading to the result of a computation at a level finer than that obtained by using the big-step semantics. A configuration is defined as a pair $\langle e, \text{Env}\rangle$, where $e \in E_A$ and $\text{Env} \in \mathbf{E}$. We define a configuration transition relation using the inference system presented in Box 3.5. Intuitively, $\langle e, \text{Env}\rangle \hookrightarrow \langle e', \text{Env}\rangle$ means that the evaluation of expression e in environment Env leads us to evaluate the "simpler" expression e' in the same environment. For example:

$$\langle 3 + (2 * 8), \text{Env}\rangle \hookrightarrow \langle 3 + 16, \text{Env}\rangle$$

$$(AS_1)\frac{}{\langle x, \text{Env}\rangle \hookrightarrow \langle \text{Env}(x), \text{Env}\rangle}(x \in \mathbf{X}) \qquad n_1, n_2 \in \mathbb{Z}$$

$$(AS_2)\frac{}{\langle n_1 + n_2, \text{Env}\rangle \hookrightarrow \langle n, \text{Env}\rangle}(n = n_1 + n_2)$$

$$(AS_3)\frac{}{\langle n_1 * n_2, \text{Env}\rangle \hookrightarrow \langle n, \text{Env}\rangle}(n = n_1 * n_2)$$

$$(AS_4)\frac{}{\langle n_1 - n_2, \text{Env}\rangle \hookrightarrow \langle n, \text{Env}\rangle}(n = n_1 - n_2)$$

$$(AS_5)\frac{\langle e_1, \text{Env}\rangle \hookrightarrow \langle e_1', \text{Env}\rangle}{\langle e_1 + e_2, \text{Env}\rangle \hookrightarrow \langle e_1' + e_2, \text{Env}\rangle} \quad (AS_6)\frac{\langle e_2, \text{Env}\rangle \hookrightarrow \langle e_2', \text{Env}\rangle}{\langle n_1 + e_2, \text{Env}\rangle \hookrightarrow \langle n_1 + e_2', \text{Env}\rangle}$$

$$(AS_7)\frac{\langle e_1, \text{Env}\rangle \hookrightarrow \langle e_1', \text{Env}\rangle}{\langle e_1 * e_2, \text{Env}\rangle \hookrightarrow \langle e_1' * e_2, \text{Env}\rangle} \quad (AS_8)\frac{\langle e_2, \text{Env}\rangle \hookrightarrow \langle e_2', \text{Env}\rangle}{\langle n_1 * e_2, \text{Env}\rangle \hookrightarrow \langle n_1 * e_2', \text{Env}\rangle}$$

$$(AS_9)\frac{\langle e_1, \text{Env}\rangle \hookrightarrow \langle e_1', \text{Env}\rangle}{\langle e_1 - e_2, \text{Env}\rangle \hookrightarrow \langle e_1' - e_2, \text{Env}\rangle} \quad (AS_{10})\frac{\langle e_2, \text{Env}\rangle \hookrightarrow \langle e_2', \text{Env}\rangle}{\langle n_1 - e_2, \text{Env}\rangle \hookrightarrow \langle n_1 - e_2', \text{Env}\rangle}$$

Box 3.5. *Small-step operational semantics*

A transition sequence is a potentially infinite sequence of the form:

$$\langle e_0, \text{Env}\rangle \hookrightarrow \langle e_1, \text{Env}\rangle \hookrightarrow \langle e_2, \text{Env}\rangle \hookrightarrow \cdots$$

such that for all $i \geq 0$, there exists an inference tree for $\langle e_i, \text{Env}\rangle \hookrightarrow \langle e_{i+1}, \text{Env}\rangle$. A configuration is said to be *terminal* if it takes the form $\langle n, \text{Env}\rangle$ where $n \in \mathbb{Z}$

(as, in this case, the result of the evaluation is available). Relation $\stackrel{\star}{\hookrightarrow}$ is then defined by $\langle e, \text{Env}\rangle \stackrel{\star}{\hookrightarrow} n$ if, and only if, there is a transition sequence of finite positive or zero length:

$$\langle e_0, \text{Env}\rangle \hookrightarrow \langle e_1, \text{Env}\rangle \hookrightarrow \langle e_2, \text{Env}\rangle \hookrightarrow \cdots \hookrightarrow \langle e_m, \text{Env}\rangle$$

such that configuration $\langle e_m, \text{Env}\rangle$ is terminal with $e_m = n$.

1) Give the transition sequence used to describe the evaluation of the expression

$$(z + (x - y)) * (x + 7) \qquad [3.14]$$

in an environment Env such that $\text{Env}(x) = 1$, $\text{Env}(y) = 3$ and $\text{Env}(z) = 2$.

2) Since an expression is "simplified" at each step of a transition sequence, we prove that all transition sequences are of finite length.

a) From the well-founded order relation[2] $\leq$ over the set of natural integers $\mathbb{N}$, define a well-founded order relation $\preceq$ over $\mathbb{N} \times \mathbb{N}$.

b) Given an expression $e \in E_A$, let $\mathcal{N}_V(e)$ be the number of occurrences of variables in e and $\mathcal{N}_O(e)$ the number of occurrences of operators appearing in e. Prove that, if $\langle e, \text{Env}\rangle \hookrightarrow \langle e', \text{Env}\rangle$, then:

$$(\mathcal{N}_V(e'), \mathcal{N}_O(e')) \prec (\mathcal{N}_V(e), \mathcal{N}_O(e))$$

where $\prec$ is the strict order associated with $\preceq$.

c) From this result, deduce that all transition sequences are finite.

d) Is this result still true when the rule below is added? Explain your answer.

$$(AS_{11})\frac{}{\langle e_1 + e_2, \text{Env}\rangle \hookrightarrow \langle e_2 + e_1, \text{Env}\rangle}$$

3) Prove that if $e \vdash \text{Env} \rightsquigarrow n$, then $\langle e, \text{Env}\rangle \stackrel{\star}{\hookrightarrow} n$.

4) Prove that if $\langle e_1, \text{Env}\rangle \hookrightarrow \langle e_2, \text{Env}\rangle$ and $\text{Env} \vdash e_2 \rightsquigarrow v$, then $\text{Env} \vdash e_1 \rightsquigarrow v$.

5) Using your result from the previous question, prove that if $\langle e, \text{Env}\rangle \stackrel{\star}{\hookrightarrow} n$, then $e \vdash \text{Env} \rightsquigarrow n$.

2 An order relation $\leq$ over a set E is well-founded if there is no strictly decreasing infinite sequence $e_1 > e_2 > \cdots$ of elements of E.

6) Now we want to describe the evaluation of expression [3.14] in the same environment Env using the following computation sequence:

$$
\begin{array}{l}
\quad\langle(z+(x-y))*(x+7),\text{Env}\rangle \hookrightarrow \langle(z+(x-y))*(1+7),\text{Env}\rangle \\
\hookrightarrow \langle(z+(x-y))*8,\text{Env}\rangle \hookrightarrow \langle(z+(x-3))*8,\text{Env}\rangle \\
\hookrightarrow \langle(z+(1-3))*8,\text{Env}\rangle \hookrightarrow \langle(z+-2)*8,\text{Env}\rangle \\
\hookrightarrow \langle(2+-2)*8,\text{Env}\rangle \hookrightarrow \langle 0*8,\text{Env}\rangle \hookrightarrow \langle 0,\text{Env}\rangle
\end{array}
$$

Propose a new inference system to define the relation $\hookrightarrow$ such that for each step in this transition sequence there is an inference tree using this system.

4

Semantics of Imperative Features

This chapter presents the three classical semantic approaches to the most common imperative features of programming languages, as they are usually treated (see [NIE 07, WIN 93]).

4.1. Syntax of a kernel of an imperative language

We consider here a language, $\mathbf{Lang}_3$, corresponding to the kernel of most imperative languages. It allows explicit handling of *references* of locations, i.e. memory addresses. This language is built from two languages: $\mathbf{Def}_3$ for definitions and $\mathbf{Exp}_3$ for expressions. Both are variations of languages $\mathbf{Def}_1$ and $\mathbf{Exp}_1$, introduced in Chapter 2. The language $\mathbf{Def}_3$, defined in Table 4.1, contains a single construct that introduces a binding into an execution environment.

$d ::= \mathsf{let}\ x = e;$ definition of a non-mutable variable $(x \in \mathbf{X},\ e \in \mathbf{Exp}_3)$

Table 4.1. *Language of defintions* $\mathbf{Def}_3$

The language $\mathbf{Exp}_3$, used for expressions, aims at manipulating values and modifying the memory by allocating locations for new values. The syntax of $\mathbf{Exp}_3$ is presented in Table 4.2.

Thus, an expression $e \in \mathbf{Exp}_3$ is either:

– a constant $k \in \mathbf{K} = \mathbb{Z} \cup \mathbb{B}$;

– an identifier $x \in \mathbf{X}$;

– an expression obtained by applying a unary (or binary) operator to an expression (or two expressions);

– an allocation $\uparrow e$ which reserves a location, stores the result of the evaluation of e at this location, and returns the reference of this location $r \in \mathbf{R}$;

– dereferencing $\downarrow e$, which evaluates the expression e and, if the value of e is a reference, returns the value stored at this location.

$e ::=$	k	Constant $(k \in \mathbf{K})$	(1)
$\mid$	x	Variable $(x \in \mathbf{X})$	(2)
$\mid$	op e_1	Application of a primitive unary operator	(3)
$\mid$	e_1 op e_2	Application of a primitive binary operator	(4)
$\mid$	$\uparrow e$	Indirection, memory allocation	(5)
$\mid$	$\downarrow e$	Dereferencing	(6)

Table 4.2. *Language of expressions* $\mathbf{Exp}_3$

We shall not describe the possible primitive operators here; depending on the chosen language, these operators may handle integers, booleans or references (pointer arithmetics).

The set of constants is defined by the union of integers, booleans and references. Using integers to represent references, the following definitions are used to represent $\mathbf{K} \cup \mathbf{R}$.

Python
```
class CInt4:
    def __init__(self,cst_int):
        self.cst_int = cst_int
class CBool4:
    def __init__(self,cst_bool):
        self.cst_bool = cst_bool
class CRef4:
    def __init__(self,cst_adr):
        self.cst_adr = cst_adr
```

OCaml
```
type const4 = CInt4 of int | CBool4 of bool | CRef4 of int
```

For our purposes, we shall use the same classical operators introduced in Chapter 3.

Python
```
class Not4:
    pass
class Minus4:
    pass
class Div4:
    pass
class Or4:
    pass
class Leq4:
    pass

class Plus4:
    pass
class Mult4:
    pass
class And4:
    pass
class Eq4:
    pass
```

OCaml
```
type unary_op4 = Not4
type bin_op4 = Plus4 | Minus4 | Mult4 | Div4 | And4 | Or4 | Eq4 | Leq4
```

Expressions of $\mathbf{Exp}_3$ are defined as follows:

Python
```
class Cste4:
    def __init__(self,cste):
        self.cste = cste
class Var4:
    def __init__(self,symb):
        self.symb = symb
class U_Op4:
    def __init__(self,op,exp1):
        self.op = op
        self.exp1 = exp1
class B_Op4:
    def __init__(self,op,exp1,exp2):
        self.op = op
        self.exp1 = exp1
        self.exp2 = exp2
class Ref4:
    def __init__(self,exp):
        self.exp = exp
class Deref4:
    def __init__(self,exp):
        self.exp = exp
```

OCaml
```
type 'a exp4 =
  | Cste4 of const4                 | Var4 of 'a
  | U_Op4 of unary_op4 * 'a exp4 | B_Op4 of bin_op4 * 'a exp4 * 'a exp4
  | Ref4 of 'a exp4                 | Deref4 of 'a exp4
```

Finally, language $\mathbf{Lang}_3$ includes assignment (already presented for language $\mathbf{Lang}_1$ in Chapter 2) along with the classical control structures (sequence, conditional, loop), occurring in most imperative languages. Syntax of $\mathbf{Lang}_3$ is defined in Table 4.3.

$c ::=$	$x := e$	Assignment $(x \in \mathbf{X},\ e \in \mathbf{Exp}_3)$	(1)
|	skip		(2)
|	$c_1 \ ; \ c_2$	Sequence	(3)
|	if e then c_1 else c_2	Conditional $(e \in \mathbf{Exp}_3)$	(4)
|	while e do c_1	Loop $(e \in \mathbf{Exp}_3)$	(5)

Table 4.3. *Language of commands* $\mathbf{Lang}_3$

At this step, the language of definitions $\mathbf{Def}_3$ and the language of programs $\mathbf{Lang}_3$ can be defined separately. However, in section 4.6, $\mathbf{Def}_3$ will be extended in order to define procedures. The body of a procedure is a program in $\mathbf{Lang}_3$, and $\mathbf{Lang}_3$ will be extended with local definitions of variables introduced by expressions in $\mathbf{Def}_3$. Languages $\mathbf{Def}_3$ and $\mathbf{Lang}_3$ are thus defined in a mutually recursive way as follows (these definitions will be extended in section 4.6).

Python
```
class Let_def4:
    def __init__(self,symb,exp):
        self.symb = symb
        self.exp = exp
class ICond4:
    def __init__(self,exp,pgm1,pgm2):
        self.exp = exp
        self.pgm1 = pgm1
```

```
class IAffect4:
    def __init__(self,symb,exp):
        self.symb = symb
        self.exp = exp
class ISkip4:
    pass
class ISeq4:
    def __init__(self,pgm1,pgm2):
        self.pgm1 = pgm1
        self.pgm2 = pgm2
```

```
        self.pgm2 = pgm2
class IWhile4:
    def __init__(self,exp,pgm):
        self.exp = exp
        self.pgm = pgm
```

```
OCaml
type 'a def4 =
  | Let_def4 of 'a * 'a exp4          | ... procedure definitions
and 'a pgm4 =
  | IAffect4 of 'a * 'a exp4          | ISkip4
  | ISeq4 of 'a pgm4 * 'a pgm4        | ICond4 of 'a exp4 * 'a pgm4 * 'a pgm4
  | IWhile4 of 'a exp4 * 'a pgm4 | ... blocks with local variables
  | ... procedure calls
```

EXAMPLE 4.1.– Program for computing GCD: the fragment of program represented in the abstract syntax tree below computes the GCD of two natural integers already stored in the memory at locations x and y:

$$(5)\dfrac{(4)\dfrac{(1)\dfrac{}{y :=\downarrow y-\downarrow x} \quad (1)\dfrac{}{x :=\downarrow x-\downarrow y}}{\text{if} \downarrow x \leq\downarrow y \text{ then } y :=\downarrow y-\downarrow x \text{ else } x :=\downarrow x-\downarrow y}}{\text{while not}(\downarrow x =\downarrow y) \text{ do if} \downarrow x \leq\downarrow y \text{ then } y :=\downarrow y-\downarrow x \text{ else } x :=\downarrow x-\downarrow y} \qquad [4.1]$$

This program is written as follows:

```
Python
ex_pgm4_1 = IWhile4(U_Op4(Not4(),
                    B_Op4(Eq4(),Deref4(Var4("x")),Deref4(Var4("y")))),
                    ICond4(B_Op4(Leq4(),Deref4(Var4("x")),Deref4(Var4("y"))),
                           IAffect4("y",B_Op4(Minus4(),Deref4(Var4("y")),
                                              Deref4(Var4("x")))),
                           IAffect4("x",B_Op4(Minus4(),Deref4(Var4("x")),
                                              Deref4(Var4("y"))))))
```

```
OCaml
let ex_pgm4_1 =
  IWhile4
    (U_Op4 (Not4, B_Op4 (Eq4, Deref4 (Var4 "x"), Deref4 (Var4 "y"))),
     ICond4 (B_Op4 (Leq4, Deref4 (Var4 "x"), Deref4 (Var4 "y")),
         IAffect4 ("y", B_Op4 (Minus4, Deref4 (Var4 "y"),
                     Deref4 (Var4 "x"))),
         IAffect4 ("x", B_Op4 (Minus4, Deref4 (Var4 "x"),
                     Deref4 (Var4 "y")))))
```

Constructs used to define functions and procedures will be introduced later. However, the following theorem characterizes the expressiveness of the language $\mathbf{Lang}_3$ defined at this stage.

THEOREM 4.1.– Böhm and Jacopini [BOE 66]: Any algorithm can be expressed as a composition of three control structures: sequence, conditional instruction and iteration.

Defining a semantics for the language $\mathbf{Lang}_3$ allows us to have a precise description of the memory modifications resulting from assignments and control instructions. Throughout this chapter, we consider that all expressions and programs over which a semantics is defined are syntactically correct and well-typed (see Chapter 5): evaluation of an expression or execution of a program never generates a typing error.

4.2. Evaluation of expressions

As in Chapter 2, an expression in $\mathbf{Exp}_3$ is evaluated in a state made up of an execution environment $\text{Env} \in \mathbf{E}$ and a memory $\text{Mem} \in \mathbf{M}$. However, the language of expressions $\mathbf{Exp}_3$ contains an operator $\uparrow$, which allocates memory, stores the value of its argument in this new location and returns its reference. Thus, while evaluating an expression never modifies the execution environment, it may trigger a modification of the memory state (known as a *side effect* of the evaluation). Evaluating an expression returns a value that may be an integer, a boolean or a reference, hence $\mathbf{V} = \mathbb{Z} \cup \mathbb{B} \cup \mathbf{R}$. Since we are only considering well-typed expressions (i.e. expressions for which evaluation does not result in a typing error), the only possible errors are definition errors Err_U (access to an identifier which is not defined in the environment) and execution errors Err_E (for example in the case of a division by 0). Hence, $\mathbb{E} = \{\text{Err}_\text{U}, \text{Err}_\text{E}\}$ and $\mathbb{V} = \mathbf{V} \cup \mathbb{E}$ is the set of all possible results of expression evaluation.

We thus define an evaluation function:

$$[\![\_]\!]_{\_}^{\_} : (\mathbf{E} \times \mathbf{M} \times \mathbf{Exp}_3) \to (\mathbb{V} \times \mathbf{M})$$

such that $[\![e]\!]_{\text{Env}}^{\text{Mem}} = (v, \text{Mem}')$ means that v is the value obtained by evaluating the expression e in the state (Env, Mem) and Mem' is the memory issued from Mem when computing v. This function is defined in Table 4.4.

We suppose that each operator op has a semantics defining how to interpret it by an operator $\text{op}_\mathbf{V}$ over the domain of values. We also suppose that during allocation (operator $\uparrow$) an external mechanism provides locations in the memory and returns their references. For expressions of the form $e_1 \,\text{op}\, e_2$, we have chosen to evaluate expression e_1 first, then – depending on the result – to evaluate e_2. Furthermore, when an error is generated during expression evaluation, the returned value corresponds to this error, and the memory is that produced by the last error-free evaluation.

e	$[\![e]\!]^{\text{Mem}}_{\text{Env}} = (v, \text{Mem}')$	
k	$v = k$ $\text{Mem}' = \text{Mem}$	$(k \in \mathbf{K})$
x	$v = \text{Env}(x)$ $\text{Mem}' = \text{Mem}$	if $x \in \text{dom}(\text{Env})$ $\quad (x \in \mathbf{X})$
	$v = \text{Err}_{\text{U}}$ $\text{Mem}' = \text{Mem}$	if $x \notin \text{dom}(\text{Env})$ $\quad (x \in \mathbf{X})$
op e_1	$v = \text{op}_{\mathbf{V}}\ v_1$ $\text{Mem}' = \text{Mem}_1$	if $[\![e_1]\!]^{\text{Mem}}_{\text{Env}} = (v_1, \text{Mem}_1)$ and $v_1 \in \mathbf{V}$
	$v = v_1$ $\text{Mem}' = \text{Mem}$	if $[\![e_1]\!]^{\text{Mem}}_{\text{Env}} = (v_1, \text{Mem}_1)$ and $v_1 \in \mathbb{E}$
e_1 op e_2	$v = v_1\ \text{op}_{\mathbf{V}}\ v_2$ $\text{Mem}' = \text{Mem}_2$	if $[\![e_1]\!]^{\text{Mem}}_{\text{Env}} = (v_1, \text{Mem}_1)$ and $v_1 \in \mathbf{V}$ and $[\![e_2]\!]^{\text{Mem}_1}_{\text{Env}} = (v_2, \text{Mem}_2)$ and $v_2 \in \mathbf{V}$
	$v = v_1$ $\text{Mem}' = \text{Mem}$	if $[\![e_1]\!]^{\text{Mem}}_{\text{Env}} = (v_1, \text{Mem}_1)$ and $v_1 \in \mathbb{E}$
	$v = v_2$ $\text{Mem}' = \text{Mem}_1$	if $[\![e_1]\!]^{\text{Mem}}_{\text{Env}} = (v_1, \text{Mem}_1)$ and $v_1 \in \mathbf{V}$ and $[\![e_2]\!]^{\text{Mem}_1}_{\text{Env}} = (v_2, \text{Mem}_2)$ and $v_2 \in \mathbb{E}$
$\uparrow e_1$	$v = r$ $\text{Mem}' = \text{Mem}_1[r := v_1]$	if $[\![e_1]\!]^{\text{Mem}}_{\text{Env}} = (v_1, \text{Mem}_1)$ and $v_1 \in \mathbf{V}$
	$v = v_1$ $\text{Mem}' = \text{Mem}$	if $[\![e_1]\!]^{\text{Mem}}_{\text{Env}} = (v_1, \text{Mem}_1)$ and $v_1 \in \mathbb{E}$
$\downarrow e_1$	$v = \text{Mem}_1(v_1)$ $\text{Mem}' = \text{Mem}_1$	if $[\![e_1]\!]^{\text{Mem}}_{\text{Env}} = (v_1, \text{Mem}_1)$ and $v_1 \in \mathbf{R}$
	$v = v_1$ $\text{Mem}' = \text{Mem}$	if $[\![e_1]\!]^{\text{Mem}}_{\text{Env}} = (v_1, \text{Mem}_1)$ and $v_1 \in \mathbb{E}$

Table 4.4. *Evaluation of expressions in* $\mathbf{Exp}_3$

In this chapter, we have chosen to use the exception mechanism (see Chapter 8) instead of introducing special values for errors. If an error is triggered by evaluating an expression, the exception `Eval_Error` is raised and the evaluation stops. The set of values $\mathbf{V}$ thus includes the constants from $\mathbf{K}$ and the references from $\mathbf{R}$. Other values will be introduced later, in addition to the definitions given below.

Python
```
class VCste4:
    def __init__(self,cst):
        self.cst = cst
```

OCaml
```
type 'a valeurs4 = | VCste4 of const4 | ...
```

Since references are represented by integers, a simple way of carrying out memory allocation is to increment the last created reference by 1. This gives us the new allocated reference r, and the binding (r, v) can be added at the head of the list representing the memory. The function below carries out this operation and returns a pair made up of the new reference and the list representing the new memory.

Python
```
def add_mem(mem,v):
    if len(mem) == 0: return (0,[(0,v)])
    else:
        n1,v1 = mem[0]
        return (n1+1,[(n1+1,v)]+mem)
```

OCaml
```
let add_mem mem v = match mem with
  | [] -> (0, [(0, v)])
  | (n1, v1) :: mem1 -> (n1 + 1, ((n1 + 1), v) :: mem)
val add_mem : (int * 'a) list -> 'a -> int * (int * 'a) list
```

Interpretation of operators is defined as in Chapter 3. Error handling is replaced by the raising of the exception `Eval_Error`.

Python
```
def apply4_unary_op(op,v):
    if isinstance(op,Not4) and isinstance(v,VCste4) \
       and isinstance(v.cst,CBool4):
        return VCste4(CBool4(not v.cst.cst_bool))
    raise ValueError

def apply4_binary_op(op,v1,v2):
    if isinstance(v1,VCste4) and isinstance(v2,VCste4):
        vc1 = v1.cst
        vc2 = v2.cst
        if isinstance(op,Plus4):
            if isinstance(vc1,CInt4) and isinstance(vc2,CInt4):
                return VCste4(CInt4(vc1.cst_int + vc2.cst_int))
            raise ValueError
        if isinstance(op,Minus4):
            if isinstance(vc1,CInt4) and isinstance(vc2,CInt4):
                return VCste4(CInt4(vc1.cst_int - vc2.cst_int))
            raise ValueError
        if isinstance(op,Mult4):
            if isinstance(vc1,CInt4) and isinstance(vc2,CInt4):
                return VCste4(CInt4(vc1.cst_int * vc2.cst_int))
            raise ValueError
        if isinstance(op,Div4):
            if isinstance(vc1,CInt4) and isinstance(vc2,CInt4)
               and vc2.cst_int != 0:
                return VCste4(CInt4(vc1.cst_int / vc2.cst_int))
            raise ValueError
```

```
        if isinstance(op,And4):
            if isinstance(vc1,CBool4) and isinstance(vc2,CBool4):
                return VCste4(CBool4(vc1.cst_bool and vc2.cst_bool))
            raise ValueError
        if isinstance(op,Or4):
            if isinstance(vc1,CBool4) and isinstance(vc2,CBool4):
                return VCste4(CBool4(vc1.cst_bool or vc2.cst_bool))
            raise ValueError
        if isinstance(op,Eq4):
            if isinstance(vc1,CInt4) and isinstance(vc2,CInt4):
                return VCste4(CBool4(vc1.cst_int == vc2.cst_int))
            raise ValueError
        if isinstance(op,Leq4):
            if isinstance(vc1,CInt4) and isinstance(vc2,CInt4):
                return VCste4(CBool4(vc1.cst_int <= vc2.cst_int))
            raise ValueError
        raise ValueError
    raise ValueError
```

OCaml
```
exception Eval_Error

let apply4_unary_op op v = match op with
  | Not4 -> (match v with
      | VCste4 (CBool4 b) -> VCste4 (CBool4 (not b))
      | _ -> raise Eval_Error)
val apply4_unary_op : unary_op4 -> 'a valeurs4 -> 'b valeurs4

let apply4_binary_op op v1 v2 =
  match op with
  | Plus4 -> (match (v1, v2) with
      | (VCste4 (CInt4 n1), VCste4 (CInt4 n2)) -> VCste4 (CInt4 (n1 + n2))
      | _ -> raise Eval_Error)
  | Minus4 -> (match (v1, v2) with
      | (VCste4 (CInt4 n1), VCste4 (CInt4 n2)) -> VCste4 (CInt4 (n1 - n2))
      | _ -> raise Eval_Error)
  | Mult4 -> (match (v1, v2) with
      | (VCste4 (CInt4 n1), VCste4 (CInt4 n2)) -> VCste4 (CInt4 (n1 * n2))
      | _ -> raise Eval_Error)
  | Div4 -> (match (v1, v2) with
      | (VCste4 (CInt4 n1), VCste4 (CInt4 n2)) ->
          if n2 = 0 then raise Eval_Error else VCste4 (CInt4 (n1 / n2))
      | _ -> raise Eval_Error)
  | And4 -> (match (v1, v2) with
      | (VCste4 (CBool4 b1), VCste4 (CBool4 b2)) -> VCste4 (CBool4 (b1 && b2))
      | _ -> raise Eval_Error)
  | Or4 -> (match (v1, v2) with
      | (VCste4 (CBool4 b1), VCste4 (CBool4 b2)) -> VCste4 (CBool4 (b1 || b2))
      | _ -> raise Eval_Error)
```

```
  | Eq4 -> (match (v1, v2) with
    | (VCste4 (CInt4 n1), VCste4 (CInt4 n2)) -> VCste4 (CBool4 (n1 = n2))
    | _ -> raise Eval_Error)
  | Leq4 -> (match (v1, v2) with
      | (VCste4 (CInt4 n1), VCste4 (CInt4 n2)) -> VCste4 (CBool4 (n1 <= n2))
      | _ -> raise Eval_Error)
val apply4_binary_op : bin_op4 -> 'a valeurs4 -> 'b valeurs4 -> 'c valeurs4
```

It is now possible to define the function that evaluates an expression. Throughout this chapter, a state will be represented by an environment binding some values of $\mathbf{V}$ to variable symbols, and by a memory binding constants in $\mathbf{K} \cup \mathbf{R}$ to references (represented by integers).

Python
```
def eval_exp4(env,mem,e):
    if isinstance(e,Cste4): return (VCste4(e.cste),mem)
    if isinstance(e,Var4):
        x = valeur_de(env,e.symb)
        if x is None: raise ValueError
        return (x,mem)
    if isinstance(e,U_Op4):
        v,memv = eval_exp4(env,mem,e.exp1)
        return (apply4_unary_op(e.op,v),memv)
    if isinstance(e,B_Op4):
        v1,mem1 = eval_exp4(env,mem,e.exp1)
        v2,mem2 = eval_exp4(env,mem1,e.exp2)
        return (apply4_binary_op(e.op,v1,v2),mem2)
    if isinstance(e,Ref4):
        v,mem1 = eval_exp4(env,mem,e.exp)
        r,mem2 = add_mem(mem1,v.cst)
        return (VCste4(CRef4(r)),mem2)
    if isinstance(e,Deref4):
        v1,mem1 = eval_exp4(env,mem,e.exp)
        if isinstance(v1,VCste4) and isinstance(v1.cst,CRef4):
            v2 = valeur_ref(mem1,v1.cst.cst_adr)
            if v2 is None: raise ValueError
            else: return (VCste4(v2),mem1)
            raise ValueError
        raise ValueError
    raise ValueError
```

OCaml
```
let rec eval_exp4 env mem e = match e with
  | Cste4 k -> (VCste4 k, mem)
  | Var4 x -> (
      match valeur_de env x with Some(v) -> (v, mem) | _ -> raise Eval_Error)
  | U_Op4 (op, e1) -> (match eval_exp4 env mem e1 with
      | (VCste4 v, mem1) -> ((apply4_unary_op op (VCste4 v)), mem1)
      | _ -> raise Eval_Error)
  | B_Op4(op,e1,e2) -> (match eval_exp4 env mem e1 with
      | (VCste4 v1, mem1) -> (match eval_exp4 env mem1 e2 with
          | (VCste4 v2, mem2) ->
              ((apply4_binary_op op (VCste4 v1) (VCste4 v2)), mem2)
```

```
        | _ -> raise Eval_Error)
      | _ -> raise Eval_Error)
  | Ref4 e1 -> (match eval_exp4 env mem e1 with
      | (VCste4 v, mem1) ->
        let (a, mem2) = (add_mem mem1 v) in (VCste4 (CRef4 a), mem2)
      | _ -> raise Eval_Error)
  | Deref4 e1 -> (match eval_exp4 env mem e1 with
      | (VCste4 (CRef4 a), mem1) -> (
        match valeur_ref mem1 a with
        | Some v -> (VCste4 v, mem1) | _ -> raise Eval_Error)
      | (VCste4 _, mem1) -> raise Eval_Error
      | _ -> raise Eval_Error)
val eval_exp4 : ('a * 'b valeurs4) list -> (int * const4) list -> 'a exp4
               -> 'b valeurs4 * (int * const4) list
```

4.3. Evaluation of definitions

Definitions introduced with language $\mathbf{Def}_3$ only modify execution environments. Their semantics is defined by:

$$(\text{Env}, \text{Mem}) \xrightarrow{\text{let } x = e;}_{\mathbf{Def}_3} ((x, v) \oplus \text{Env}, \text{Mem}') \quad \text{if} \quad [\![e]\!]_{\text{Env}}^{\text{Mem}} = (v, \text{Mem}')$$

This definition adds the binding (x, v) to the environment Env and sets the memory to Mem' when $[\![e]\!]_{\text{Env}}^{\text{Mem}} = (v, \text{Mem}')$. Using a finite sequence of definitions $\mathbf{d} = [d_1, \cdots, d_n]$ and an initial state $(\text{Env}_0, \text{Mem}_0)$, this relation computes the state $(\text{Env}_n, \text{Mem}_n)$:

$$(\text{Env}_0, \text{Mem}_0) \xrightarrow{d_1}_{\mathbf{Def}_3} (\text{Env}_1, \text{Mem}_1) \xrightarrow{d_2}_{\mathbf{Def}_3} \cdots \xrightarrow{d_n}_{\mathbf{Def}_3} (\text{Env}_n, \text{Mem}_n)$$

This sequence of transitions is denoted by: $(\text{Env}_0, \text{Mem}_0) \xrightarrow{\mathbf{d}}{}^{\star}_{\mathbf{Def}_3} (\text{Env}_n, \text{Mem}_n)$.

State transformations generated by a definition, respectively, a list of definitions, are defined as follows. These definitions will be completed further as the language $\mathbf{Def}_3$ is extended.

Python
```python
def trans_def4(e,d):
    env,mem = e
    if isinstance(d,Let_def4):
        v,mem1 = eval_exp4(env,mem,d.exp)
        return (ajout_liaison_env(env,d.symb,v),mem1)
    if ...
    raise ValueError

def trans_def4_exec(e,ld):
    if len(ld) == 0: return e
    else: return trans_def4_exec(trans_def4(e,ld[0]),ld[1:])
```

OCaml
```
let trans_def4 (env, mem) d = match d with
  | Let_def4 (x, e) -> (match eval_exp4 env mem e with
        (VCste4(v),mem1) -> ((ajout_liaison_env env x v),mem1))
  | ...
val trans_def4 : ('a * 'a valeurs4) list * (int * const4) list -> 'a def4
                 -> ('a * 'a valeurs4) list * (int * const4) list

let trans_def4_exec (env, mem) ld = (List.fold_left trans_def4 (env, mem) ld)
val trans_def4_exec : ('a * 'a valeurs4) list * (int * const4) list
  -> 'a def4 list -> ('a * 'a valeurs4) list * (int * const4) list
```

Unlike the language of definitions $\mathbf{Def}_1$ presented in Chapter 2, $\mathbf{Def}_3$ includes no commands of the form var $x = e$; to define mutable variables. From a semantic perspective, however, the same effect can be obtained using the $\uparrow$ operator. Considering an expression e, of which the evaluation produces a value v and does not modify the memory, i.e. such that $[\![e]\!]_{\text{Env}}^{\text{Mem}} = (v, \text{Mem})$, then we have $[\![\uparrow e]\!]_{\text{Env}}^{\text{Mem}} = (r, \text{Mem}[r := v])$, and hence:

$$(\text{Env}, \text{Mem}) \xrightarrow{\mathsf{var}\ x = e;}_{\mathbf{Def}_1} ((x, r) \oplus \text{Env}, \text{Mem}[r := v])$$
$$(\text{Env}, \text{Mem}) \xrightarrow{\mathsf{let}\ x =\uparrow e;}_{\mathbf{Def}_3} ((x, r) \oplus \text{Env}, \text{Mem}[r := v])$$

EXAMPLE 4.2.– The transition sequence in Box 4.1 shows how, from the following sequence of definitions:

$$\mathsf{let}\ x_1 =\uparrow 4;\ \mathsf{let}\ x_2 = x_1;\ \mathsf{let}\ x_3 = 2 + \downarrow x_1;\ \mathsf{let}\ x_4 =\uparrow x_1;\ \mathsf{let}\ x_5 =\uparrow\uparrow x_1; \quad [4.2]$$

and a state (Env, Mem), we obtain the state:

$$\begin{pmatrix} (x_5, r_{x_5}) \oplus (x_4, r_{x_4}) \oplus (x_3, 6) \oplus (x_2, r_{x_1}) \oplus (x_1, r_{x_1}) \oplus \text{Env} \\ \text{Mem}[r_{x_1} := 4][r_{x_4} := r_{x_1}][r_a := r_{x_1}][r_{x_5} := r_a] \end{pmatrix} \quad [4.3]$$

The sequence defined in [4.2] is coded as:

Python
```
ex_ld4_1 = [Let_def4("x1",Ref4(Cste4(CInt4(4)))),
            Let_def4("x2",Var4("x1")),
            Let_def4("x3",B_Op4(Plus4(),Cste4(CInt4(2)),Deref4(Var4("x1")))),
            Let_def4("x4",Ref4(Var4("x1"))),
            Let_def4("x5",Ref4(Ref4(Var4("x1"))))]
```

OCaml
```
let ex_ld4_1 = [
  Let_def4 ("x1", Ref4 (Cste4 (CInt4 4))) ;
  Let_def4 ("x2", Var4 "x1") ;
  Let_def4 ("x3", B_Op4 (Plus4, Cste4 (CInt4 2), Deref4 (Var4 "x1"))) ;
  Let_def4 ("x4", Ref4 (Var4"x1")) ;
  Let_def4 ("x5", Ref4 (Ref4 (Var4 "x1")))]
```

And its execution started in a state with an empty environment is obtained as follows:

```
Python
>>> print_state(trans_def4_exec(([],[]),ex_ld4_1))
[ ( x5 , VCste4 ('CRef4', 3) ), ( x4 , VCste4 ('CRef4', 1) ),
  ( x3 , VCste4 ('CInt4', 6) ), ( x2 , VCste4 ('CRef4', 0) ),
  ( x1 , VCste4 ('CRef4', 0) ), ]
[ ( 3 , ('CRef4', 2) ), ( 2 , ('CRef4', 0) ), ( 1 , ('CRef4', 0) ),
  ( 0 , ('CInt4', 4) ), ]
```

```
OCaml
# trans_def4_exec ([], []) ex_ld4_1 ;;
- : (string * string valeurs4) list * (int * const4) list =
([("x5", VCste4 (CRef4 3)); ("x4", VCste4 (CRef4 1));
  ("x3", VCste4 (CInt4 6)); ("x2", VCste4 (CRef4 0));
  ("x1", VCste4 (CRef4 0))],
 [(3, CRef4 2); (2, CRef4 0); (1, CRef4 0); (0, CInt4 4)])
```

Taking $r_{x_1} = 0$, $r_{x_4} = 1$, $r_{x_5} = 3$ and $r_a = 0$, we thus obtain the state defined in [4.3].

$$(\text{Env}, \text{Mem})$$

$$\xrightarrow{\text{let } x_1 = \uparrow 4;}_{\text{Def}_3} \left(\underbrace{(x_1, r_{x_1}) \oplus \text{Env}}_{\text{Env}_1}, \underbrace{\text{Mem}[r_{x_1} := 4]}_{\text{Mem}_1} \right) \quad (i)$$

$$\xrightarrow{\text{let } x_2 = x_1;}_{\text{Def}_3} \left(\underbrace{(x_2, r_{x_1}) \oplus \text{Env}_1}_{\text{Env}_2}, \text{Mem}_1 \right) \quad (ii)$$

$$\xrightarrow{\text{let } x_3 = 2 + \downarrow x_1;}_{\text{Def}_3} \left(\underbrace{(x_3, 6) \oplus \text{Env}_2}_{\text{Env}_3}, \text{Mem}_1 \right) \quad (iii)$$

$$\xrightarrow{\text{let } x_4 = \uparrow x_1;}_{\text{Def}_3} \left(\underbrace{(x_4, r_{x_4}) \oplus \text{Env}_3}_{\text{Env}_4}, \underbrace{\text{Mem}_1[r_{x_4} := r_{x_1}]}_{\text{Mem}_2} \right) \quad (iv)$$

$$\xrightarrow{\text{let } x_5 = \uparrow\uparrow x_1;}_{\text{Def}_3} \left(\underbrace{(x_5, r_{x_5}) \oplus \text{Env}_4}_{\text{Env}_5} \underbrace{\text{Mem}_2[r_a := r_{x_1}][r_{x_5} := r_a]}_{\text{Mem}_3} \right) \quad (v)$$

(i) $[\![4]\!]^{\text{Mem}}_{\text{Env}} = (4, \text{Mem})$

(ii) $[\![x_1]\!]^{\text{Mem}_1}_{\text{Env}_1} = (\text{Env}_1(x_1), \text{Mem}_1) = (r_{x_1}, \text{Mem}_1)$

(iii) $[\![2 + \downarrow x_1]\!]^{\text{Mem}_1}_{\text{Env}_2} = (2 + \text{Mem}_1(\text{Env}_2(x_1)), \text{Mem}_1) = (6, \text{Mem}_1)$

(iv) $[\![\uparrow x_1]\!]^{\text{Mem}_1}_{\text{Env}_3} = (r_{x_4}, \text{Mem}_1[r_{x_4} := \text{Env}_3(x_1)]) = (r_{x_4}, \text{Mem}_1[r_{x_4} := r_{x_1}])$

(v) $[\![\uparrow\uparrow x_1]\!]^{\text{Mem}_2}_{\text{Env}_4} = (r_{x_5}, \text{Mem}_2[r_a := r_{x_1}][r_{x_5} := r_a])$

since $[\![\uparrow x_1]\!]^{\text{Mem}_2}_{\text{Env}_4} = (r_a, \text{Mem}_2[r_a := \text{Env}_4(x_1)]) = (r_a, \text{Mem}_2[r_a := r_{x_1}])$

Box 4.1. *Example: execution of a sequence of definitions*

4.4. Operational semantics

The instruction at the heart of all imperative languages is *assignment*, which changes the state of the memory, i.e. modifies its contents. Execution of an assignment $x := e$ in a state (Env, Mem) consists of evaluating the expression e in this state to obtain a value v and a new memory state Mem', searching the value $r = \text{Env}(x) \in \mathbf{R}$ of x in environment Env, and then, finally, writing the value v at the location r to obtain memory $\text{Mem}'[r := v]$.

REMARK.– Few high-level languages include a syntactic construct to distinguish between the reference of the location of a variable and the value stored at this location. These two distinct values are often denoted by the variable name alone; it is left to users to determine which value is to be used for each occurrence of the variable name in an expression. As we saw at the end of section 2.3, in these languages, the meaning of variable names occurring in assignments depends on their position to the left or to the right of the assignment symbol, on the left variable names denote a location, and on the right, they denote a value.

The execution of an imperative program consists of executing a sequence of assignments and tests. The control structures included in a program specify the assignments to execute, the order of execution and the number of times they should be executed. Given an execution environment, the semantics of $\mathbf{Lang}_3$ describes the memory transformations associated with each instruction in $\mathbf{Lang}_3$. These transformations are described by a relation, following the approach first introduced in [PLO 04].

4.4.1. *Big-step semantics*

Given an execution environment Env, the big-step operational semantics of language $\mathbf{Lang}_3$ defines a transition relation:

$$\twoheadrightarrow_{\text{Env}} \subseteq \mathbf{M} \times \mathbf{Lang}_3 \times \mathbf{M}$$

such that $\langle \text{Mem}, c \rangle \twoheadrightarrow_{\text{Env}} \text{Mem}'$ means that the execution of instruction c in state (Env, Mem) leads to the memory state Mem'. This relation is defined by induction, using the inference system presented in Box 4.2.

Rule C_1 defines the semantics of the assignment $x := e$: the evaluation of e in state (Env, Mem) must not raise an error, and evaluation of the variable x in the environment Env must produce a reference, otherwise the execution is aborted. As the premises for this rule do not take the form $\langle \text{Mem}, c \rangle \twoheadrightarrow_{\text{Env}} \text{Mem}'$, they must be considered as conditions for applying the rule.

$$(C_1)\frac{\text{Env}(x) = r \in \mathbf{R} \qquad [\![e]\!]_{\text{Env}}^{\text{Mem}_1} = (v, \text{Mem}_2) \in \mathbf{V} \times \mathbf{M}}{\langle \text{Mem}_1, x := e\rangle \twoheadrightarrow_{\text{Env}} \text{Mem}_2[r := v]}$$

$$(C_2)\frac{}{\langle \text{Mem}, \mathsf{skip}\rangle \twoheadrightarrow_{\text{Env}} \text{Mem}}$$

$$(C_3)\frac{\langle \text{Mem}, c_1\rangle \twoheadrightarrow_{\text{Env}} \text{Mem}_1 \qquad \langle \text{Mem}_1, c_2\rangle \twoheadrightarrow_{\text{Env}} \text{Mem}_2}{\langle \text{Mem}, c_1 \,;\, c_2\rangle \twoheadrightarrow_{\text{Env}} \text{Mem}_2}$$

$$(C_4)\frac{[\![e]\!]_{\text{Env}}^{\text{Mem}} = (\mathit{true}, \text{Mem}_1) \qquad \langle \text{Mem}_1, c_1\rangle \twoheadrightarrow_{\text{Env}} \text{Mem}_2}{\langle \text{Mem}, \mathsf{if}\ e\ \mathsf{then}\ c_1\ \mathsf{else}\ c_2\rangle \twoheadrightarrow_{\text{Env}} \text{Mem}_2}$$

$$(C_5)\frac{[\![e]\!]_{\text{Env}}^{\text{Mem}} = (\mathit{false}, \text{Mem}_1) \qquad \langle \text{Mem}_1, c_2\rangle \twoheadrightarrow_{\text{Env}} \text{Mem}_2}{\langle \text{Mem}, \mathsf{if}\ e\ \mathsf{then}\ c_1\ \mathsf{else}\ c_2\rangle \twoheadrightarrow_{\text{Env}} \text{Mem}_2}$$

$$(C_6)\frac{[\![e]\!]_{\text{Env}}^{\text{Mem}} = (\mathit{true}, \text{Mem}_1) \qquad \begin{array}{c}\langle \text{Mem}_1, c\rangle \twoheadrightarrow_{\text{Env}} \text{Mem}_2 \\ \langle \text{Mem}_2, \mathsf{while}\ e\ \mathsf{do}\ c\rangle \twoheadrightarrow_{\text{Env}} \text{Mem}_3\end{array}}{\langle \text{Mem}, \mathsf{while}\ e\ \mathsf{do}\ c\rangle \twoheadrightarrow_{\text{Env}} \text{Mem}_3}$$

$$(C_7)\frac{[\![e]\!]_{\text{Env}}^{\text{Mem}} = (\mathit{false}, \text{Mem}_1)}{\langle \text{Mem}, \mathsf{while}\ e\ \mathsf{do}\ c\rangle \twoheadrightarrow_{\text{Env}} \text{Mem}_1}$$

Box 4.2. *Big-step operational semantics of* $\mathbf{Lang}_3$

Rule C_2 simply indicates that executing the instruction skip does not affect the state. Rule C_3 describes the evolution of the memory during the execution of a sequence of two instructions. Rules C_4 and C_5 specify the memory transformation associated with a conditional; they contain a condition to apply them which depends on the result of the evaluation of expression e.

Finally, rules C_6 and C_7 relate to the while e do c loop, and specify how many times the execution of the body c of the loop is to be repeated, according to the value of the conditional expression e. Unlike the other instructions, while e do c allows us to write programs where execution does not terminate, and which, consequently, cannot be described by an inference tree since such a tree is a finite structure. For such non-terminating programs c, there is no memory state Mem' such that $\langle \text{Mem}, c\rangle \twoheadrightarrow_{\text{Env}} \text{Mem}'$. For example, taking the program $foo = \mathsf{while}\ \mathit{true}\ \mathsf{do}\ c$, there is no memory state Mem' such that $\langle \text{Mem}, foo\rangle \twoheadrightarrow_{\text{Env}} \text{Mem}'$. In fact, this program contains an infinite loop, hence it is impossible to build a finite execution tree for it:

$$(C_6)\frac{\begin{array}{c}[\![\mathit{true}]\!]_{\text{Env}}^{\text{Mem}} \\ = (\mathit{true}, \text{Mem})\end{array} \quad (C_2)\frac{}{\langle \text{Mem}, \mathsf{skip}\rangle \twoheadrightarrow_{\text{Env}} \text{Mem}} \quad (C_6)\frac{\vdots}{\langle \text{Mem}, foo\rangle \twoheadrightarrow_{\text{Env}} ?}}{\langle \text{Mem}, foo\rangle \twoheadrightarrow_{\text{Env}} ?}$$

since $\langle \text{Mem}, foo\rangle \twoheadrightarrow_{\text{Env}} ?$ occurs both in the root and in the premise. Big-step operational semantics uses inference trees – which are finite – and cannot, therefore, take account of the execution of non-terminating programs. A finer approach, suitable for describing the execution of non-terminating programs, will be presented in section 4.4.2.

Let us implement a function that, given a state (Env, Mem) and a program c, computes the new memory Mem' such that $\langle\text{Mem}, c\rangle \twoheadrightarrow_{\text{Env}} \text{Mem}'$. Note however that, just like the evaluation function defined in Chapter 3, this does not correspond to a mathematical function. Indeed, this function may not terminate if it is applied to a program whose execution in the given state does not terminate. This is the case when execution leads to an infinite loop (as we saw in Chapter 3 for expressions corresponding to the application of a recursive function, where evaluation resulted in an infinite number of recursive calls). This function reproduces the execution mechanism described by the semantic rules in Box 4.2. As such, it is not compositional: the execution of a program does not simply correspond to the execution of each of its components. Rule C_6 describes the execution of the instruction while e do c using the execution of the c component, but also the execution of the loop itself. We shall return to this point in section 4.7.1. As we assume that executions of programs do not generate typing errors, we just have to introduce here a single exception `Exec_Error`, which will be raised in case of error of execution. Once again, the definition of this function will be extended as language $\mathbf{Lang}_3$ is enriched.

Python
```
def exec_pgm4(env,mem,c):
    if isinstance(c,IAffect4):
        r = valeur_de(env,c.symb)
        if r is None: raise ValueError
        if isinstance(r,VCste4) and isinstance(r.cst,CRef4):
            v,mem1 = eval_exp4(env,mem,c.exp)
            if isinstance(v,VCste4):
                return write_mem(mem1,r.cst.cst_adr,v.cst)
            raise ValueError
        raise ValueError
    if isinstance(c,ISkip4): return mem
    if isinstance(c,ISeq4):
        return exec_pgm4(env,exec_pgm4(env,mem,c.pgm1),c.pgm2)
    if isinstance(c,ICond4):
        b,mem1 = eval_exp4(env,mem,c.exp)
        if isinstance(b,VCste4) and isinstance(b.cst,CBool4):
            if b.cst.cst_bool: return exec_pgm4(env,mem1,c.pgm1)
            else: return exec_pgm4(env,mem1,c.pgm2)
        raise ValueError
    if isinstance(c,IWhile4):
        b,mem1 = eval_exp4(env,mem,c.exp)
        if isinstance(b,VCste4) and isinstance(b.cst,CBool4):
            if b.cst.cst_bool:
                return exec_pgm4(env,exec_pgm4(env,mem1,c.pgm),c)
            else: return mem1
        raise ValueError
    if ...
    raise ValueError
```

OCaml
```
exception Exec_Error

let rec exec_pgm4 env mem c = match c with
  | IAffect4 (x, e) -> (
     match ((valeur_de env x), (eval_exp4 env mem e)) with
      | (Some (VCste4 (CRef4 a)), (VCste4 v, mem1)) -> write_mem mem1 a v
    | _ -> raise Exec_Error)
  | ISkip4 -> mem
  | ISeq4 (c1, c2) -> exec_pgm4 env (exec_pgm4 env mem c1) c2
  | ICond4(e,c1,c2) -> (match eval_exp4 env mem e with
      | (VCste4 (CBool4 true), mem1) -> exec_pgm4 env mem1 c1
      | (VCste4 (CBool4 false), mem1) -> exec_pgm4 env mem1 c2
      | _ -> raise Exec_Error)
  | IWhile4 (e, c1) -> (match eval_exp4 env mem e with
      | (VCste4 (CBool4 true), mem1) -> exec_pgm4 env (exec_pgm4 env mem1 c1) c
      | (VCste4 (CBool4 false), mem1) -> mem1
      | _ -> raise Exec_Error)
  | ...
val exec_pgm4 : ('a * 'a valeurs4) list -> (int * const4) list -> 'a pgm4
                -> (int * const4) list
```

EXAMPLE 4.3.– The execution of the program from example 4.1:

$$\text{while } \underbrace{\text{not}(\downarrow x = \downarrow y)}_{e_w} \text{ do } \underbrace{\text{if } \underbrace{\downarrow x \leq \downarrow y}_{e} \text{ then } \underbrace{y := \downarrow y - \downarrow x}_{c_1} \text{ else } \underbrace{x := \downarrow x - \downarrow y}_{c_2}}_{c_w}$$

in a state $(\text{Env}, \text{Mem}) = ((y, r_y) \oplus (x, r_x) \oplus \text{Env}_0, \text{Mem}_0[r_x := 4][r_y := 2])$ is described by the following inference tree:

$$(C_6)\frac{[\![e_w]\!]^{\text{Mem}}_{\text{Env}} = (true, \text{Mem}) \quad \underset{1}{\nabla} \quad \underset{2}{\nabla}}{\langle \text{Mem}, \text{while } e_w \text{ do } c_w \rangle \twoheadrightarrow_{\text{Env}} \text{Mem}[r_x := 2]}$$

where ∇_1 is the tree:

$$(C_5)\frac{[\![e]\!]^{\text{Mem}}_{\text{Env}} = (false, \text{Mem}) \quad (C_1)\dfrac{\text{Env}(x) = r_x \quad [\![\downarrow x - \downarrow y]\!]^{\text{Mem}}_{\text{Env}} = (2, \text{Mem})}{\langle \text{Mem}, c_2 \rangle \twoheadrightarrow_{\text{Env}} \text{Mem}[r_x := 2]}}{\langle \text{Mem}, c_w \rangle \twoheadrightarrow_{\text{Env}} \text{Mem}[r_x := 2]}$$

and ∇_2 is the tree:

$$(C_5)\frac{[\![e_w]\!]^{\text{Mem}[r_x:=2]}_{\text{Env}} = (false, \text{Mem}[r_x := 2])}{\langle \text{Mem}[r_x := 2], \text{while } e_w \text{ do } c_w \rangle \twoheadrightarrow_{\text{Env}} \text{Mem}[r_x := 2]}$$

Returning to program ex_pgm4_1, introduced in example 4.1, this execution is obtained as follows:

```
Python
ex_ld4_2 = [Let_def4("x",Ref4(Cste4(CInt4(4)))),
            Let_def4("y",Ref4(Cste4(CInt4(2))))]
(env_ex_ld4_2,mem_ex_ld4_2) = trans_def4_exec(([],[]),ex_ld4_2)
>>> print_state((env_ex_ld4_2,exec_pgm4(env_ex_ld4_2,mem_ex_ld4_2,ex_pgm4_1)))
[ ( y , VCste4 ('CRef4', 1) ), ( x , VCste4 ('CRef4', 0) ), ]
[ ( 1 , ('CInt4', 2) ), ( 0 , ('CInt4', 2) ), ]
```

```
OCaml
let ex_ld4_2 = [ Let_def4 ("x", Ref4 (Cste4 (CInt4 4)));
                 Let_def4 ("y", Ref4 (Cste4 (CInt4 2)))]
# let (env, mem) = trans_def4_exec ([],[]) ex_ld4_2 in exec_pgm4
                                  env mem ex_pgm4_1 ;;
  - : (int * const4) list = [(1, CInt4 2); (0, CInt4 2)]
```

4.4.2. *Small-step semantics*

The big-step operational semantics presented in section 4.4.1 handles elements of the form $\langle \text{Mem}_1, c\rangle \twoheadrightarrow_{\text{Env}} \text{Mem}_2$, which are only able to specify the final state Mem_2 of the memory obtained after the execution of a program c, without providing any precise information concerning the steps, which led up to it (this information can be retrieved in the tree for $\langle \text{Mem}_1, c\rangle \twoheadrightarrow_{\text{Env}} \text{Mem}_2$). Furthermore, as inference trees are, by convention, finite objects, this semantics is unable to account for the execution of non-terminating programs, as we saw earlier. Small-step operational semantics provides finer, more detailed information concerning program execution. To present it, we define a *configuration* as a pair $\langle c, \text{Mem}\rangle \in \mathbf{Lang}_3 \times \mathbf{M}$ of a program c and a memory state Mem. We then define a *transition relation* between configurations using the inference system defined in Box 4.3.

Intuitively, $\langle c_1, \text{Mem}_1\rangle \hookrightarrow_{\text{Env}} \langle c_2, \text{Mem}_2\rangle$ means that executing one step of the program c_1 in a state $(\text{Env}, \text{Mem}_1)$ – which corresponds to the configuration $\langle c_1, \text{Mem}_1\rangle$ – leads to a state $(\text{Env}, \text{Mem}_2)$ in which c_2 is the rest of the program to be executed: this corresponds to the configuration $\langle c_2, \text{Mem}_2\rangle$. A configuration is said to be *terminal* if it takes the form $\langle \mathsf{skip}, \text{Mem}\rangle$, and in this case, the memory Mem will no longer be modified. Formally, the only two rules leading to a terminal configuration correspond to an assignment (S_1) and to an exit of loop (S_7).

$$(S_1)\frac{\text{Env}(x) = r \in \mathbf{R} \quad [\![e]\!]^{\text{Mem}_1}_{\text{Env}} = (v, \text{Mem}_2) \in \mathbf{V} \times \mathbf{M}}{\langle x := e, \text{Mem}_1\rangle \hookrightarrow_{\text{Env}} \langle \textsf{skip}, \text{Mem}_2[r := v]\rangle}$$

$$(S_2)\frac{\langle c_1, \text{Mem}_1\rangle \hookrightarrow_{\text{Env}} \langle c_1', \text{Mem}_2\rangle}{\langle c_1 \,;\, c_2, \text{Mem}_1\rangle \hookrightarrow_{\text{Env}} \langle c_1' \,;\, c_2, \text{Mem}_2\rangle} \qquad (S_3)\frac{}{\langle \textsf{skip} \,;\, c, \text{Mem}\rangle \hookrightarrow_{\text{Env}} \langle c, \text{Mem}\rangle}$$

$$(S_4)\frac{[\![e]\!]^{\text{Mem}_1}_{\text{Env}} = (\mathit{true}, \text{Mem}_2)}{\langle \textsf{if } e \textsf{ then } c_1 \textsf{ else } c_2, \text{Mem}_1\rangle \hookrightarrow_{\text{Env}} \langle c_1, \text{Mem}_2\rangle}$$

$$(S_5)\frac{[\![e]\!]^{\text{Mem}_1}_{\text{Env}} = (\mathit{false}, \text{Mem}_2)}{\langle \textsf{if } e \textsf{ then } c_1 \textsf{ else } c_2, \text{Mem}_1\rangle \hookrightarrow_{\text{Env}} \langle c_2, \text{Mem}_2\rangle}$$

$$(S_6)\frac{[\![e]\!]^{\text{Mem}_1}_{\text{Env}} = (\mathit{true}, \text{Mem}_2)}{\langle \textsf{while } e \textsf{ do } c, \text{Mem}_1\rangle \hookrightarrow_{\text{Env}} \langle c ; \textsf{while } e \textsf{ do } c, \text{Mem}_2\rangle}$$

$$(S_7)\frac{[\![e]\!]^{\text{Mem}_1}_{\text{Env}} = (\mathit{false}, \text{Mem}_2)}{\langle \textsf{while } e \textsf{ do } c, \text{Mem}_1\rangle \hookrightarrow_{\text{Env}} \langle \textsf{skip}, \text{Mem}_2\rangle}$$

Box 4.3. *Small-step operational semantics of* $\mathbf{Lang}_3$

A *sequence of computations* is a sequence, possibly infinite, of the form:

$$\langle c_0, \text{Mem}_0\rangle \hookrightarrow_{\text{Env}} \langle c_1, \text{Mem}_1\rangle \hookrightarrow_{\text{Env}} \langle c_2, \text{Mem}_2\rangle \hookrightarrow_{\text{Env}} \cdots$$

such that there exists an inference tree for $\langle c_i, \text{Mem}_i\rangle \hookrightarrow_{\text{Env}} \langle c_{i+1}, \text{Mem}_{i+1}\rangle$ for all $i \geq 0$. The relation $\hookrightarrow^{\star}_{\text{Env}}$ can then be defined by $\langle c, \text{Mem}\rangle \hookrightarrow^{\star}_{\text{Env}} \text{Mem}'$ if, and only if, there exists a sequence of computations, containing only a finite number of transitions (or zero transitions)

$$\langle c_0, \text{Mem}_0\rangle \hookrightarrow_{\text{Env}} \langle c_1, \text{Mem}_1\rangle \hookrightarrow_{\text{Env}} \langle c_2, \text{Mem}_2\rangle \hookrightarrow_{\text{Env}} \cdots \hookrightarrow_{\text{Env}} \langle c_k, \text{Mem}_k\rangle$$

such that $c = c_0$, $\text{Mem} = \text{Mem}_0$ and $\text{Mem}' = \text{Mem}_k$ and such that the configuration $\langle c_k, \text{Mem}_k\rangle$ is terminal (i.e. $c_k = \textsf{skip}$). If we know the length of the sequence, as in this case, we write $\langle c, \text{Mem}\rangle \hookrightarrow^{k}_{\text{Env}} \text{Mem}'$.

EXAMPLE 4.4.– Consider the execution of the program fragment shown in example 4.1:

$$\underbrace{\textsf{while } \underbrace{\textsf{not}(\downarrow x = \downarrow y)}_{e_w} \textsf{ do } \underbrace{\textsf{if } \underbrace{\downarrow x \leq \downarrow y}_{e} \textsf{ then } \underbrace{y := \downarrow y - \downarrow x}_{c_1} \textsf{ else } \underbrace{x := \downarrow x - \downarrow y}_{c_2}}_{c_w}}_{c}$$

in a state $(\text{Env}, \text{Mem}) = ((y, r_y) \oplus (x, r_x) \oplus \text{Env}_0, \text{Mem}_0[r_x := 4][r_y := 2])$. The small-step semantics of this program builds the following sequence:

$$\begin{array}{llll}
 & \langle c, \text{Mem}\rangle & (S_6) & [\![e_w]\!]_{\text{Env}}^{\text{Mem}} = (true, \text{Mem}) \\
\hookrightarrow_{\text{Env}} & \langle c_w \,;\, c, \text{Mem}\rangle & (S_2, S_5) & [\![e]\!]_{\text{Env}}^{\text{Mem}} = (false, \text{Mem}) \\
\hookrightarrow_{\text{Env}} & \langle c_2 \,;\, c, \text{Mem}\rangle & (S_2, S_1) & \text{Env}(x) = r_x \text{ and} \\
 & & & [\![\downarrow x - \downarrow y]\!]_{\text{Env}}^{\text{Mem}} = (2, \text{Mem}) \\
\hookrightarrow_{\text{Env}} & \langle \mathsf{skip} \,;\, c, \text{Mem}[r_x := 2]\rangle & (S_3) & \\
\hookrightarrow_{\text{Env}} & \langle c, \text{Mem}[r_x := 2]\rangle & (S_7) & [\![e_w]\!]_{\text{Env}}^{\text{Mem}[r_x:=2]} \\
 & & & \quad = (false, \text{Mem}[r_x := 2]) \\
\hookrightarrow_{\text{Env}} & \langle \mathsf{skip}, \text{Mem}[r_x := 2]\rangle & &
\end{array}$$

So we get: $\langle c, \text{Mem}\rangle \hookrightarrow^{\star}_{\text{Env}} \text{Mem}[r_x := 2]$.

4.4.3. *Expressiveness of operational semantics*

As shown by examples 4.3 and 4.4, when used to describe the execution of terminating programs in $\mathbf{Lang}_3$, big-step and small-step operational semantics produce the same result. This semantic equivalence is proved in section 4.5.4. Nevertheless, as it provides a finer description of executions of programs, small-step semantics can handle certain features of some programming languages, which cannot be described using big-step semantics. Parallelism is one such feature. Consider the construction $c_1 \parallel c_2$ which denotes the parallel execution of two programs c_1 and c_2. Big-step semantics only allows us to describe memory states after the execution of programs c_1 and c_2; in this approach, the only possible ways of executing $c_1 \parallel c_2$ are to execute c_1 then execute c_2, or to execute c_2 then execute c_1:

$$(C_P^1)\frac{\langle \text{Mem}, c_1\rangle \twoheadrightarrow_{\text{Env}} \text{Mem}' \quad \langle \text{Mem}', c_2\rangle \twoheadrightarrow_{\text{Env}} \text{Mem}''}{\langle \text{Mem}, c_1 \parallel c_2\rangle \twoheadrightarrow_{\text{Env}} \text{Mem}''}$$

$$(C_P^2)\frac{\langle \text{Mem}, c_2\rangle \twoheadrightarrow_{\text{Env}} \text{Mem}' \quad \langle \text{Mem}', c_1\rangle \twoheadrightarrow_{\text{Env}} \text{Mem}''}{\langle \text{Mem}, c_1 \parallel c_2\rangle \twoheadrightarrow_{\text{Env}} \text{Mem}''}$$

For example, using these rules, there are only two possible executions of program c:

$$x := 2 \parallel (x := 3 \,;\, x :=\downarrow x * \downarrow x)$$

giving us:

$$\langle \text{Mem}, c\rangle \twoheadrightarrow_{\text{Env}} \text{Mem}[\text{Env}(x) := 2] \text{ and } \langle \text{Mem}, c\rangle \twoheadrightarrow_{\text{Env}} \text{Mem}[\text{Env}(x) := 9]$$

This semantics is unable to take account of possible interleavings between the executions of c_1 and c_2. Small-step semantics, on the other hand, allows us to describe the execution of just one part of a program:

$$(S_P^1)\frac{\langle c_1, \mathrm{Mem}\rangle \hookrightarrow_{\mathrm{Env}} \langle c_1', \mathrm{Mem}'\rangle}{\langle c_1 \parallel c_2, \mathrm{Mem}\rangle \hookrightarrow_{\mathrm{Env}} \langle c_1' \parallel c_2, \mathrm{Mem}'\rangle}$$

$$(S_P^2)\frac{\langle c_2, \mathrm{Mem}\rangle \hookrightarrow_{\mathrm{Env}} \langle c_2', \mathrm{Mem}'\rangle}{\langle c_1 \parallel c_2, \mathrm{Mem}\rangle \hookrightarrow_{\mathrm{Env}} \langle c_1 \parallel c_2', \mathrm{Mem}'\rangle}$$

$$(S_P^3)\frac{\langle c_1, \mathrm{Mem}\rangle \hookrightarrow_{\mathrm{Env}} \langle \mathsf{skip}, \mathrm{Mem}'\rangle}{\langle c_1 \parallel c_2, \mathrm{Mem}\rangle \hookrightarrow_{\mathrm{Env}} \langle c_2, \mathrm{Mem}'\rangle} \quad (S_P^4)\frac{\langle c_2, \mathrm{Mem}\rangle \hookrightarrow_{\mathrm{Env}} \langle \mathsf{skip}, \mathrm{Mem}'\rangle}{\langle c_1 \parallel c_2, \mathrm{Mem}\rangle \hookrightarrow_{\mathrm{Env}} \langle c_1, \mathrm{Mem}'\rangle}$$

These rules can be used to describe the following three executions of the program c:

$$\begin{aligned}\langle c, \mathrm{Mem}\rangle &\hookrightarrow_{\mathrm{Env}} \langle x := 3\,;\, x :=\downarrow x * \downarrow x, \mathrm{Mem}[\mathrm{Env}(x) := 2]\rangle \\ &\hookrightarrow_{\mathrm{Env}} \langle \mathsf{skip}\,;\, x :=\downarrow x * \downarrow x, \mathrm{Mem}[\mathrm{Env}(x) := 3]\rangle \\ &\hookrightarrow_{\mathrm{Env}} \langle x :=\downarrow x * \downarrow x, \mathrm{Mem}[\mathrm{Env}(x) := 3]\rangle \\ &\hookrightarrow_{\mathrm{Env}} \langle \mathsf{skip}, \mathrm{Mem}[\mathrm{Env}(x) := 9]\rangle\end{aligned}$$

$$\begin{aligned}\langle c, \mathrm{Mem}\rangle &\hookrightarrow_{\mathrm{Env}} \langle x := 2 \parallel (\mathsf{skip}\,;\, x :=\downarrow x * \downarrow x), \mathrm{Mem}[\mathrm{Env}(x) := 3]\rangle \\ &\hookrightarrow_{\mathrm{Env}} \langle x := 2 \parallel x :=\downarrow x * \downarrow x, \mathrm{Mem}[\mathrm{Env}(x) := 3]\rangle \\ &\hookrightarrow_{\mathrm{Env}} \langle x := 2, \mathrm{Mem}[\mathrm{Env}(x) := 9]\rangle \\ &\hookrightarrow_{\mathrm{Env}} \langle \mathsf{skip}, \mathrm{Mem}[\mathrm{Env}(x) := 2]\rangle\end{aligned}$$

$$\begin{aligned}\langle c, \mathrm{Mem}\rangle &\hookrightarrow_{\mathrm{Env}} \langle x := 2 \parallel (\mathsf{skip}\,;\, x :=\downarrow x * \downarrow x), \mathrm{Mem}[\mathrm{Env}(x) := 3]\rangle \\ &\hookrightarrow_{\mathrm{Env}} \langle x := 2 \parallel x :=\downarrow x * \downarrow x, \mathrm{Mem}[\mathrm{Env}(x) := 3]\rangle \\ &\hookrightarrow_{\mathrm{Env}} \langle x :=\downarrow x * \downarrow x, \mathrm{Mem}[\mathrm{Env}(x) := 2]\rangle \\ &\hookrightarrow_{\mathrm{Env}} \langle \mathsf{skip}, \mathrm{Mem}[\mathrm{Env}(x) := 4]\rangle\end{aligned}$$

4.5. Semantic properties

In this section, we present applications of the semantic definitions given earlier, along with a few classical proof techniques using these formalisms.

4.5.1. *Equivalent programs*

As an operational semantics describes program behavior, it is possible to identify "equivalent" programs, i.e. programs whose executions terminate and lead to the same memory state. Using big-step operational semantics, this equivalence relation can be defined formally as follows:

DEFINITION 4.1.– *Equivalent programs: two programs c_1 and c_2 are equivalent, noted $c_1 \equiv c_2$, if and only if:*

$$\forall \text{Env} \in \mathbf{E}\ \forall \text{Mem}, \text{Mem}' \in \mathbf{M} \quad \langle \text{Mem}, c_1 \rangle \twoheadrightarrow_{\text{Env}} \text{Mem}' \Leftrightarrow \langle \text{Mem}, c_2 \rangle \twoheadrightarrow_{\text{Env}} \text{Mem}'$$

EXAMPLE 4.5.– Consider the following two programs:

$$c_1 : (\textsf{if } e \textsf{ then } c \textsf{ else } c')\,;\, c'' \qquad c_2 : \textsf{if } e \textsf{ then } c\,;\, c'' \textsf{ else } c'\,;\, c''$$

To prove that $c_1 \equiv c_2$, we build an inference tree for $\langle \text{Mem}, c_1 \rangle \twoheadrightarrow_{\text{Env}} \text{Mem}'$ from an inference tree for $\langle \text{Mem}, c_2 \rangle \twoheadrightarrow_{\text{Env}} \text{Mem}'$ and vice versa. To this end, we examine all the possible rules for constructing such a tree, distinguishing between two possible cases for the evaluation of e. We must then simply "combine" the inference trees. If e is evaluated as $true$, then, from the tree

$$(C_3)\dfrac{(C_4)\dfrac{\begin{array}{c}[\![e]\!]^{\text{Mem}}_{\text{Env}} \\ = (true, \text{Mem}_1)\end{array} \quad (C_i)\dfrac{\underset{1}{\nabla}}{\langle \text{Mem}_1, c \rangle \twoheadrightarrow_{\text{Env}} \text{Mem}_2}}{\langle \text{Mem}, \textsf{if } e \textsf{ then } c \textsf{ else } c' \rangle \twoheadrightarrow_{\text{Env}} \text{Mem}_2} \quad (C_j)\dfrac{\underset{2}{\nabla}}{\langle \text{Mem}_2, c'' \rangle \twoheadrightarrow_{\text{Env}} \text{Mem}'}}{\langle \text{Mem}, (\textsf{if } e \textsf{ then } c \textsf{ else } c')\,;\, c'' \rangle \twoheadrightarrow_{\text{Env}} \text{Mem}'}$$

we can obtain the tree

$$(C_4)\dfrac{\begin{array}{c}[\![e]\!]^{\text{Mem}}_{\text{Env}} \\ = (true, \text{Mem}_1)\end{array} \quad (C_3)\dfrac{(C_i)\dfrac{\underset{1}{\nabla}}{\langle \text{Mem}_1, c \rangle \twoheadrightarrow_{\text{Env}} \text{Mem}_2} \quad (C_j)\dfrac{\underset{2}{\nabla}}{\langle \text{Mem}_2, c'' \rangle \twoheadrightarrow_{\text{Env}} \text{Mem}'}}{\langle \text{Mem}_1, c\,;\, c'' \rangle \twoheadrightarrow_{\text{Env}} \text{Mem}'}}{\langle \text{Mem}, \textsf{if } e \textsf{ then } c\,;\, c'' \textsf{ else } c'\,;\, c'' \rangle \twoheadrightarrow_{\text{Env}} \text{Mem}'}$$

and vice versa. A similar reasoning applies when e is evaluated as $false$.

EXAMPLE 4.6.– Consider the following two programs:

$$c_1 : c\,;\, (\textsf{if } e \textsf{ then } c' \textsf{ else } c'') \qquad c_2 : \textsf{if } e \textsf{ then } c\,;\, c' \textsf{ else } c\,;\, c''$$

To show that $c_1 \not\equiv c_2$, it is enough to build a counterexample where c, c' and c'' are instructions $x := 0$, $x :=\downarrow x + 1$ and $x :=\downarrow x + 5$, respectively, and where e is the expression $1 \leq \downarrow x$. Starting from a state (Env, Mem) such that $\text{Env}(x) = r_x \in \mathbf{R}$ and $\text{Mem}(r_x) = 2$, we obtain:

$$\begin{array}{ll}\langle \text{Mem}, c\,;\, (\textsf{if } e \textsf{ then } c' \textsf{ else } c'') \rangle \twoheadrightarrow_{\text{Env}} \text{Mem}_1 & \text{with } \text{Mem}_1(x) = 5 \\ \langle \text{Mem}, \textsf{if } e \textsf{ then } c\,;\, c' \textsf{ else } c\,;\, c'' \rangle \twoheadrightarrow_{\text{Env}} \text{Mem}_2 & \text{with } \text{Mem}_2(x) = 1\end{array}$$

4.5.2. *Program termination*

The existence of an inference tree describing the execution of a program guarantees that this execution is finite. As all inference trees are finite, showing that a program c terminates from a starting state (Env, Mem) using big-step operational semantics comes down to proving the existence of a memory state Mem' such that $\langle \text{Mem}, c\rangle \twoheadrightarrow_{\text{Env}} \text{Mem}'$.

EXAMPLE 4.7.– Returning to the program fragment used in example 4.1:

$$\underbrace{\text{while } \underbrace{\text{not}(\downarrow x = \downarrow y)}_{e_w} \text{ do } \underbrace{\text{if } \underbrace{\downarrow x \leq \downarrow y}_{e} \text{ then } \underbrace{y := \downarrow y - \downarrow x}_{c_1} \text{ else } \underbrace{x := \downarrow x - \downarrow y}_{c_2}}_{c_w}}_{c}$$

let us show that the execution of this program terminates when starting from any memory state storing strictly positive integer values at locations x and y. Formally, we must prove that:

$$\forall \text{Env} \in \mathbf{E}\, \forall \text{Mem} \in \mathbf{M}$$
$$(\text{Mem}(\text{Env}(x)) \geq 1 \text{ and } \text{Mem}(\text{Env}(y)) \geq 1) \Rightarrow \exists \text{Mem}' \in \mathbf{M}\ \langle \text{Mem}, c\rangle \twoheadrightarrow_{\text{Env}} \text{Mem}'$$

To prove this property, given an environment $\text{Env} \in \mathbf{E}$, we define the set:

$$M_{\text{Env}} = \{\text{Mem} \mid \text{Mem}(\text{Env}(x)) \geq 1 \text{ and } \text{Mem}(\text{Env}(y)) \geq 1\} \subseteq \mathbf{M}$$

and we must prove that:

$$\forall \text{Mem} \in M_{\text{Env}}\ \underbrace{\exists \text{Mem}' \in \mathbf{M}\ \langle \text{Mem}, c\rangle \twoheadrightarrow_{\text{Env}} \text{Mem}'}_{P_{\text{Env}}(\text{Mem})}$$

The proof is obtained by well-founded induction. First, we define a relation $\preceq_{\text{Env}}$ over the elements of M_{Env}. Then we show that this relation is a well-founded order relation, that is, there is no strictly decreasing infinite sequence of values according to $\preceq_{\text{Env}}$. Finally, we prove that for any element $\text{Mem} \in M_{\text{Env}}$, under the hypothesis $P_{\text{Env}}(\text{Mem}')$ for all $\text{Mem}' \prec_{\text{Env}} \text{Mem}$, we can get $P_{\text{Env}}(\text{Mem})$.

Since at least one value associated with the two variables x and y decreases for each execution of the loop body, the relation $\preceq_{\text{Env}}$ over M_{Env} can be defined as follows:

$$\text{Mem}_1 \preceq_{\text{Env}} \text{Mem}_2 \Leftrightarrow \left(\begin{array}{l} \quad\ \text{Mem}_1(\text{Env}(x)) \leq \text{Mem}_2(\text{Env}(x)) \\ \text{and } \text{Mem}_1(\text{Env}(y)) \leq \text{Mem}_2(\text{Env}(y)) \\ \text{and } \forall z \in \mathbf{X} \setminus \{x, y\}\ \text{Mem}_1(\text{Env}(z)) = \text{Mem}_2(\text{Env}(z)) \end{array}\right)$$

The strict order $\prec_{\text{Env}}$ associated with $\preceq_{\text{Env}}$ is thus defined by:

$$\text{Mem}_1 \prec_{\text{Env}} \text{Mem}_2 \Leftrightarrow \left(\begin{array}{l} \left(\begin{array}{c} \text{Mem}_1(\text{Env}(x)) \neq \text{Mem}_2(\text{Env}(x)) \\ \text{or } \text{Mem}_1(\text{Env}(y)) \neq \text{Mem}_2(\text{Env}(y)) \end{array} \right) \\ \text{and } \text{Mem}_1(\text{Env}(x)) \leq \text{Mem}_2(\text{Env}(x)) \\ \text{and } \text{Mem}_1(\text{Env}(y)) \leq \text{Mem}_2(\text{Env}(y)) \\ \text{and } \forall z \in \mathbf{X} \setminus \{x, y\}\ \text{Mem}_1(\text{Env}(z)) = \text{Mem}_2(\text{Env}(z)) \end{array} \right)$$

In the below proof, it is helpful to note that if $\text{Mem}_1 \prec_{\text{Env}} \text{Mem}_2$, then $\text{Mem}_1(\text{Env}(x)) < \text{Mem}_2(\text{Env}(x))$ or $\text{Mem}_1(\text{Env}(y)) < \text{Mem}_2(\text{Env}(y))$. Relation $\preceq_{\text{Env}}$ clearly defines an order relation (i.e. it is reflexive, anti-symmetric and transitive).

Let us prove that $\preceq_{\text{Env}}$ is well founded.

Suppose that there exists a strictly decreasing infinite sequence:

$$\text{Mem}_1 \succ_{\text{Env}} \text{Mem}_2 \succ_{\text{Env}} \cdots$$

The memories in M_{Env} only associate strictly positive values to variables x and y and these values themselves decrease or remain the same. As there is no infinite decreasing sequence of strictly positive integers, there exists some rank k such that all elements of this infinite sequence of rank greater than k are equal. Thus, we would have:

$$\forall j \geq k \quad \text{Mem}_k(\text{Env}(x)) = \text{Mem}_j(\text{Env}(x)) \text{ and } \text{Mem}_k(\text{Env}(y)) = \text{Mem}_j(\text{Env}(y))$$

But, by definition, if $\text{Mem}_k \succ_{\text{Env}} \text{Mem}_{k+1}$, then:

$$\text{Mem}_k(\text{Env}(x)) \neq \text{Mem}_{k+1}(\text{Env}(x)) \text{ or } \text{Mem}_k(\text{Env}(y)) \neq \text{Mem}_{k+1}(\text{Env}(y))$$

This is a contradiction: our hypothesis is false.

Now, let us prove $P_{\text{Env}}(\text{Mem})$ by well-founded induction. Take $\text{Mem} \in M_{\text{Env}}$. By induction hypothesis, we have $\forall \text{Mem}' \prec_{\text{Env}} \text{Mem}$, $P_{\text{Env}}(\text{Mem}')$ and we must prove the property $P_{\text{Env}}(\text{Mem})$. Let $\text{Mem}(\text{Env}(x)) = m$ and $\text{Mem}(\text{Env}(y)) = n$. Two cases are possible:

1) if $m = n$, then we get $P_{\text{Env}}(\text{Mem})$ since we can construct the following tree:

$$(C_7)\frac{[\![e_w]\!]^{\text{Mem}}_{\text{Env}} = (false, \text{Mem})}{\langle \text{Mem}, c \rangle \twoheadrightarrow_{\text{Env}} \text{Mem}}$$

2) if $m \neq n$, then there are two possible subcases:

a) if $m \leq n$, then we can construct the tree ∇_1 :

$$(C_4)\frac{[\![e]\!]^{\text{Mem}}_{\text{Env}} = (true, \text{Mem}) \quad (C_1)\dfrac{[\![\downarrow y - \downarrow x]\!]^{\text{Mem}}_{\text{Env}} = (n - m, \text{Mem})}{\langle \text{Mem}, c_1 \rangle \twoheadrightarrow_{\text{Env}} \text{Mem}[\text{Env}(y) := n - m]}}{\langle \text{Mem}, c_w \rangle \twoheadrightarrow_{\text{Env}} \text{Mem}[\text{Env}(y) := n - m]}$$

But, by definition, $\text{Mem}[\text{Env}(y) := n - m] \prec_{\text{Env}} \text{Mem}$, and by induction hypothesis, we obtain the tree ∇_2 :

$$(C_i)\frac{\vdots}{\langle \text{Mem}[\text{Env}(y) := n - m], c_w\rangle \twoheadrightarrow_{\text{Env}} \text{Mem}'}$$

b) if $m > n$, then we can build the tree ∇_1:

$$(C_5)\frac{[\![e]\!]_{\text{Env}}^{\text{Mem}} = (false, \text{Mem}) \quad (C_1)\dfrac{[\![\downarrow x - \downarrow y]\!]_{\text{Env}}^{\text{Mem}} = (m - n, \text{Mem})}{\langle \text{Mem}, c_2\rangle \twoheadrightarrow_{\text{Env}} \text{Mem}[\text{Env}(x) := m - n]}}{\langle \text{Mem}, c_w\rangle \twoheadrightarrow_{\text{Env}} \text{Mem}[\text{Env}(x) := m - n]}$$

Once again, we obtain $\text{Mem}[\text{Env}(x) := m - n] \prec_{\text{Env}} \text{Mem}$, and using the induction hypothesis, we obtain the tree ∇_2:

$$(C_j)\frac{\vdots}{\langle \text{Mem}[\text{Env}(x) := m - n], c_w\rangle \twoheadrightarrow_{\text{Env}} \text{Mem}'}$$

In all cases, we get the result since we can construct the following tree:

$$(C_6)\frac{[\![e_w]\!]_{\text{Env}}^{\text{Mem}} = (true, \text{Mem}) \quad \underset{1}{\nabla} \quad \underset{2}{\nabla}}{\langle \text{Mem}, c\rangle \twoheadrightarrow_{\text{Env}} \text{Mem}'}$$

This proof is typically used to show that a program terminates: first, a well-founded order must be built (a point which is not always immediately obvious), then the proof can be done by well-founded induction.

4.5.3. *Determinism of program execution*

To ensure that the same program executed from the same state will always produce the same result, we prove that the execution of instructions is deterministic.

PROPOSITION 4.1.– Determinism of instruction execution:

$$\forall c \in \mathbf{Lang}_3\ \forall \text{Env} \in \mathbf{E}\ \forall \text{Mem}, \text{Mem}_1, \text{Mem}_2 \in \mathbf{M}$$
$$((\langle \text{Mem}, c\rangle \twoheadrightarrow_{\text{Env}} \text{Mem}_1 \text{ and } \langle \text{Mem}, c\rangle \twoheadrightarrow_{\text{Env}} \text{Mem}_2) \Rightarrow \text{Mem}_1 = \text{Mem}_2$$

We first attempt to prove this proposition by induction over the program structure. For the program while e do c, by induction hypothesis, we may suppose that the property is verified for c. Using the following two trees

$$\left(\underset{1}{\nabla}\right) \quad (C_j)\frac{\vdots}{\langle \text{Mem}', c\rangle \twoheadrightarrow_{\text{Env}} \text{Mem}_1} \qquad \left(\underset{2}{\nabla}\right) \quad (C_{j'})\frac{\vdots}{\langle \text{Mem}', c\rangle \twoheadrightarrow_{\text{Env}} \text{Mem}_1'}$$

we therefore obtain $\text{Mem}_1 = \text{Mem}_1'$.

In the case where e is evaluated as $true$, we must prove the equality between two memory states Mem_2 and Mem'_2, obtained in the following manner (note that the evaluation of expressions is deterministic):

$$
(C_6)\dfrac{\begin{array}{c}[\![e]\!]_{\mathrm{Env}}^{\mathrm{Mem}}\\ =(true,\mathrm{Mem}')\end{array} \quad \underset{1}{\nabla} \quad (C_k)\dfrac{\vdots}{\langle \mathrm{Mem}_1, \textsf{while } e \textsf{ do } c\rangle \twoheadrightarrow_{\mathrm{Env}} \mathrm{Mem}_2}}{\langle \mathrm{Mem}, \textsf{while } e \textsf{ do } c\rangle \twoheadrightarrow_{\mathrm{Env}} \mathrm{Mem}_2}
$$

$$
(C_6)\dfrac{\begin{array}{c}[\![e]\!]_{\mathrm{Env}}^{\mathrm{Mem}}\\ =(true,\mathrm{Mem}')\end{array} \quad \underset{2}{\nabla} \quad (C_{k'})\dfrac{\vdots}{\langle \mathrm{Mem}'_1, \textsf{while } e \textsf{ do } c\rangle \twoheadrightarrow_{\mathrm{Env}} \mathrm{Mem}'_2}}{\langle \mathrm{Mem}, \textsf{while } e \textsf{ do } c\rangle \twoheadrightarrow_{\mathrm{Env}} \mathrm{Mem}'_2}
$$

The induction hypothesis over c establishes the equality $\mathrm{Mem}_1 = \mathrm{Mem}'_1$. To deduce $\mathrm{Mem}_2 = \mathrm{Mem}'_2$, an induction hypothesis over the instruction while e do c would also be needed but there is no way to get it.

Hence, reasoning by induction over the structure of programs is not sufficient to prove the determinism of program execution. In this case, instead of using the induction scheme generated by the inductive definition of $\mathbf{Lang}_3$, we need to use the induction scheme generated by the inference system defining the relation $\langle \mathrm{Mem}, c\rangle \twoheadrightarrow_{\mathrm{Env}} \mathrm{Mem}'$. This leads to reason by induction over executions instead of programs. A formal definition of this induction scheme is shown in Box 4.4.

$$
\begin{array}{ll}
\text{if} & \forall \mathrm{Mem} \in \mathbf{M}\ \mathrm{Env}(x) = r \in \mathbf{R} \text{ and } [\![e]\!]_{\mathrm{Env}}^{\mathrm{Mem}} = (v, \mathrm{Mem}') \in \mathbf{V} \times \mathbf{M} \\
& \Rightarrow P(\langle \mathrm{Mem}, x := e\rangle \twoheadrightarrow_{\mathrm{Env}} \mathrm{Mem}'[r := v]) \\
\text{and} & \forall \mathrm{Mem} \in \mathbf{M}\ P(\langle \mathrm{Mem}, \textsf{skip}\rangle \twoheadrightarrow_{\mathrm{Env}} \mathrm{Mem}) \\
\text{and} & \forall c_1, c_2 \in \mathbf{Lang}_3\ \forall \mathrm{Mem}, \mathrm{Mem}_1, \mathrm{Mem}_2 \in \mathbf{M} \\
& (P(\langle \mathrm{Mem}, c_1\rangle \twoheadrightarrow_{\mathrm{Env}} \mathrm{Mem}_1) \text{ and } P(\langle \mathrm{Mem}_1, c_2\rangle \twoheadrightarrow_{\mathrm{Env}} \mathrm{Mem}_2)) \\
& \Rightarrow P(\langle \mathrm{Mem}, c_1\, ;\, c_2\rangle \twoheadrightarrow_{\mathrm{Env}} \mathrm{Mem}_2) \\
\text{and} & \forall c_1, c_2 \in \mathbf{Lang}_3\ \forall \mathrm{Mem}, \mathrm{Mem}_1, \mathrm{Mem}_2 \in \mathbf{M}\ \forall e \in \mathbf{Exp}_3 \\
& ([\![e]\!]_{\mathrm{Env}}^{\mathrm{Mem}} = (true, \mathrm{Mem}_1) \text{ and } P(\langle \mathrm{Mem}_1, c_1\rangle \twoheadrightarrow_{\mathrm{Env}} \mathrm{Mem}_2)) \\
& \Rightarrow P(\langle \mathrm{Mem}, \textsf{if } e \textsf{ then } c_1 \textsf{ else } c_2\rangle \twoheadrightarrow_{\mathrm{Env}} \mathrm{Mem}_2) \\
\text{and} & \forall c_1, c_2 \in \mathbf{Lang}_3\ \forall \mathrm{Mem}, \mathrm{Mem}_1, \mathrm{Mem}_2 \in \mathbf{M}\ \forall e \in \mathbf{Exp}_3 \\
& ([\![e]\!]_{\mathrm{Env}}^{\mathrm{Mem}} = (false, \mathrm{Mem}_1) \text{ and } P(\langle \mathrm{Mem}_1, c_2\rangle \twoheadrightarrow_{\mathrm{Env}} \mathrm{Mem}_2)) \\
& \Rightarrow P(\langle \mathrm{Mem}, \textsf{if } e \textsf{ then } c_1 \textsf{ else } c_2\rangle \twoheadrightarrow_{\mathrm{Env}} \mathrm{Mem}_2) \\
\text{and} & \forall c \in \mathbf{Lang}_3\ \forall \mathrm{Mem}, \mathrm{Mem}_1, \mathrm{Mem}_2, \mathrm{Mem}_3 \in \mathbf{M}\ \forall e \in \mathbf{Exp}_3 \\
& ([\![e]\!]_{\mathrm{Env}}^{\mathrm{Mem}} = (true, \mathrm{Mem}_1) \text{ and } P(\langle \mathrm{Mem}_1, c\rangle \twoheadrightarrow_{\mathrm{Env}} \mathrm{Mem}_2) \\
& \quad \text{and } P(\langle \mathrm{Mem}_2, \textsf{while } e \textsf{ do } c\rangle \twoheadrightarrow_{\mathrm{Env}} \mathrm{Mem}_3)) \\
& \Rightarrow P(\langle \mathrm{Mem}, \textsf{while } e \textsf{ do } c\rangle \twoheadrightarrow_{\mathrm{Env}} \mathrm{Mem}_3) \\
\text{and} & \forall c \in \mathbf{Lang}_3\ \forall \mathrm{Mem}, \mathrm{Mem}_1 \in \mathbf{M}\ \forall e \in \mathbf{Exp}_3 \\
& [\![e]\!]_{\mathrm{Env}}^{\mathrm{Mem}} = (true, \mathrm{Mem}_1) \Rightarrow P(\langle \mathrm{Mem}, \textsf{while } e \textsf{ do } c\rangle \twoheadrightarrow_{\mathrm{Env}} \mathrm{Mem}_1) \\
\text{then} & \forall c \in \mathbf{Lang}_3\ \forall \mathrm{Mem}, \mathrm{Mem}' \in \mathbf{M}\ P(\langle \mathrm{Mem}, c\rangle \twoheadrightarrow_{\mathrm{Env}} \mathrm{Mem}')
\end{array}
$$

Box 4.4. *Reasoning scheme by structural induction over* $\langle \mathrm{Mem}, c\rangle \twoheadrightarrow_{\mathrm{Env}} \mathrm{Mem}'$

PROOF.– (This is the proof of proposition 4.1)

The property to prove $P(\langle \text{Mem}, c\rangle \twoheadrightarrow_{\text{Env}} \text{Mem}_1)$ is defined by:

$$\forall \text{Mem}_2 \in \mathbf{M}\ \langle \text{Mem}, c\rangle \twoheadrightarrow_{\text{Env}} \text{Mem}_2 \Rightarrow \text{Mem}_1 = \text{Mem}_2$$

We prove the following points.

– The form of the tree for $\langle \text{Mem}, x := e\rangle \twoheadrightarrow_{\text{Env}} \text{Mem}_1$ can only be:

$$(C_1)\frac{\text{Env}(x) = r \in \mathbf{R} \qquad [\![e]\!]_{\text{Env}}^{\text{Mem}} = (v, \text{Mem}') \in \mathbf{V} \times \mathbf{M}}{\langle \text{Mem}, x := e\rangle \twoheadrightarrow_{\text{Env}} \text{Mem}'[r := v]}$$

hence $\text{Mem}_1 = \text{Mem}'[r := v]$. As the evaluation of expressions is deterministic, this is the only possible tree for $\langle \text{Mem}, x := e\rangle \twoheadrightarrow_{\text{Env}} \text{Mem}_2$, so $\text{Mem}_1 = \text{Mem}_2$.

– The form of the tree for $\langle \text{Mem}, \textsf{skip}\rangle \twoheadrightarrow_{\text{Env}} \text{Mem}_1$ can only be:

$$(C_2)\frac{}{\langle \text{Mem}, \textsf{skip}\rangle \twoheadrightarrow_{\text{Env}} \text{Mem}}$$

This is also the only possible tree for $\langle \text{Mem}, \textsf{skip}\rangle \twoheadrightarrow_{\text{Env}} \text{Mem}_2$, so $\text{Mem}_1 = \text{Mem} = \text{Mem}_2$.

– The tree for $\langle \text{Mem}, c_1 ; c_2\rangle \twoheadrightarrow_{\text{Env}} \text{Mem}_1$ must take the form:

$$(C_3)\frac{(C_i)\dfrac{\vdots}{\langle \text{Mem}, c_1\rangle \twoheadrightarrow_{\text{Env}} \text{Mem}'} \qquad (C_j)\dfrac{\vdots}{\langle \text{Mem}', c_2\rangle \twoheadrightarrow_{\text{Env}} \text{Mem}_1}}{\langle \text{Mem}, c_1 ; c_2\rangle \twoheadrightarrow_{\text{Env}} \text{Mem}_1}$$

Similarly, the tree for $\langle \text{Mem}, c_1 ; c_2\rangle \twoheadrightarrow_{\text{Env}} \text{Mem}_2$ must take the form:

$$(C_3)\frac{(C_{i'})\dfrac{\vdots}{\langle \text{Mem}, c_1\rangle \twoheadrightarrow_{\text{Env}} \text{Mem}''} \qquad (C_{j'})\dfrac{\vdots}{\langle \text{Mem}'', c_2\rangle \twoheadrightarrow_{\text{Env}} \text{Mem}_2}}{\langle \text{Mem}, c_1 ; c_2\rangle \twoheadrightarrow_{\text{Env}} \text{Mem}_2}$$

By induction hypothesis, we have $P(\langle \text{Mem}, c_1\rangle \twoheadrightarrow_{\text{Env}} \text{Mem}')$ and thus $\text{Mem}' = \text{Mem}''$ and since we have, still from the induction hypothesis, $P(\langle \text{Mem}', c_2\rangle \twoheadrightarrow_{\text{Env}} \text{Mem}_1)$, we finally obtain $\text{Mem}_1 = \text{Mem}_2$.

– When $[\![e]\!]_{\text{Env}}^{\text{Mem}} = (true, \text{Mem}')$, the tree for $\langle \text{Mem}, \textsf{if}\ e\ \textsf{then}\ c_1\ \textsf{else}\ c_2\rangle \twoheadrightarrow_{\text{Env}} \text{Mem}_1$ must take the form:

$$(C_4)\frac{[\![e]\!]_{\text{Env}}^{\text{Mem}} = (true, \text{Mem}') \qquad (C_i)\dfrac{\vdots}{\langle \text{Mem}', c_1\rangle \twoheadrightarrow_{\text{Env}} \text{Mem}_1}}{\langle \text{Mem}, \textsf{if}\ e\ \textsf{then}\ c_1\ \textsf{else}\ c_2\rangle \twoheadrightarrow_{\text{Env}} \text{Mem}_1}$$

Similarly, the tree for $\langle \text{Mem}, \textsf{if } e \textsf{ then } c_1 \textsf{ else } c_2 \rangle \twoheadrightarrow_{\text{Env}} \text{Mem}_2$ must take the form:

$$(C_4)\dfrac{[\![e]\!]_{\text{Env}}^{\text{Mem}} = (true, \text{Mem}') \qquad (C_j)\dfrac{\vdots}{\langle \text{Mem}', c_1 \rangle \twoheadrightarrow_{\text{Env}} \text{Mem}_2}}{\langle \text{Mem}, \textsf{if } e \textsf{ then } c_1 \textsf{ else } c_2 \rangle \twoheadrightarrow_{\text{Env}} \text{Mem}_2}$$

By induction hypothesis, we have $P(\langle \text{Mem}', c_1 \rangle \twoheadrightarrow_{\text{Env}} \text{Mem}_1)$ and thus $\text{Mem}_1 = \text{Mem}_2$. The reasoning used when $[\![e]\!]_{\text{Env}}^{\text{Mem}} = (false, \text{Mem}')$ is similar.

– When $[\![e]\!]_{\text{Env}}^{\text{Mem}} = (true, \text{Mem}')$, the tree for $\langle \text{Mem}, \textsf{while } e \textsf{ do } c \rangle \twoheadrightarrow_{\text{Env}} \text{Mem}_1$ must take the form:

$$(C_6)\dfrac{\begin{array}{c}[\![e]\!]_{\text{Env}}^{\text{Mem}} \\ = \\ (true, \text{Mem}')\end{array} \quad (C_i)\dfrac{\vdots}{\langle \text{Mem}', c \rangle \twoheadrightarrow_{\text{Env}} \text{Mem}_1'} \quad (C_j)\dfrac{\vdots}{\langle \text{Mem}_1', \textsf{while } e \textsf{ do } c \rangle \twoheadrightarrow_{\text{Env}} \text{Mem}_1}}{\langle \text{Mem}, \textsf{while } e \textsf{ do } c \rangle \twoheadrightarrow_{\text{Env}} \text{Mem}_1}$$

Similarly, the tree for $\langle \text{Mem}, \textsf{while } e \textsf{ do } c \rangle \twoheadrightarrow_{\text{Env}} \text{Mem}_2$ must take the form:

$$(C_6)\dfrac{\begin{array}{c}[\![e]\!]_{\text{Env}}^{\text{Mem}} \\ = \\ (true, \text{Mem}')\end{array} \quad (C_{i'})\dfrac{\vdots}{\langle \text{Mem}', c \rangle \twoheadrightarrow_{\text{Env}} \text{Mem}_2'} \quad (C_{j'})\dfrac{\vdots}{\langle \text{Mem}_2', \textsf{while } e \textsf{ do } c \rangle \twoheadrightarrow_{\text{Env}} \text{Mem}_2}}{\langle \text{Mem}, \textsf{while } e \textsf{ do } c \rangle \twoheadrightarrow_{\text{Env}} \text{Mem}_2}$$

By induction hypothesis, we have $P(\langle \text{Mem}', c \rangle \twoheadrightarrow_{\text{Env}} \text{Mem}_1')$ and thus $\text{Mem}_1' = \text{Mem}_2'$, and since we have, here again by induction hypothesis, $P(\langle \text{Mem}_1', \textsf{while } e \textsf{ do } c \rangle \twoheadrightarrow_{\text{Env}} \text{Mem}_1)$, we finally obtain $\text{Mem}_1 = \text{Mem}_2$.

– When $[\![e]\!]_{\text{Env}}^{\text{Mem}} = (false, \text{Mem}')$, the tree for $\langle \text{Mem}, \textsf{while } e \textsf{ do } c \rangle \twoheadrightarrow_{\text{Env}} \text{Mem}_1$ must take the form:

$$(C_7)\dfrac{[\![e]\!]_{\text{Env}}^{\text{Mem}} = (false, \text{Mem}')}{\langle \text{Mem}, \textsf{while } e \textsf{ do } c \rangle \twoheadrightarrow_{\text{Env}} \text{Mem}'}$$

This is also the only possible tree for $\langle \text{Mem}, \textsf{while } e \textsf{ do } c \rangle \twoheadrightarrow_{\text{Env}} \text{Mem}_2$ and we therefore obtain $\text{Mem}_1 = \text{Mem}' = \text{Mem}_2$. □

4.5.4. *Big steps versus small steps*

Big-step and small-step operational semantics provide two ways to describe the execution of programs in **Lang**$_3$. These descriptions are based on different formalisms, but are identical from a semantic perspective. As we can see from examples 4.3 and 4.4, the same finite execution of a program may be described by a tree, in the case of big-step semantics, or by a transition sequence in the case of small-step semantics; the results obtained using the two approaches are identical. We now formally prove this semantic equivalence. We begin by proving the following lemma.

LEMMA 4.1.– If $\langle c_1 \,;\, c_2, \text{Mem}\rangle \hookrightarrow^{k}_{\text{Env}} \text{Mem}''$, then there exists $\text{Mem}' \in \mathbf{M}$ and two integers k_1 and k_2 such that $\langle c_1, \text{Mem}\rangle \hookrightarrow^{k_1}_{\text{Env}} \text{Mem}'$ and $\langle c_2, \text{Mem}'\rangle \hookrightarrow^{k_2}_{\text{Env}} \text{Mem}''$ with $k = k_1 + k_2 + 1$.

PROOF.– We proceed by induction over k. If $k = 0$, the property is trivial (since in this case the hypothesis $\langle c_1 \,;\, c_2, \text{Mem}\rangle \hookrightarrow^{0}_{\text{Env}} \text{Mem}$ cannot be verified, since $\langle c_1 \,;\, c_2, \text{Mem}\rangle$ is not a terminal configuration). If $k = k_0 + 1$, two cases are possible:

1) If the transition sequence in the hypothesis is:

$$\langle c_1 \,;\, c_2, \text{Mem}\rangle \hookrightarrow_{\text{Env}} \underbrace{\langle c_1' \,;\, c_2, \text{Mem}_1\rangle \hookrightarrow_{\text{Env}} \cdots \hookrightarrow_{\text{Env}} \langle \textsf{skip}, \text{Mem}''\rangle}_{\text{sequence of length } k_0}$$

with $\langle c_1, \text{Mem}\rangle \hookrightarrow_{\text{Env}} \langle c_1', \text{Mem}_1\rangle$, then, by induction hypothesis, there exists two integers k_1' and k_2' such that $\langle c_1', \text{Mem}_1\rangle \hookrightarrow^{k_1'}_{\text{Env}} \text{Mem}_1'$ and $\langle c_2, \text{Mem}_1'\rangle \hookrightarrow^{k_2'}_{\text{Env}} \text{Mem}''$ with $k_0 = k_1' + k_2' + 1$. We construct the following sequence of length $k_1' + 1$:

$$\langle c_1, \text{Mem}\rangle \hookrightarrow_{\text{Env}} \langle c_1', \text{Mem}_1\rangle \hookrightarrow^{k_1'}_{\text{Env}} \text{Mem}_1'$$

which allows us to conclude with $k_1 = k_1' + 1$ and $k_2 = k_2'$.

2) If the transition sequence in the hypothesis is:

$$\langle \textsf{skip} \,;\, c_2, \text{Mem}\rangle \hookrightarrow_{\text{Env}} \underbrace{\langle c_2, \text{Mem}\rangle \hookrightarrow_{\text{Env}} \cdots \hookrightarrow_{\text{Env}} \langle \textsf{skip}, \text{Mem}''\rangle}_{\text{sequence of length } k_0}$$

then we can conclude directly, with $k_1 = 0$ and $k_2 = k_0$. □

The equivalence between the big-step and small-step operational semantics can now be proved.

PROPOSITION 4.2.– $\forall c \in \mathbf{Lang}_3\ \forall \text{Env} \in \mathbf{E}\ \forall \text{Mem}_1, \text{Mem}_2 \in \mathbf{M}$:

$$\langle \text{Mem}_1, c\rangle \twoheadrightarrow_{\text{Env}} \text{Mem}_2 \Leftrightarrow \langle c, \text{Mem}_1\rangle \hookrightarrow^{\star}_{\text{Env}} \text{Mem}_2$$

PROOF.–

($\Rightarrow$) We proceed by induction over $\langle \text{Mem}_1, c\rangle \twoheadrightarrow_{\text{Env}} \text{Mem}_2$ (applying the induction scheme presented in Box 4.4).

If $\langle \text{Mem}_1, c\rangle \twoheadrightarrow_{\text{Env}} \text{Mem}_2$ has been obtained using rule C_1:

$$(C_1)\frac{\text{Env}(x) = r \in \mathbf{R} \qquad [\![e]\!]_{\text{Env}}^{\text{Mem}_1} = (v, \text{Mem}) \in \mathbf{V} \times \mathbf{M}}{\langle \text{Mem}_1, x := e\rangle \twoheadrightarrow_{\text{Env}} \text{Mem}[r := v]}$$

then rule S_1 gives us $\langle x := e, \text{Mem}_1\rangle \hookrightarrow_{\text{Env}} \langle \mathsf{skip}, \text{Mem}[r := v]\rangle$ and since $\langle \mathsf{skip}, \text{Mem}[r := v]\rangle$ is a terminal configuration, we get:

$$\langle x := e, \text{Mem}_1\rangle \hookrightarrow_{\text{Env}}^{\star} \text{Mem}[r := v]$$

If $\langle \text{Mem}_1, c\rangle \twoheadrightarrow_{\text{Env}} \text{Mem}_2$ has been obtained using rule C_2, then c is the instruction skip and $\text{Mem}_1 = \text{Mem}_2$, and since $\langle \mathsf{skip}, \text{Mem}_2\rangle$ is a terminal configuration, by considering a sequence of length zero we get:

$$\langle \mathsf{skip}, \text{Mem}_1\rangle \hookrightarrow_{\text{Env}}^{\star} \text{Mem}_2$$

If $\langle \text{Mem}_1, c\rangle \twoheadrightarrow_{\text{Env}} \text{Mem}_2$ has been obtained using rule C_3:

$$(C_3)\frac{(C_i)\dfrac{\vdots}{\langle \text{Mem}_1, c_1\rangle \twoheadrightarrow_{\text{Env}} \text{Mem}} \qquad (C_j)\dfrac{\vdots}{\langle \text{Mem}, c_2\rangle \twoheadrightarrow_{\text{Env}} \text{Mem}_2}}{\langle \text{Mem}_1, c_1 \,;\, c_2\rangle \twoheadrightarrow_{\text{Env}} \text{Mem}_2}$$

by induction hypothesis, we have $\langle c_1, \text{Mem}_1\rangle \hookrightarrow_{\text{Env}}^{\star} \text{Mem}$ and $\langle c_2, \text{Mem}\rangle \hookrightarrow_{\text{Env}}^{\star} \text{Mem}_2$. Hence, we have two transition sequences:

$$\langle c_1, \text{Mem}_1\rangle \hookrightarrow_{\text{Env}} \cdots \hookrightarrow_{\text{Env}} \langle \mathsf{skip}, \text{Mem}\rangle \qquad [4.4]$$

$$\langle c_2, \text{Mem}\rangle \hookrightarrow_{\text{Env}} \cdots \hookrightarrow_{\text{Env}} \langle \mathsf{skip}, \text{Mem}_2\rangle \qquad [4.5]$$

Applying rule S_2 to each of the transitions in sequence [4.4], we can build the following sequence:

$$\langle c_1 \,;\, c_2, \text{Mem}_1\rangle \hookrightarrow_{\text{Env}} \cdots \hookrightarrow_{\text{Env}} \langle \mathsf{skip} \,;\, c_2, \text{Mem}\rangle \qquad [4.6]$$

and since rule S_3 allows us to construct the transition:

$$\langle \mathsf{skip} \,;\, c_2, \text{Mem}\rangle \hookrightarrow_{\text{Env}} \langle c_2, \text{Mem}\rangle \qquad [4.7]$$

by concatenating [4.6], [4.7] and [4.5], we get the following sequence:

$$\langle c_1 \,;\, c_2, \text{Mem}_1\rangle \hookrightarrow_{\text{Env}} \cdots \hookrightarrow_{\text{Env}} \langle \mathsf{skip}, \text{Mem}_2\rangle$$

and thus $\langle c_1 \,;\, c_2, \text{Mem}_1\rangle \hookrightarrow_{\text{Env}}^{\star} \text{Mem}_2$.

– If $\langle \text{Mem}_1, c\rangle \twoheadrightarrow_{\text{Env}} \text{Mem}_2$ has been obtained using rule C_4:

$$(C_4)\frac{[\![e]\!]_{\text{Env}}^{\text{Mem}_1} = (true, \text{Mem}) \qquad (C_i)\dfrac{\vdots}{\langle \text{Mem}, c_1\rangle \twoheadrightarrow_{\text{Env}} \text{Mem}_2}}{\langle \text{Mem}_1, \mathsf{if}\ e\ \mathsf{then}\ c_1\ \mathsf{else}\ c_2\rangle \twoheadrightarrow_{\text{Env}} \text{Mem}_2}$$

then, by induction hypothesis, we have $\langle c_1, \text{Mem}_1\rangle \hookrightarrow^\star_{\text{Env}} \text{Mem}$ and there exists a transition sequence:

$$\langle c_1, \text{Mem}_1\rangle \hookrightarrow_{\text{Env}} \cdots \hookrightarrow_{\text{Env}} \langle \mathsf{skip}, \text{Mem}\rangle$$

from which, by applying rule S_4, we can obtain:

$$\langle \mathsf{if}\ e\ \mathsf{then}\ c_1\ \mathsf{else}\ c_2, \text{Mem}_1\rangle \hookrightarrow_{\text{Env}} \langle c_1, \text{Mem}_1\rangle \hookrightarrow_{\text{Env}} \cdots \hookrightarrow_{\text{Env}} \langle \mathsf{skip}, \text{Mem}\rangle$$

which gives us $\langle \mathsf{if}\ e\ \mathsf{then}\ c_1\ \mathsf{else}\ c_2, \text{Mem}_1\rangle \hookrightarrow^\star_{\text{Env}} \text{Mem}_2$.

– If $\langle \text{Mem}_1, c\rangle \twoheadrightarrow_{\text{Env}} \text{Mem}_2$ has been obtained using rule C_5, the reasoning is similar to that used in the previous case.

– If $\langle \text{Mem}_1, c\rangle \twoheadrightarrow_{\text{Env}} \text{Mem}_2$ has been obtained using rule C_6:

$$(C_6)\dfrac{\begin{array}{c}[\![e]\!]^{\text{Mem}_1}_{\text{Env}}\\ =\\ (true, \text{Mem})\end{array} \quad (C_i)\dfrac{\vdots}{\langle \text{Mem}, c_0\rangle \twoheadrightarrow_{\text{Env}} \text{Mem}'} \quad (C_j)\dfrac{\vdots}{\langle \text{Mem}', \mathsf{while}\ e\ \mathsf{do}\ c_0\rangle \twoheadrightarrow_{\text{Env}} \text{Mem}_2}}{\langle \text{Mem}_1, \mathsf{while}\ e\ \mathsf{do}\ c_0\rangle \twoheadrightarrow_{\text{Env}} \text{Mem}_2}$$

then, by induction hypothesis, we get $\langle c_0, \text{Mem}\rangle \hookrightarrow^\star_{\text{Env}} \text{Mem}'$ and $\langle \mathsf{while}\ e\ \mathsf{do}\ c_0, \text{Mem}'\rangle \hookrightarrow^\star_{\text{Env}} \text{Mem}_2$, and we thus have two transition sequences:

$$\langle c_0, \text{Mem}\rangle \hookrightarrow_{\text{Env}} \cdots \hookrightarrow_{\text{Env}} \langle \mathsf{skip}, \text{Mem}'\rangle \qquad [4.8]$$

$$\langle \mathsf{while}\ e\ \mathsf{do}\ c_0, \text{Mem}'\rangle \hookrightarrow_{\text{Env}} \cdots \hookrightarrow_{\text{Env}} \langle \mathsf{skip}, \text{Mem}_2\rangle \qquad [4.9]$$

Applying rule S_2 to each of the transitions in sequence [4.8], the following sequence can be constructed:

$$\langle c_0\ ;\ \mathsf{while}\ e\ \mathsf{do}\ c_0, \text{Mem}\rangle \hookrightarrow_{\text{Env}} \cdots \hookrightarrow_{\text{Env}} \langle \mathsf{skip}\ ;\ \mathsf{while}\ e\ \mathsf{do}\ c_0, \text{Mem}'\rangle \qquad [4.10]$$

and since rules S_3 and S_6 allow us to construct transitions:

$$\langle \mathsf{skip}\ ;\ \mathsf{while}\ e\ \mathsf{do}\ c_0, \text{Mem}'\rangle \hookrightarrow_{\text{Env}} \langle \mathsf{while}\ e\ \mathsf{do}\ c_0, \text{Mem}'\rangle \qquad [4.11]$$

$$\langle \mathsf{while}\ e\ \mathsf{do}\ c_0, \text{Mem}_1\rangle \hookrightarrow_{\text{Env}} \langle c_0\ ;\ \mathsf{while}\ e\ \mathsf{do}\ c_0, \text{Mem}\rangle \qquad [4.12]$$

by concatenating [4.12], [4.10], [4.11] and [4.9], we obtain the following sequence:

$$\langle \mathsf{while}\ e\ \mathsf{do}\ c_0, \text{Mem}_1\rangle \hookrightarrow_{\text{Env}} \cdots \hookrightarrow_{\text{Env}} \langle \mathsf{skip}, \text{Mem}_2\rangle$$

hence $\langle \mathsf{while}\ e\ \mathsf{do}\ c_0, \text{Mem}_1\rangle \hookrightarrow^\star_{\text{Env}} \text{Mem}_2$.

– If $\langle \text{Mem}_1, c\rangle \twoheadrightarrow_{\text{Env}} \text{Mem}_2$ has been obtained using rule C_7, then c is the instruction $\mathsf{while}\ e\ \mathsf{do}\ c_0$ and $[\![e]\!]^{\text{Mem}_1}_{\text{Env}} = (false, \text{Mem}_2)$. Rule S_7 gives us the transition $\langle \textsf{while } e \textsf{ do } c, \text{Mem}_1\rangle \hookrightarrow_{\text{Env}} \langle \textsf{skip}, \text{Mem}_2\rangle$ and since $\langle \textsf{skip}, \text{Mem}_2\rangle$ is a terminal configuration, we obtain $\langle \textsf{while } e \textsf{ do } c_0, \text{Mem}_1\rangle \hookrightarrow^{\star}_{\text{Env}} \text{Mem}_2$.

($\Leftarrow$) Since, by hypothesis $\langle c, \text{Mem}_1\rangle \hookrightarrow^{\star}_{\text{Env}} \text{Mem}_2$, there exists a sequence of transitions of length k:

$$\langle c, \text{Mem}_1\rangle \hookrightarrow_{\text{Env}} \langle c'_1, \text{Mem}'_1\rangle \hookrightarrow_{\text{Env}} \cdots \hookrightarrow_{\text{Env}} \langle c'_k, \text{Mem}'_k\rangle = \langle \textsf{skip}, \text{Mem}_2\rangle$$

Proof is obtained by well-founded induction over k then by case over the first transition in this sequence. If $k = 0$, then the starting configuration is terminal, $\text{Mem}_1 = \text{Mem}_2$ and rule C_2 allows us to conclude. Otherwise, there are several possible cases.

– If the first transition is $\langle x := e, \text{Mem}_1\rangle \hookrightarrow_{\text{Env}} \langle \textsf{skip}, \text{Mem}[r := v]\rangle$ (obtained using rule S_1), then $k = 1$ since $\langle \textsf{skip}, \text{Mem}[r := v]\rangle$ is a terminal configuration, and using rule C_1 we get:

$$\langle \text{Mem}_1, x := e\rangle \twoheadrightarrow_{\text{Env}} \text{Mem}[r := v]$$

– If the first transition is obtained using rule S_2:

$$\langle c_1 \,;\, c_2, \text{Mem}_1\rangle \hookrightarrow_{\text{Env}} \langle \underbrace{c''_1 \,;\, c_2}_{c'_1}, \text{Mem}'_1\rangle \hookrightarrow_{\text{Env}} \cdots \hookrightarrow_{\text{Env}} \underbrace{\langle c'_k, \text{Mem}'_k\rangle}_{=\langle \textsf{skip}, \text{Mem}_2\rangle}$$

with $\langle c_1, \text{Mem}_1\rangle \hookrightarrow_{\text{Env}} \langle c''_1, \text{Mem}'_1\rangle$, then, by lemma 4.1, there exists two integers k_1 and k_2 such that $\langle c_1, \text{Mem}_1\rangle \hookrightarrow^{k_1}_{\text{Env}} \text{Mem}'$ and $\langle c_2, \text{Mem}'\rangle \hookrightarrow^{k_2}_{\text{Env}} \text{Mem}_2$ with $k_1 + k_2 + 1 = k$. Hence $k_1 < k$ and $k_2 < k$ and, by induction hypothesis, we get $\langle \text{Mem}_1, c_1\rangle \twoheadrightarrow_{\text{Env}} \text{Mem}'$ and $\langle \text{Mem}', c_2\rangle \twoheadrightarrow_{\text{Env}} \text{Mem}_2$ and we can conclude by building the tree:

$$(C_3)\dfrac{(C_i)\dfrac{\vdots}{\langle \text{Mem}_1, c_1\rangle \twoheadrightarrow_{\text{Env}} \text{Mem}'} \qquad (C_j)\dfrac{\vdots}{\langle \text{Mem}', c_2\rangle \twoheadrightarrow_{\text{Env}} \text{Mem}_2}}{\langle \text{Mem}_1, c_1 \,;\, c_2\rangle \twoheadrightarrow_{\text{Env}} \text{Mem}_2}$$

– If the first transition is obtained using rule S_3:

$$\langle \textsf{skip} \,;\, c'_1, \text{Mem}_1\rangle \hookrightarrow_{\text{Env}} \langle c'_1, \text{Mem}'_1\rangle \hookrightarrow_{\text{Env}} \cdots \hookrightarrow_{\text{Env}} \underbrace{\langle c'_k, \text{Mem}'_k\rangle}_{=\langle \textsf{skip}, \text{Mem}_2\rangle}$$

with $\text{Mem}_1 = \text{Mem}'_1$, then we get $\langle \text{Mem}_1, c'_1\rangle \twoheadrightarrow_{\text{Env}} \text{Mem}_2$ by induction hypothesis and we can conclude by constructing the tree:

$$(C_3)\dfrac{(C_2)\dfrac{\vdots}{\langle \text{Mem}_1, \textsf{skip}\rangle \twoheadrightarrow_{\text{Env}} \text{Mem}_1} \qquad (C_i)\dfrac{\vdots}{\langle \text{Mem}_1, c'_1\rangle \twoheadrightarrow_{\text{Env}} \text{Mem}_2}}{\langle \text{Mem}_1, \textsf{skip} \,;\, c'_1\rangle \twoheadrightarrow_{\text{Env}} \text{Mem}_2}$$

– If the first transition is obtained using rule S_4:

$$\langle \textsf{if } e \textsf{ then } c_1' \textsf{ else } c_2, \text{Mem}_1\rangle \hookrightarrow_{\text{Env}} \langle c_1', \text{Mem}_1'\rangle \hookrightarrow_{\text{Env}} \cdots \hookrightarrow_{\text{Env}} \underbrace{\langle c_k', \text{Mem}_k'\rangle}_{=\langle \textsf{skip}, \text{Mem}_2\rangle}$$

with $[\![e]\!]_{\text{Env}}^{\text{Mem}_1} = (\mathit{true}, \text{Mem}_1')$, then $\langle \text{Mem}_1, c_1'\rangle \twoheadrightarrow_{\text{Env}} \text{Mem}_2$ by induction hypothesis, and we conclude by constructing the tree:

$$(C_4)\frac{[\![e]\!]_{\text{Env}}^{\text{Mem}_1} = (\mathit{true}, \text{Mem}_1') \qquad (C_i)\dfrac{\vdots}{\langle \text{Mem}_1, c_1'\rangle \twoheadrightarrow_{\text{Env}} \text{Mem}_2}}{\langle \text{Mem}_1, \textsf{if } e \textsf{ then } c_1' \textsf{ else } c_2\rangle \twoheadrightarrow_{\text{Env}} \text{Mem}_2}$$

– Similar reasoning is used if the first transition is obtained using rule S_5.

– If the first transition is obtained using rule S_6:

$$\begin{aligned}&\langle \textsf{while } e \textsf{ do } c_0, \text{Mem}_1\rangle \\ &\hookrightarrow_{\text{Env}} \langle \underbrace{c_0 \,;\, \textsf{while } e \textsf{ do } c_0}_{c_1'}, \text{Mem}_1'\rangle \hookrightarrow_{\text{Env}} \cdots \hookrightarrow_{\text{Env}} \underbrace{\langle c_k', \text{Mem}_k'\rangle}_{=\langle \textsf{skip}, \text{Mem}_2\rangle}\end{aligned}$$

with $[\![e]\!]_{\text{Env}}^{\text{Mem}_1} = (\mathit{true}, \text{Mem}_1')$, then $\langle \text{Mem}_1, c_1'\rangle \twoheadrightarrow_{\text{Env}} \text{Mem}_2$ by induction hypothesis, and we can construct the tree:

$$(C_4)\frac{[\![e]\!]_{\text{Env}}^{\text{Mem}_1} = (\mathit{true}, \text{Mem}_1') \qquad (C_i)\dfrac{\vdots}{\langle \text{Mem}_1', c_1'\rangle \twoheadrightarrow_{\text{Env}} \text{Mem}_2}}{\langle \text{Mem}_1, \textsf{if } e \textsf{ then } (c_0 \,;\, \textsf{while } e \textsf{ do } c_0) \textsf{ else skip}\rangle \twoheadrightarrow_{\text{Env}} \text{Mem}_2}$$

and since we can show (see exercise 4.1) that:

$$\textsf{while } e \textsf{ do } c_0 \equiv \textsf{if } e \textsf{ then } (c_0 \,;\, \textsf{while } e \textsf{ do } c_0) \textsf{ else skip}$$

we obtain $\langle \text{Mem}_1, \textsf{while } e \textsf{ do } c_0\rangle \twoheadrightarrow_{\text{Env}} \text{Mem}_2$.

– If the first transition is obtained using rule S_7:

$$\langle \textsf{while } e \textsf{ do } c_0, \text{Mem}_1\rangle \hookrightarrow_{\text{Env}} \langle \textsf{skip}, \text{Mem}_2\rangle$$

with $[\![e]\!]_{\text{Env}}^{\text{Mem}_1} = (\mathit{false}, \text{Mem}_2)$ (and thus $k = 1$), we can construct the tree:

$$(C_7)\frac{[\![e]\!]_{\text{Env}}^{\text{Mem}_1} = (\mathit{false}, \text{Mem}_2)}{\langle \text{Mem}_1, \textsf{while } e \textsf{ do } c_0\rangle \twoheadrightarrow_{\text{Env}} \text{Mem}_2}$$

□

4.6. Procedures

Procedures are used to structure the code of a program, isolating certain parts and making them accessible from any point in the program (by means of a procedure call instruction). We shall begin by introducing the notion of a block, which confines definitions of variables and procedures in a portion of code. We will then build on this notion to introduce procedures, which trigger the execution of a block, and outline different possible modes of parameter transmission. The concepts of functions and procedures give the basis of program architecture and will be extended with modular and object features in Volume 2. The big-step operational semantics formalism will be used throughout this section as a basis for presenting new constructs.

4.6.1. *Blocks*

The syntax of $\mathbf{Lang}_3$ is extended by adding the construct:

begin $\mathbf{d}$ c end

where $\mathbf{d}$ is a sequence (which may be empty) of definitions in language $\mathbf{Def}_3$ (which may be extended in order to allow local definition of procedures) and c is a program in $\mathbf{Lang}_3$. This new construct ensures that the variables and procedures defined by $\mathbf{d}$ will be accessible during an execution of c. The execution of begin $\mathbf{d}$ c end in a state $(\mathrm{Env}, \mathrm{Mem})$ is a two-step process:

1) construction of the state in which program c is to be executed by evaluating the definitions in $\mathbf{d}$:

$$(\mathrm{Env}, \mathrm{Mem}) \xrightarrow{\mathbf{d}}{}^{\star}_{\mathbf{Def}_3} (\mathrm{Env}_1, \mathrm{Mem}_1)$$

where $\rightarrow^{\star}_{\mathbf{Def}_3}$ is the relation defined in section 4.2 (adapted for language $\mathbf{Def}_3$);

2) execution of the program c in state $(\mathrm{Env}_1, \mathrm{Mem}_1)$:

$$\langle \mathrm{Mem}_1, c\rangle \twoheadrightarrow_{\mathrm{Env}_1} \mathrm{Mem}_2$$

To take account of this new construct, the rule C_8 given in Box 4.5 is added to the rules in Box 4.2.

$$(C_8)\,\frac{(\mathrm{Env}, \mathrm{Mem}) \xrightarrow{\mathbf{d}}{}^{\star}_{\mathbf{Def}_3} (\mathrm{Env}_1, \mathrm{Mem}_1) \qquad \langle \mathrm{Mem}_1, c\rangle \twoheadrightarrow_{\mathrm{Env}_1} \mathrm{Mem}_2}{\langle \mathrm{Mem}, \text{begin } \mathbf{d}\ c \text{ end}\rangle \twoheadrightarrow_{\mathrm{Env}} \mathrm{Mem}_2}$$

Box 4.5. *Big-step operational semantics: blocks*

Once again, only the memory can be modified by executing a block: the evaluation of the declarations **d** transforms Mem into Mem_1, leading to the state where the execution of the instruction c starts; this execution leads to Mem_2. The starting environment is only temporarily modified during execution of c.

We introduce the syntactic construct that adds blocks to language $\mathbf{Lang}_3$. First, with Python, we define the following class:

Python
```
class IBloc4:
    def __init__(self,list_def,pgm):
        self.list_def = list_def
        self.pgm = pgm
```

With OCaml, the following constructor is added to the definition of the sum type `'a pgm4` (see section 4.1):

OCaml
```
| IBloc4 of ('a def4) list * 'a pgm4
```

The function `exec_pgm4`, which executes programs in $\mathbf{Lang}_3$ (see section 4.1), is extended to handle this new construct. With Python, we simply insert the following instruction into the function code:

Python
```
    if isinstance(c,IBloc4):
        env1,mem1 = trans_def4_exec((env,mem),c.list_def)
        return exec_pgm4(env1,mem1,c.pgm)
```

With OCaml, the following pattern is added to the function:

OCaml
```
| IBloc4 (d, c1) ->
    let (env1, mem1) = trans_def4_exec (env, mem) d in exec_pgm4 env1 mem1 c1
```

EXAMPLE 4.8.– In a state (Env, Mem) defined by $\text{Env} = (y, r_y) \oplus (x, r_x) \oplus \text{Env}_0$ and $\text{Mem} = \text{Mem}_0[r_x := 3][r_y := 0]$ where $(\text{Env}_0, \text{Mem}_0)$ is any given state, the execution of a block c defined by:

$$\text{begin } \underbrace{(\text{let } z = x; \text{let } w = \uparrow\downarrow x; \text{let } x = 5;)}_{\mathbf{d}} \underbrace{z := x\, ;\, y := w}_{c_B} \text{ end}$$

is obtained by, first, constructing the state in which the block instructions are to be executed:

$$\begin{array}{ll} & (\text{Env}, \text{Mem}) \\ \xrightarrow[\textbf{Def}_3]{\text{let } z = x;} & ((z, r_x) \oplus \text{Env}, \text{Mem}) \\ \xrightarrow[\textbf{Def}_3]{\text{let } w = \uparrow \downarrow x;} & ((w, r_w) \oplus (z, r_x) \oplus \text{Env}, \text{Mem}[r_w := 3]) \\ \xrightarrow[\textbf{Def}_3]{\text{let } x = 5;} & \left(\underbrace{(x, 5) \oplus (w, r_w) \oplus (z, r_x) \oplus \text{Env}}_{\text{Env}_1}, \underbrace{\text{Mem}[r_w := 3]}_{\text{Mem}_1} \right) \end{array}$$

Using rule C_8, we obtain the following execution tree:

$$(C_8)\frac{(\text{Env}, \text{Mem}) \xrightarrow{\mathbf{d}}{}^{\star}_{\textbf{Def}_3} (\text{Env}_1, \text{Mem}_1) \quad (C_3)\dfrac{\nabla_1 \quad \nabla_2}{\langle \text{Mem}_1, c_B \rangle \twoheadrightarrow_{\text{Env}_1} \text{Mem}'}}{\langle \text{Mem}, c \rangle \twoheadrightarrow_{\text{Env}} \underbrace{\text{Mem}_1[r_x := 5][r_y := r_w]}_{\text{Mem}'}}$$

where ∇_1 and ∇_2 are the trees:

$$\nabla_1 : (C_1)\frac{\text{Env}_1(z) = r_x \quad [\![x]\!]_{\text{Env}_1}^{\text{Mem}_1} = (5, \text{Mem}_1)}{\langle \text{Mem}_1, z := x \rangle \twoheadrightarrow_{\text{Env}_1} \text{Mem}_1[r_x := 5]}$$

$$\nabla_2 : (C_1)\frac{\text{Env}_1(y) = r_y \quad [\![w]\!]_{\text{Env}_1}^{\text{Mem}_1[r_x := 5]} = (r_w, \text{Mem}_1[r_x := 5])}{\langle \text{Mem}_1[r_x := 5], y := w \rangle \twoheadrightarrow_{\text{Env}_1} \text{Mem}_1[r_x := 5][r_y := r_w]}$$

During execution of block c, the environment Env_1 binds the variable x to the value 5, and the variable z to the reference r_x. Hence, assignment $z := x$ stores the value 5 in the location denoted by r_x. Allocation also takes place during execution of the block: the environment Env_1 binds variable w to a reference r_w. At the end of execution of this block, environment Env_1 is no longer accessible and this binding ceases to be available. Nevertheless, the assignment $y := w$ stored the reference r_w into the location denoted by r_y, and the contents of the location denoted by r_w thus remain accessible by evaluating the expression $\downarrow \downarrow y$.

The definitions of state (Env, Mem) and block c (both introduced at the beginning of this example), and the construction of state $(\text{Env}, \text{Mem}')$ such that $\langle \text{Mem}, c \rangle \twoheadrightarrow_{\text{Env}} \text{Mem}'$, are obtained as follows:

```
Python
(env_ex_ld4_3,mem_ex_ld4_3) \
  = trans_def4_exec(([],[]),
                        [Let_def4("x",Ref4(Cste4(CInt4(3)))),
                         Let_def4("y",Ref4(Cste4(CInt4(0))))])
ex_pgm4_2 = IBloc4([Let_def4("z",Var4("x")),
                         Let_def4("w",Ref4(Deref4(Var4("x")))),
                         Let_def4("x",Cste4(CInt4(5)))],
                        ISeq4(IAffect4("z",Var4("x")),IAffect4("y",Var4("w"))))
>>> print_state((env_ex_ld4_3,exec_pgm4(env_ex_ld4_3,mem_ex_ld4_3,ex_pgm4_2)))
[ ( y , VCste4 ('CRef4', 1) ), ( x , VCste4 ('CRef4', 0) ), ]
[ ( 2 , ('CInt4', 3) ), ( 1 , ('CRef4', 2) ), ( 0 , ('CInt4', 5) ), ]
```

```
OCaml
# let (env,mem) = (trans_def4_exec ([],[])
                        [Let_def4("x",Ref4(Cste4(CInt4(3))));
                         Let_def4("y",Ref4(Cste4(CInt4(0))))])
in let ex_pgm4_2 =
 IBloc4([Let_def4("z",Var4("x"));
         Let_def4("w",Ref4(Deref4(Var4("x"))));
         Let_def4("x",Cste4(CInt4(5)))],
        ISeq4(IAffect4("z",Var4("x")),IAffect4("y",Var4("w"))))
in (env,(exec_pgm4 env mem ex_pgm4_2));;
                     - : (string * string valeurs4) list * (int * const4) list =
([("y", VCste4 (CRef4 1)); ("x", VCste4 (CRef4 0))],
 [(2, CInt4 3); (1, CRef4 2); (0, CInt4 5)])
```

4.6.2. *Procedures*

4.6.2.1. *Language of definitions*

A new syntactic construct is added to $\mathbf{Def}_3$: proc $p(x)\{c\}$. This construct defines a procedure named p (which is an identifier), having a parameter denoted by a variable x and a body which is a program c in $\mathbf{Lang}_3$. For the sake of simplicity, we shall limit our discussion to procedures with a single parameter. The descriptions below can easily be extended to procedures with multiple parameters, for example by replacing the single parameter with an n-uple. For procedures without arguments, the argument is a value of the "empty" type.

We implement the syntactic construct used to define procedures in $\mathbf{Def}_3$. With Python, we define the following class:

```
Python
class Let_proc4:
    def __init__(self,name,param,pgm):
        self.name = name
        self.param = param
        self.pgm = pgm
```

With OCaml, the definition of the sum type `'a def4` (see section 4.1) is extended with the following constructor:

```
OCaml
| Let_proc4 of 'a * 'a * 'a pgm4
```

The execution of a procedure is triggered by a call (also called an invocation) to it. As in Chapter 3, where we describe the application of functions, there are several possible choices for determining the state in which a procedure should be executed.

Using a static (or lexical) scope mechanism, a procedure is executed in the state in which it was defined (augmented with the binding for its parameter), and the value associated with the procedure is therefore a closure containing its definition environment.

Using a dynamic scope mechanism, a procedure is executed in the state in which the call is done, and the value associated with the procedure is therefore a closure containing only the name of the parameter and the procedure body.

In both cases, the binding of the procedure name to its value is added to the execution environment, allowing definitions of recursive procedures. Table 4.5 describes how relation $\rightarrow_{\mathbf{Def}_3}$ is extended in the two cases.

Static scope	$(\text{Env}, \text{Mem}) \xrightarrow{\mathsf{proc}\ p(x)\{c\}}_{\mathbf{Def}_3} ((p, \langle x, c, \text{Env}\rangle) \oplus \text{Env}, \text{Mem})$
Dynamic scope	$(\text{Env}, \text{Mem}) \xrightarrow{\mathsf{proc}\ p(x)\{c\}}_{\mathbf{Def}_3} ((p, \langle x, c\rangle) \oplus \text{Env}, \text{Mem})$

Table 4.5. *Extension of relation $\rightarrow_{\mathbf{Def}_3}$: procedures*

Allowing the definition of procedures requires defining a new value corresponding to a closure. Let us first consider a static scope mechanism. With Python, we simply add the following class:

```
Python
class Ferm4:
    def __init__(self,param,pgm,env):
        self.param = param
        self.pgm = pgm
        self.env = env
```

With OCaml, the sum type `'a valeurs4` (see section 4.2) must be extended with the following constructor:

```
OCaml
| Ferm4 of 'a * 'a pgm4 * ('a * 'a valeurs4) list
```

Using a dynamic scope mechanism, with Python, we add the following class:

```
Python
class Ferm4:
    def __init__(self,param,pgm):
        self.param = param
        self.pgm = pgm
```

while with OCaml, the definition of the sum type `'a valeurs4` is extended with the constructor:

```
OCaml
| Ferm4 of 'a * 'a pgm4
```

To extend the relation $\rightarrow_{\mathbf{Def}_3}$ whilst taking into account procedure definitions, we need to modify the function `trans_def4` defined in section 4.3.

With Python, in case of static scope, this is done by inserting the following instruction into the function code:

```
Python
    if isinstance(d,Let_proc4):
        return (ajout_liaison_env(env,d.name,Ferm4(d.symb,d.pgm,d.env)),mem)
```

and, in case of dynamic scope, the following instruction is inserted:

```
Python
    if isinstance(d,Let_proc4):
        return (ajout_liaison_env(env,d.name,Ferm4(d.symb,d.pgm)),mem)
```

With OCaml, with static scope, we extend the function `trans_def4` with the following pattern:

```
OCaml
  | Let_proc4(p,x,c) -> ((ajout_liaison_env env p (Ferm4(x,c,env))),mem)
```

with dynamic scope, we add the following pattern:

```
OCaml
  | Let_proc4(p,x,c) -> ((ajout_liaison_env env p (Ferm4(x,c))),mem)
```

4.6.2.2. *Procedure call mechanisms*

4.6.2.2.1. Call by value

When calling a procedure, the call by value mechanism transmits a value, which is the result of evaluating an expression e in the current environment of the call. Using this mechanism, the syntactic construct of a procedure call in $\mathbf{Lang}_3$ is given as:

call $p(e)$

To add the syntactic construct for a procedure call with Python, we define the following class:

```
Python
class ICall4:
    def __init__(self,name,exp):
        self.name = name
        self.exp = exp
```

With OCaml, the definition of the sum type `'a pgm4` (see section 4.1) is extended with the following constructor:

```
OCaml
  | ICall4 of 'a * 'a exp4
```

Box 4.6 presents different semantics of this call mechanism. In the first version, the transmitted value is seen as a constant: the binding of the parameter to the value is directly present in the environment (and thus this value cannot be modified).

In the second version, the parameter of the procedure is considered as a mutable variable local to the procedure, which is initialized using the transmitted value. In this case, an allocation is done, the parameter references this location and the contents of this location may be modified by the instructions in the procedure body. As in the case of blocks, this reference may remain accessible once the procedure has been executed, if it was assigned to a variable "external" to the procedure. Both versions are presented for static and dynamic scope mechanisms.

Call by value: constant parameter

$$\text{Static scope} \quad (C_9)\frac{\begin{array}{l}\text{Env}(p) = \langle x, c, \text{Env}_p\rangle \quad [\![e]\!]^{\text{Mem}}_{\text{Env}} = (v, \text{Mem}') \\ \text{Env}' = (p, \langle x, c, \text{Env}_p\rangle) \oplus (x, v) \oplus \text{Env}_p \\ \langle \text{Mem}', c\rangle \twoheadrightarrow_{\text{Env}'} \text{Mem}''\end{array}}{\langle \text{Mem}, \mathsf{call}\ p(e)\rangle \twoheadrightarrow_{\text{Env}} \text{Mem}''}$$

$$\text{Dynamic scope} \quad (C_9)\frac{\begin{array}{l}\text{Env}(p) = \langle x, c\rangle \quad [\![e]\!]^{\text{Mem}}_{\text{Env}} = (v, \text{Mem}') \\ \text{Env}' = (p, \langle x, c\rangle) \oplus (x, v) \oplus \text{Env} \\ \langle \text{Mem}', c\rangle \twoheadrightarrow_{\text{Env}'} \text{Mem}''\end{array}}{\langle \text{Mem}, \mathsf{call}\ p(e)\rangle \twoheadrightarrow_{\text{Env}} \text{Mem}''}$$

Call by value: mutable parameter

$$\text{Static scope} \quad (C_9)\frac{\begin{array}{l}\text{Env}(p) = \langle x, c, \text{Env}_p\rangle \quad [\![e]\!]^{\text{Mem}}_{\text{Env}} = (v, \text{Mem}') \\ \text{Env}' = (p, \langle x, c, \text{Env}_p\rangle) \oplus (x, r_x) \oplus \text{Env}_p \\ \langle \text{Mem}'[r_x := v], c\rangle \twoheadrightarrow_{\text{Env}'} \text{Mem}''\end{array}}{\langle \text{Mem}, \mathsf{call}\ p(e)\rangle \twoheadrightarrow_{\text{Env}} \text{Mem}''}$$

$$\text{Dynamic scope} \quad (C_9)\frac{\begin{array}{l}\text{Env}(p) = \langle x, c\rangle \quad [\![e]\!]^{\text{Mem}}_{\text{Env}} = (v, \text{Mem}') \\ \text{Env}' = (p, \langle x, c\rangle) \oplus (x, r_x) \oplus \text{Env} \\ \langle \text{Mem}'[r_x := v], c\rangle \twoheadrightarrow_{\text{Env}'} \text{Mem}''\end{array}}{\langle \text{Mem}, \mathsf{call}\ p(e)\rangle \twoheadrightarrow_{\text{Env}} \text{Mem}''}$$

Call by reference

$$\text{Static scope} \quad (C_9)\frac{\begin{array}{l}\text{Env}(p) = \langle x_p, c, \text{Env}_p\rangle \quad \text{Env}(x) = r_x \in \mathbf{R} \\ \text{Env}' = (p, \langle x_p, c, \text{Env}_p\rangle) \oplus (x_p, r_x) \oplus \text{Env}_p \\ \langle \text{Mem}, c\rangle \twoheadrightarrow_{\text{Env}'} \text{Mem}'\end{array}}{\langle \text{Mem}, \mathsf{call}\ p(x)\rangle \twoheadrightarrow_{\text{Env}} \text{Mem}'}$$

$$\text{Dynamic scope} \quad (C_9)\frac{\begin{array}{l}\text{Env}(p) = \langle x_p, c\rangle \quad \text{Env}(x) = r_x \in \mathbf{R} \\ \text{Env}' = (p, \langle x_p, c\rangle) \oplus (x_p, r_x) \oplus \text{Env} \\ \langle \text{Mem}, c\rangle \twoheadrightarrow_{\text{Env}'} \text{Mem}'\end{array}}{\langle \text{Mem}, \mathsf{call}\ p(x)\rangle \twoheadrightarrow_{\text{Env}} \text{Mem}'}$$

Box 4.6. *Big-step operational semantics: calls to a procedure*

Using a static scope mechanism and call by value with a constant parameter, we add the following instruction to the code of the Python function `exec_pgm4` (see section 4.4.1):

Python
```
    if isinstance(c,ICall4):
        v,mem1 = eval_exp4(env,mem,c.exp)
        fp = valeur_de(env,c.name)
        if fp is None: raise ValueError
        if isinstance(fp,Ferm4) and isinstance(v,VCste4):
            enve = ajout_liaison(ajout_liaison_env(fp.env,fp.param,v),
                                  c.name,fp)
            return exec_pgm4(enve,mem1,fp.pgm)
        raise ValueError
    raise ValueError
```

In this case, with OCaml, the definition of function `exec_pgm4` is extended to include the following pattern:

OCaml
```
| ICall4(p,e) -> (match ((valeur_de env p),(eval_exp4 env mem e)) with
                    (Some(Ferm4(xp,cp,envp)),(VCste4(v),mem1))
                    -> let enve = (ajout_liaison_env
                                    (ajout_liaison_env envp xp (VCste4(v)))
                                    p (Ferm4(xp,cp,envp)))
                       in (exec_pgm4 enve mem1 cp)
              | _ -> raise Exec_Error)
```

Using a dynamic scope mechanism and call by value with a constant parameter, we add the following instruction to the code of the Python function `exec_pgm4`:

Python
```
    if isinstance(c,ICall4):
        v,mem1 = eval_exp4(env,mem,c.exp)
        fp = valeur_de(env,c.name)
        if fp is None:
            raise ValueError
        if isinstance(fp,Ferm4) and isinstance(v,VCste4):
            enve = ajout_liaison(ajout_liaison_env(env,fp.param,v),
                                  c.name,fp)
            return exec_pgm4(enve,mem1,fp.pgm)
        raise ValueError
```

In this case, with OCaml, the definition of function `exec_pgm4` is extended to include the following pattern:

OCaml
```
| ICall4(p,e) -> (match ((valeur_de env p),(eval_exp4 env mem e)) with
                    (Some(Ferm4(xp,cp)),(VCste4(v),mem1))
```

```
                    -> let enve = (ajout_liaison_env
                                   (ajout_liaison_env env xp (VCste4(v)))
                                   p (Ferm4(xp,cp)))
                       in (exec_pgm4 enve mem1 cp)
         | _ -> raise Exec_Error)
```

We consider only the implementation of the two cases described above. The other cases (mutable parameter and call by reference) are obtained in similar ways.

4.6.2.2.2. Call by reference

The call by reference mechanism transmits to the procedure a value, which is the reference of a mutable variable present in the current environment of the call. Using this mechanism, the syntactic construct of the procedure call in $\mathbf{Lang}_3$ is:

call $p(x)$

where x is a mutable variable in the calling environment. Its contents may be directly modified when executing the body of p.

The formal definition of this mechanism is presented in Box 4.6. Unlike the call by value mechanism, where the parameter is considered as a variable that is local to the procedure, there is no allocation in this case. This mechanism will be discussed in more detail in section 7.4, where we explain its implementation using pointers.

4.7. Other approaches

Only small-step and big-step operational semantics have been considered up to this point. Both cases involve the definition of a transition relation, which characterizes the transformation of states resulting from program execution (and from the evaluation of expressions occurring in the program, as the selected expression language permits side-effects). In this section, we present two other classical approaches to the semantics of imperative languages. For the sake of simplicity, we suppose that the evaluation of expressions does not modify the states of the memory, and we write $[\![e]\!]^{\mathrm{Mem}}_{\mathrm{Env}} = v$ to express that the evaluation of e in a state $(\mathrm{Env}, \mathrm{Mem})$ returns the value v. We use the language $\mathbf{Lang}_3$ (without blocks and procedures), whose syntax is defined in Table 4.3.

4.7.1. *Denotational semantics*

4.7.1.1. *Partial functions associated with a program*

The function `exec_pgm4`, defined in section 4.4.1, computes the state of the memory Mem' obtained after executing a program c in a state $(\mathrm{Env}, \mathrm{Mem})$. However, as we said, `exec_pgm4` does not correspond to a mathematical function as computation of Mem' does not necessarily terminate. We now wish to define this semantics by a true mathematical function:

$$[\![\_]\!]_{\_} : (\mathbf{E} \times \mathbf{Lang}_3) \rightarrow (\mathbf{M} \rightharpoonup \mathbf{M})$$

This is the purpose of denotational semantics. Some properties over programs and their executions are easier to prove within this framework, developed in detail in [STO 77].

Of course, this function $[\![\_]\!]_{\_}$ and the operational semantics must be semantically equivalent. In formal terms:

$$\begin{array}{l}\forall \text{Env} \in \mathbf{E}\ \forall c \in \mathbf{Lang}_3 \\ \forall \text{Mem}, \text{Mem}' \in \mathbf{M}\end{array} \quad [\![c]\!]_{\text{Env}}(\text{Mem}) = \text{Mem}' \Leftrightarrow \langle \text{Mem}, c\rangle \twoheadrightarrow_{\text{Env}} \text{Mem}' \qquad [4.13]$$

Thus, $[\![c]\!]_{\text{Env}}$ is a partial function from $\mathbf{M}$ to $\mathbf{M}$ (noted $\mathbf{M} \rightharpoonup \mathbf{M}$) which, given a memory state Mem, computes – in a finite amount of time – the memory state Mem' obtained after executing a program c in an environment Env. If the execution in state (Env, Mem) of the program c does not terminate, then the function $[\![c]\!]_{\text{Env}}$ is not defined for Mem (we denote this case by $[\![c]\!]_{\text{Env}}(\text{Mem}) = \text{undef}$). Hence, the denotational semantics defines a function $[\![\_]\!]_{\text{Env}}$ providing for every program c a function describing the memory state transformations, which occur when c is executed. This function $[\![\_]\!]_{\text{Env}}$ is partial and also compositional: function $[\![c]\!]_{\text{Env}}$ is obtained by composing the functions corresponding to the components of c. This is not the case for `exec_pgm4`, which defines the semantics of while e do c in a certain state by recursively using the semantics of the same instruction in a different state, the one obtained after one execution of the body of the loop.

$$[\![x := e]\!]_{\text{Env}} = \text{Mem} \mapsto \begin{cases} \text{Mem}[r := v] & \text{if } \text{Env}(x) = r \in \mathbf{R} \text{ and } v = [\![e]\!]^{\text{Mem}}_{\text{Env}} \in \mathbf{V} \\ \text{undef} & \text{otherwise} \end{cases}$$

$$[\![\mathsf{skip}]\!]_{\text{Env}} = id = \text{Mem} \mapsto \text{Mem}$$

$$[\![c_1\ ;\ c_2]\!]_{\text{Env}} = [\![c_2]\!]_{\text{Env}} \circ [\![c_1]\!]_{\text{Env}} = \text{Mem} \mapsto [\![c_2]\!]_{\text{Env}}([\![c_1]\!]_{\text{Env}}(\text{Mem}))$$

$$[\![\mathsf{if}\ e\ \mathsf{then}\ c_1\ \mathsf{else}\ c_2]\!]_{\text{Env}} = \mathsf{cond}([\![e]\!]_{\text{Env}}, [\![c_1]\!]_{\text{Env}}, [\![c_2]\!]_{\text{Env}})$$

$$\mathsf{cond} : (\mathbf{M} \rightharpoonup \mathbb{B}) \times (\mathbf{M} \rightharpoonup \mathbf{M}) \times (\mathbf{M} \rightharpoonup \mathbf{M}) \rightarrow (\mathbf{M} \rightharpoonup \mathbf{M})$$

$$\mathsf{cond}(f_b, f_1, f_2) = \text{Mem} \mapsto \begin{cases} f_1(\text{Mem}) & \text{if } f_b(\text{Mem}) = 1 \\ f_2(\text{Mem}) & \text{if } f_b(\text{Mem}) = 0 \\ \text{undef} & \text{if } f_b(\text{Mem}) = \text{undef} \end{cases}$$

$$[\![\mathsf{while}\ e\ \mathsf{do}\ c]\!]_{\text{Env}} = \text{lfp}(F) = \bigsqcup \{F^n(\bot) \mid n \geq 0\}$$

$$\begin{array}{l} F : (\mathbf{M} \rightharpoonup \mathbf{M}) \rightarrow (\mathbf{M} \rightharpoonup \mathbf{M}) \\ F : f \mapsto \mathsf{cond}([\![e]\!]_{\text{Env}}, f \circ [\![c]\!]_{\text{Env}}, id) \end{array} \qquad \bot : \text{Mem} \mapsto \text{undef}$$

Box 4.7. *Denotational semantics*

The definition of $[\![c]\!]_{\text{Env}}$ is given in Box 4.7. The assignment instruction $x := e$ is associated with the function that transforms a memory state by storing the result of evaluating the expression e (if this does not raise an error) at the location denoted by x (if x denotes a reference). The function associated with the instruction skip is the identity function. For the sequence c_1 ; c_2, we compose the two functions associated with c_1 and c_2; moreover, if one of these is applied to a state for which it is undefined, then the result of applying the composition is also undefined. The function associated with if e then c_1 else c_2 uses the function $\mathsf{cond}(f_b, f_1, f_2)$ which, given a memory state, selects which function to apply to this state (f_1 or f_2) according to the result of the boolean function (f_b) in this state. Finally, consider the instruction while e do c. Using the big-step operational semantics formalism, it is easy to show (see exercise 4.1) that:

$$\text{while } e \text{ do } c \equiv \text{if } e \text{ then } (c\,;\ \text{while } e \text{ do } c) \text{ else skip}$$

and thus, using the definitions given above, the function associated with the loop while e do c could be:

$$[\![\text{while } e \text{ do } c]\!]_{\text{Env}} = \mathsf{cond}([\![e]\!]_{\text{Env}}, [\![\text{while } e \text{ do } c]\!]_{\text{Env}} \circ [\![c]\!]_{\text{Env}}, id)$$

However, such a definition using $[\![\text{while } e \text{ do } c]\!]_{\text{Env}}$ is not compositional. One solution to this problem is to abstract the function $[\![\text{while } e \text{ do } c]\!]_{\text{Env}}$ to get the function:

$$F : f \mapsto \mathsf{cond}([\![e]\!]_{\text{Env}}, f \circ [\![c]\!]_{\text{Env}}, id)$$

This function F is known as a functional as its argument is itself a function. Its body uses only components of while e do c, F is therefore compositional. Furthermore, it verifies:

$$F([\![\text{while } e \text{ do } c]\!]_{\text{Env}}) = [\![\text{while } e \text{ do } c]\!]_{\text{Env}}$$

In other words, $[\![\text{while } e \text{ do } c]\!]_{\text{Env}}$ is a fixpoint of F. To adopt such an approach, we must establish the existence of at least one fixpoint for F and, when several fixpoints exist, determine which to use.

For example, consider the program while $\downarrow x = 0$ do skip, which may be associated with the functional $F : f \mapsto \mathsf{cond}([\![\downarrow x = 0]\!]_{\text{Env}}, f, id)$ as $[\![\text{skip}]\!]_{\text{Env}} = id$ and $f \circ id = f$. According to the definition of cond, we therefore have:

$$F : f \mapsto \left(\text{Mem} \mapsto \begin{cases} f(\text{Mem}) & \text{if Mem(Env}(x)) = 0 \text{ and Env}(x) \in \mathbf{R} \\ \text{Mem} & \text{if Mem(Env}(x)) \neq 0 \text{ and Env}(x) \in \mathbf{R} \\ \text{undef} & \text{else} \end{cases} \right)$$

Given a function $h : \mathbf{M} \rightharpoonup \mathbf{M}$, we define the following function $g_h : \mathbf{M} \rightharpoonup \mathbf{M}$:

$$g_h : \text{Mem} \mapsto \begin{cases} h(\text{Mem}) & \text{if Mem(Env}(x)) = 0 \text{ and Env}(x) \in \mathbf{R} \\ \text{Mem} & \text{if Mem(Env}(x)) \neq 0 \text{ and Env}(x) \in \mathbf{R} \\ \text{undef} & \text{else} \end{cases}$$

According to the definition of F, we therefore have:

$$F(g_h) : \text{Mem} \mapsto \begin{cases} g_h(\text{Mem}) & \text{if } \text{Mem}(\text{Env}(x)) = 0 \text{ and } \text{Env}(x) \in \mathbf{R} \\ \text{Mem} & \text{if } \text{Mem}(\text{Env}(x)) \neq 0 \text{ and } \text{Env}(x) \in \mathbf{R} \\ \text{undef} & \text{else} \end{cases}$$

and from the definition of g_h we obtain:

$$F(g_h) : \text{Mem} \mapsto \begin{cases} h(\text{Mem}) & \text{if } \text{Mem}(\text{Env}(x)) = 0 \text{ and } \text{Env}(x) \in \mathbf{R} \\ \text{Mem} & \text{if } \text{Mem}(\text{Env}(x)) \neq 0 \text{ and } \text{Env}(x) \in \mathbf{R} \\ \text{undef} & \text{else} \end{cases}$$

Hence $F(g_h) = g_h$ and so for every function $h : \mathbf{M} \rightharpoonup \mathbf{M}$, g_h is a fixpoint of F. This example shows that there exist several fixpoints of the functional F.

We now need to determine which of these fixpoints characterize exactly the same states as the operational semantics (see property [4.13]). For a memory state Mem such that $\text{Mem}(\text{Env}(x)) = 2$, the body of the loop is not executed, and the memory is not modified. For a memory state Mem such that $\text{Mem}(\text{Env}(x)) = 0$, the body of the loop is executed an infinite number of times and the program does not terminate, so we wish that $[\![\text{while} \downarrow x = 0 \text{ do skip}]\!]_{\text{Env}}(\text{Mem}) = \text{undef}$. From the function k : $\text{Mem} \mapsto \text{undef}$, we define g_k:

$$g_k : \text{Mem} \mapsto \begin{cases} \text{Mem} & \text{if } \text{Mem}(\text{Env}(x)) \neq 0 \text{ and } \text{Env}(x) \in \mathbf{R} \\ \text{undef} & \text{else} \end{cases}$$

We see that g_k is a fixpoint of F, and this is the fixpoint that we wish to associate to $[\![\text{while} \downarrow x = 0 \text{ do skip}]\!]_{\text{Env}}$.

For any other fixpoint g_h in F, if $g_k(\text{Mem}) \neq \text{undef}$, then $g_k(\text{Mem}) = g_h(\text{Mem})$. As we will see, by defining a partial order relation over partial functions from $\mathbf{M}$ to $\mathbf{M}$, this fixpoint corresponds to the least fixpoint of the functional F, denoted as $\text{lfp}(F)$. The proofs of the existence and uniqueness of this fixpoint, along with a procedure used to compute it, are presented in the following.

4.7.1.2. *Existence, uniqueness and computation of the least fixpoint of a functional*

4.7.1.2.1. Partial order over partial functions

To define the notion of the least fixpoint of a functional, we introduce the order relation $\sqsubseteq$ over the set noted $[A \rightharpoonup B]$ of partial functions from A to B:

$$f_1 \sqsubseteq f_2 \Leftrightarrow (\forall x \in A \; f_1(x) \neq \text{undef} \Rightarrow f_1(x) = f_2(x))$$

Intuitively, $f_1 \sqsubseteq f_2$ if and only if f_2 extends f_1, i.e. f_2 keeps all the results of f_1 and possibly provides some new ones. We can easily prove that this relation defines a partial order noted $(\sqsubseteq, [A \rightharpoonup B])$: the relation is reflexive, anti-symmetric and transitive.

We recall that, given a partial order $(D, \sqsubseteq)$, the upper bounds and lower bounds of a subset $Y \subseteq D$ are defined by:

$$Maj(Y) = \{d \in D \mid \forall y \in Y\ y \sqsubseteq d\} \qquad Min(Y) = \{d \in D \mid \forall y \in Y\ d \sqsubseteq y\}$$

The supremum $\sqcup Y$ of $Y \subseteq D$ is the least upper bound of Y, if there exists:

$$\forall d \in D\ (\forall y \in Y\ y \sqsubseteq d) \Rightarrow \sqcup Y \sqsubseteq d$$

The least element (minimum) of $Y \subseteq D$, if it exists, is the only element in $Min(Y) \cap Y$. The proposition below can be proved easily.

PROPOSITION 4.3.– $\perp$: Mem $\mapsto$ undef is the least element of $[\mathbf{M} \rightharpoonup \mathbf{M}]$.

Given a partial order $(D, \sqsubseteq)$ and a function $f : D \to D$, an element $d \in D$ is a fixpoint of f if and only if $f(d) = d$. If it exists, the least fixpoint of f, denoted as $\text{lfp}(f)$, is a fixpoint of f such that:

$$\forall d \in D\ (f(d) = d \Rightarrow \text{lfp}(f) \sqsubseteq d)$$

4.7.1.2.2. Complete partial orders

The least fixpoint of a functional is obtained in an incremental way by constructing a sequence of increasingly precise approximations. The least fixpoint is the least upper bound of this sequence of approximations. Complete partial orders provide a framework for establishing the existence of a least upper bound in these situations. Given a partial order $(D, \sqsubseteq)$, a subset $Y \subseteq D$ is a *chain* if its elements are totally ordered:

$$\forall y_1, y_2 \in Y\ y_1 \sqsubseteq y_2 \text{ or } y_2 \sqsubseteq y_1$$

A ccpo – *chain-complete partial order* – is a partial order $(D, \sqsubseteq)$ such that, for any chain $Y \subseteq D$, $\sqcup Y$ exists. If $\sqcup Y$ exists for every subset $Y \subseteq D$, the partial order $(D, \sqsubseteq)$ is a complete lattice (and consequently a ccpo).

LEMMA 4.2.– Every ccpo $(D, \sqsubseteq)$ has a least element $\perp = \sqcup\emptyset$.

PROOF.– $\emptyset$ is clearly a chain, and since $(D, \sqsubseteq)$ is a ccpo, $\sqcup\emptyset$ exists. Since $\sqcup\emptyset \in D$, to show that $\sqcup\emptyset$ is the least element in D, it suffices to prove that it is a lower bound of D. Using the definition of $\sqcup\emptyset$, we have:

$$\forall d \in D\ (\forall x \in \emptyset\ x \sqsubseteq d) \Rightarrow \sqcup\emptyset \sqsubseteq d$$

and since $\forall x \in \emptyset\ x \sqsubseteq d$ is always true, we have $\forall d \in D\ \sqcup\emptyset \sqsubseteq d$ and thus $\sqcup\emptyset$ is a lower bound of D. □

The following lemma, which is an immediate consequence of definitions, will be used as follows.

LEMMA 4.3.– Given a ccpo $(D, \sqsubseteq)$ and a chain $Y \subseteq D$, we have:

$$\sqcup(Y \cup \{\bot\}) = \sqcup Y$$

The following proposition characterizes the least upper bound of a chain of partial functions from $\mathbf{M}$ to $\mathbf{M}$.

PROPOSITION 4.4.– $([\mathbf{M} \rightharpoonup \mathbf{M}], \sqsubseteq)$ is a ccpo such that for any chain Y:

$$\sqcup Y : \text{Mem} \mapsto \begin{cases} f(\text{Mem}) & \text{if there exists } f \in Y \text{ such that } f(\text{Mem}) \neq \text{undef} \\ \text{undef} & \text{else} \end{cases}$$

PROOF.– Let Y be a chain in $([\mathbf{M} \rightharpoonup \mathbf{M}], \sqsubseteq)$.

First, let us prove that $\sqcup Y$ defines a function. If a function $f \in Y$ exists such that $f(\text{Mem}) \neq$ undef, then for any function $f' \in Y$ such that $f'(\text{Mem}) \neq$ undef, we have $f'(\text{Mem}) = f(\text{Mem})$. Since Y is a chain, by definition, either $f \sqsubseteq f'$ and thus $f'(\text{Mem}) = f(\text{Mem})$ since $f(\text{Mem}) \neq$ undef, or $f' \sqsubseteq f$ and since $f'(\text{Mem}) \neq$ undef we have $f(\text{Mem}) = f'(\text{Mem})$.

Now, let us show that $\sqcup Y$ defines the least upper bound of Y. Given $f \in Y$ and $\text{Mem} \in \mathbf{M}$, if $f(\text{Mem}) \neq$ undef, then by construction we have $(\sqcup Y)(\text{Mem}) = f(\text{Mem})$ and $f \sqsubseteq \sqcup Y$. So $\sqcup Y$ defines an upper bound of Y. Now we show that this is the least upper bound of Y: let g be an upper bound of Y, we have to prove that $\sqcup Y \sqsubseteq g$. Take $\text{Mem} \in \mathbf{M}$ such that $(\sqcup Y)(\text{Mem}) \neq$ undef; by definition, there exists $f \in Y$ such that $f(\text{Mem}) = (\sqcup Y)(\text{Mem})$ and since $f \sqsubseteq g$ (since g is an upper bound of Y), we obtain $(\sqcup Y)(\text{Mem}) = f(\text{Mem}) = g(\text{Mem})$ and thus the conclusion. □

4.7.1.2.3. Continuous functions

The computation of the least fixpoint of a functional F is based on the continuity of this function. Given two partial orders $(D_1, \sqsubseteq_1)$ and $(D_2, \sqsubseteq_2)$, a function $f : D_1 \rightarrow D_2$ is monotonic when it preserves the order, i.e. if and only if:

$$\forall x, y \in D_1 \quad x \sqsubseteq_1 y \Rightarrow f(x) \sqsubseteq_2 f(y)$$

The lemma below is easy to prove since $\sqsubseteq$ is a transitive relation.

LEMMA 4.4.– Let $(D_1, \sqsubseteq_1)$, $(D_2, \sqsubseteq_2)$ and $(D_3, \sqsubseteq_3)$ be three partial orders. If the functions $f : D_1 \rightarrow D_2$ and $f' : D_2 \rightarrow D_3$ are monotonic, then $f' \circ f : D_1 \rightarrow D_3$ is monotonic.

The following lemma establishes a property over the least upper bounds of a chain and of its image by a monotonic function.

LEMMA 4.5.– Let $(D_1, \sqsubseteq_1)$, $(D_2, \sqsubseteq_2)$ be two ccpo and $f : D_1 \to D_2$ a monotonic function. If Y is a chain of $(D_1, \sqsubseteq_1)$, then $f(Y)$ is a chain of $(D_2, \sqsubseteq_2)$ such that $\sqcup_2 f(Y) \sqsubseteq_2 f(\sqcup_1 Y)$.

PROOF.– If $Y = \emptyset$, then $f(Y) = \emptyset$ is a chain of $(D_2, \sqsubseteq_2)$ and since $(D_2, \sqsubseteq_2)$ is a ccpo, following lemma 4.2, $\sqcup_2\emptyset$ is the least element in D_2 and thus $\sqcup_2\emptyset \sqsubseteq_2 f(\sqcup_1\emptyset)$. If $Y \neq \emptyset$, then if $x_2, y_2 \in f(Y)$, by definition, there exists $x_1, y_1 \in Y$ such that $x_2 = f(x_1)$ and $y_2 = f(y_1)$. Since Y is a chain of $(D_1, \sqsubseteq_1)$, either $x_1 \sqsubseteq_1 y_1$ and so $x_2 = f(x_1) \sqsubseteq_2 f(y_1) = y_2$, or $y_1 \sqsubseteq_1 x_1$ and so $y_2 = f(y_1) \sqsubseteq_2 f(x_1) = x_2$ (as f is monotonic). $f(Y)$ is therefore a chain of $(D_2, \sqsubseteq_2)$.

Let us prove that $\sqcup_2 f(Y) \sqsubseteq_2 f(\sqcup_1 Y)$. By definition, $\forall y \in Y\ y \sqsubseteq_1 \sqcup_1 Y$, and since f is monotonic, $\forall y \in Y\ f(y) \sqsubseteq_2 f(\sqcup_1 Y)$, hence $f(\sqcup_1 Y)$ is an upper bound of $f(Y)$. We know that $\sqcup_2 f(Y)$ is the least upper bound of Y and so $\sqcup_2 f(Y) \sqsubseteq_2 f(\sqcup_1 Y)$. □

Given two partial orders $(D_1, \sqsubseteq_1)$ and $(D_2, \sqsubseteq_2)$, a function $f : D_1 \to D_2$ is said to be *continuous* if and only if it is monotonic and keeps the least upper bounds of the chains, i.e. if and only if f is monotonic and $\sqcup_2 f(Y) = f(\sqcup_1 Y)$ for any non-empty chain Y in $(D_1, \sqsubseteq_1)$. The function f is said to be *strict* if the property is also verified for an empty chain:

$$\sqcup_2 f(\emptyset) = \sqcup_2\emptyset = \bot_2 = f(\bot_1) = f(\sqcup_1\emptyset)$$

Once again, the composition of two continuous functions is itself a continuous function.

LEMMA 4.6.– Take three ccpo $(D_1, \sqsubseteq_1)$, $(D_2, \sqsubseteq_2)$ and $(D_3, \sqsubseteq_3)$. If $f : D_1 \to D_2$ and $f' : D_2 \to D_3$ are continuous, then $f' \circ f : D_1 \to D_3$ is continuous.

PROOF.– Functions f and f' are continuous, so we can show that $f' \circ f$ is monotonic, as stated in lemma 4.4. Let Y be a non-empty chain in $(D_1, \sqsubseteq_1)$. Since f is continuous, by definition, $\sqcup_2 f(Y) = f(\sqcup_1 Y)$ and f is monotonic. According to lemma 4.5, $f(Y)$ is a non-empty chain in $(D_2, \sqsubseteq_2)$ and since f' is continuous, we obtain $\sqcup_3 f'(f(Y)) = f'(\sqcup_2 f(Y))$; since $\sqcup_2 f(Y) = f(\sqcup_1 Y)$, we obtain $\sqcup_3 (f' \circ f)(Y) = f'(f(\sqcup_1 Y))$ and thus $\sqcup_3 (f' \circ f)(Y) = (f' \circ f)(\sqcup_1 Y)$, which proves that $f' \circ f$ is continuous. □

It is now possible to establish that the following functional is continuous:

$$F : f \mapsto \mathsf{cond}([\![e]\!]_{\text{Env}}, f \circ [\![c]\!]_{\text{Env}}, id)$$

To do this, we consider the functional as a composition of two functions, whose continuity is proved as follows:

PROPOSITION 4.5.– Take two partial functions $f_b : \mathbf{M} \rightharpoonup \mathbb{B}$ and $f_0 : \mathbf{M} \rightharpoonup \mathbf{M}$. Function $F : f \mapsto \mathsf{cond}(f_b, f, f_0)$ is continuous.

PROOF.– First, let us prove that the function F is monotonic.

Let f_1 and f_2 be two functions in $[\mathbf{M} \rightharpoonup \mathbf{M}]$ such that $f_1 \sqsubseteq f_2$, and let us show that $F(f_1) \sqsubseteq F(f_2)$. Assume $\text{Mem} \in \mathbf{M}$ such that $(F(f_1))(\text{Mem}) \neq \text{undef}$; we show that $(F(f_1))\ (\text{Mem}) = (F(f_2))(\text{Mem})$.

If $f_b(\text{Mem}) = 1$, then:

$$(F(f_1))(\text{Mem}) = f_1(\text{Mem}) \quad \text{and} \quad (F(f_2))(\text{Mem}) = f_2(\text{Mem})$$

Since $f_1 \sqsubseteq f_2$ and $f_1(\text{Mem}) = (F(f_1))(\text{Mem}) \neq \text{undef}$, we have $f_1(\text{Mem}) = f_2(\text{Mem})$ and therefore $(F(f_1))(\text{Mem}) = (F(f_2))(\text{Mem})$, establishing that $F(f_1) \sqsubseteq F(f_2)$. If $f_b(\text{Mem}) = 0$, then $(F(f_1))(\text{Mem}) = f_0(\text{Mem})$ and $(F(f_2))$ $(\text{Mem}) = f_0(\text{Mem})$, hence $(F(f_1))(\text{Mem}) = (F(f_2))(\text{Mem})$, which allows us to establish that $F(f_1) \sqsubseteq F(f_2)$. If $f_b(\text{Mem}) = \text{undef}$, then $(F(f_1))(\text{Mem}) = \text{undef}$, which is contradictory.

Let us show that for any non-empty chain Y in $([\mathbf{M} \rightharpoonup \mathbf{M}], \sqsubseteq)$, $\sqcup F(Y) = F(\sqcup Y)$.

Since we know that F is monotonic, by lemma 4.5, we get $\sqcup F(Y) \sqsubseteq F(\sqcup Y)$. It remains to prove that $F(\sqcup Y) \sqsubseteq \sqcup F(Y)$, i.e. $F(\sqcup Y) \sqsubseteq \sqcup\{F(f) \mid f \in Y\}$. Take $\text{Mem}_1 \in \mathbf{M}$. Let us prove that if $(F(\sqcup Y))(\text{Mem}_1) = \text{Mem}_2 \neq \text{undef}$, then $(\sqcup F(Y)) = \text{Mem}_2$. If $f_b(\text{Mem}_1) = 1$, then $(F(\sqcup Y))\ (\text{Mem}_1) = (\sqcup Y)$ $(\text{Mem}_1) = \text{Mem}_2$ and, by proposition 4.4, there exists $f \in Y$ such that $f(\text{Mem}_1) = \text{Mem}_2$. By definition, for all $f \in Y$, we have $F(f) \sqsubseteq \sqcup F(Y)$ and we therefore obtain $(\sqcup F(Y))(\text{Mem}_1) = \text{Mem}_2$. Otherwise, if $f_b(\text{Mem}_1) = 0$, then $(F(\sqcup Y))(\text{Mem}_1) = f_0(\text{Mem}_1) = \text{Mem}_2$ and for any function $f \in Y$:

$$(F(f))(\text{Mem}_1) = f_0(\text{Mem}_1) = \text{Mem}_2 \neq \text{undef}$$

Once again, by definition, for all $f \in Y$, we have $F(f) \sqsubseteq \sqcup F(Y)$ and we thus obtain $(\sqcup F(Y))(\text{Mem}_1) = \text{Mem}_2$. Finally, if $f_b(\text{Mem}_1) = \text{undef}$, then $(F(\sqcup Y))$ $(\text{Mem}_1) = \text{undef}$, which is contradictory. □

PROPOSITION 4.6.– Take $f_0 : \mathbf{M} \rightharpoonup \mathbf{M}$. The function $F : f \mapsto f \circ f_0$ is continuous.

PROOF.– First, let us prove that F is monotonic. Let f_1 and f_2 be two functions of $[\mathbf{M} \rightharpoonup \mathbf{M}]$ such that $f_1 \sqsubseteq f_2$, and let us prove that $F(f_1) \sqsubseteq F(f_2)$. Take $\text{Mem} \in \mathbf{M}$ such that $(F(f_1))(\text{Mem}) \neq \text{undef}$: we can show that $(F(f_1))(\text{Mem}) = (F(f_2))(\text{Mem})$. By definition, $(F(f_1))(\text{Mem}) = f_1(f_0(\text{Mem})) \neq \text{undef}$, hence

$f_0(\text{Mem}) \neq$ undef and since $f_1 \sqsubseteq f_2$, we obtain $f_1(f_0(\text{Mem})) = f_2(f_0(\text{Mem}))$, allowing us to conclude. Now, we must prove that F is continuous:

$$(F(\sqcup Y))(\text{Mem}_1) = (\sqcup Y)(f_0(\text{Mem}_1)) = \text{Mem}_2$$
$$\Leftrightarrow \exists f \in Y \; f(f_0(\text{Mem}_1)) = \text{Mem}_2 \quad \text{(see proposition 4.4)}$$

$$\Leftrightarrow \exists f \in Y \; (F(f))(\text{Mem}_1) = \text{Mem}_2$$
$$\Leftrightarrow (\sqcup F(Y))(\text{Mem}_1) = \text{Mem}_2 \quad \text{(see proposition 4.4)}$$

□

PROPOSITION 4.7.– The functional $F : f \mapsto \mathsf{cond}([\![e]\!]_{\text{Env}}, f \circ [\![c]\!]_{\text{Env}}, id)$ is a continuous function.

PROOF.– Taking $F_1 : f \mapsto \mathsf{cond}([\![e]\!]_{\text{Env}}, f, id)$ and $F_2 : f \mapsto f \circ [\![c]\!]_{\text{Env}}$, we have $F(f) = F_1(F_2(f)) = F_1(f \circ [\![c]\!]_{\text{Env}}) = \mathsf{cond}([\![e]\!]_{\text{Env}}, f \circ [\![c]\!]_{\text{Env}}, id)$ hence $F = F_1 \circ F_2$. According to proposition 4.5, function F_1 is continuous, and according to proposition 4.6, function F_2 is continuous. Lemma 4.6 thus gives the conclusion. □

4.7.1.2.4. Least fixpoint of a continuous function over a ccpo

Now let us show how the least fixpoint of the functional F can be computed incrementally. This construction is based on the theorem as follows.

THEOREM 4.2.– Let $(D, \sqsubseteq)$ be a ccpo and $f : D \to D$ a continuous function:

$$\text{lfp}(f) = \sqcup(\{f^n(\bot) \mid n \geq 0\})$$

PROOF.– First, let us prove the existence of $\sqcup(\{f^n(\bot) \mid n \geq 0\})$. It is easily done by a proof by induction over n:

$$\forall n \in \mathbb{N} \; \forall m \in \mathbb{N} \; m \geq n \Rightarrow f^n(\bot) \sqsubseteq f^m(\bot)$$

(since, according to lemma 4.2, $\bot$ is the least element of D and f is continuous, and thus monotonic). $\{f^n(\bot) \mid n \geq 0\}$ is therefore a chain, and since $(D, \sqsubseteq)$ is a ccpo, $\sqcup(\{f^n(\bot) \mid n \geq 0\})$ exists. Now, we can show that $\sqcup(\{f^n(\bot) \mid n \geq 0\})$ is a fixpoint of f:

$$f(\sqcup(\{f^n(\bot) \mid n \geq 0\})) = \sqcup(f(\{f^n(\bot) \mid n \geq 0\})) \text{ as } f \text{ is continuous}$$
$$= \sqcup(\{f^n(\bot) \mid n \geq 1\}) = \sqcup(\{f^n(\bot) \mid n \geq 1\} \cup \{\bot\}) = \sqcup(\{f^n(\bot) \mid n \geq 0\})$$

It remains to prove that $\sqcup(\{f^n(\bot) \mid n \geq 0\})$ is the least fixpoint of f. Let $d \in D$ be a fixpoint of f. According to lemma 4.2, $\bot \sqsubseteq d$ and thus for any natural number n, we have $f^n(\bot) \sqsubseteq f^n(d)$ (which can be proved by induction over n since f is continuous and thus monotonic). Furthermore, we get $\forall n \in \mathbb{N} \; f^n(d) = d$, by induction over n since d is a fixpoint of f. d is therefore an upper bound of

$\{f^n(\bot) \mid n \geq 0\}$ and since $\sqcup(\{f^n(\bot) \mid n \geq 0\})$ is the least upper bound of $\{f^n(\bot) \mid n \geq 0\}$, we obtain $\sqcup(\{f^n(\bot) \mid n \geq 0\}) \sqsubseteq d$. □

4.7.1.2.5. Example

Consider the following program, which computes $\downarrow y = \downarrow x!$ where $\downarrow x$ is a strictly positive integer:

$$\underbrace{\underbrace{y := 1}_{c_1};\underbrace{\texttt{while}\ \underbrace{\texttt{not}(\downarrow x = 1)}_{e}\ \texttt{do}\ \underbrace{(\underbrace{y :=\downarrow y \times \downarrow x}_{c_4};\underbrace{x :=\downarrow x - 1}_{c_5})}_{c_3}}_{c_2}}_{c}$$

Let us compute the result of executing this program in a state $(\text{Env}_0, \text{Mem}_0)$ such that $\text{Mem}_0\ (\text{Env}_0(x)) = 5$ and $\text{Env}_0(y) \in \mathbf{R}$. In what follows, we suppose that $\text{Env}(x)$ and $\text{Env}(y)$ are elements of $\mathbf{R}$ noted r_x and r_y respectively. By definition, the functions associated with each of the components of this program are as follows:

$$[\![c_1]\!]_{\text{Env}} : \text{Mem} \mapsto \text{Mem}[r_y := 1]$$

$$[\![e]\!]_{\text{Env}} : \text{Mem} \mapsto \begin{cases} 1 \text{ if } \text{Mem}(r_x) \neq 1 \\ 0 \text{ if } \text{Mem}(r_x) = 1 \end{cases}$$

$$\begin{aligned} &[\![c_3]\!]_{\text{Env}} = [\![c_5]\!]_{\text{Env}} \circ [\![c_4]\!]_{\text{Env}} \\ &[\![c_3]\!]_{\text{Env}} : \text{Mem} \mapsto [\![c_5]\!]_{\text{Env}}(\text{Mem}[r_y := \text{Mem}(r_y) \times \text{Mem}(r_x)]) \\ &\qquad = \text{Mem}[r_y := \text{Mem}(r_y) \times \text{Mem}(r_x)][r_x := \text{Mem}(r_x) - 1] \end{aligned}$$

$$\begin{aligned} &[\![c_2]\!]_{\text{Env}} = \text{lfp}(F) \text{ where } F : f \mapsto \textsf{cond}([\![e]\!]_{\text{Env}}, f \circ [\![c_3]\!]_{\text{Env}}, id) \\ &F(f) : \text{Mem} \mapsto \begin{cases} f\left([\![c_3]\!]_{\text{Env}}(\text{Mem})\right) & \text{if } \text{Mem}(r_x) \neq 1 \\ \text{Mem} & \text{if } \text{Mem}(r_x) = 1 \end{cases} \end{aligned}$$

The function $[\![c_3]\!]_{\text{Env}}$ describes the state obtained after executing the body of the loop. When the body of the loop is executed twice, function $[\![c_3]\!]^2_{\text{Env}} = [\![c_3; c_3]\!]_{\text{Env}} = [\![c_3]\!]_{\text{Env}} \circ [\![c_3]\!]_{\text{Env}}$ is defined by:

$$\begin{aligned} \text{Mem} \mapsto &\left(\begin{array}{l} \text{Mem } [r_y := \text{Mem}(r_y) \times \text{Mem}(r_x)] \\ \qquad [r_x := \text{Mem}(r_x) - 1] \\ \qquad [r_y := \text{Mem}(r_y) \times \text{Mem}(r_x) \times (\text{Mem}(r_x) - 1)] \\ \qquad [r_x := \text{Mem}(r_x) - 1 - 1] \end{array}\right) \\ = &\left(\begin{array}{l} \text{Mem } [r_y := \text{Mem}(r_y) \times \text{Mem}(r_x) \times (\text{Mem}(r_x) - 1)] \\ \qquad [r_x := \text{Mem}(r_x) - 2] \end{array}\right) \end{aligned}$$

More generally, when the body of the loop is executed n times, the function $[\![c3]\!]^n_{\text{Env}}$ is defined by:

$$\text{Mem} \mapsto \left(\begin{array}{l} \text{Mem} \left[\begin{array}{l} r_y := \text{Mem}(r_y) \times \text{Mem}(r_x) \times (\text{Mem}(r_x) - 1) \\ \qquad \times \cdots \times (\text{Mem}(r_x) - (n-1)) \end{array} \right] \\ [r_x := \text{Mem}(r_x) - n] \end{array} \right)$$

Thus, in a state such that $\text{Mem}(r_x) = p$, we have:

$$\begin{aligned} [\![c3]\!]^{p-1}_{\text{Env}}(\text{Mem}) &= \left(\begin{array}{l} \text{Mem } [r_y := \text{Mem}(r_y) \times p \times (p-1) \times \cdots \times 2] \\ \qquad [r_x := \text{Mem}(r_x) - 1] \end{array} \right) \\ &= \text{Mem}[r_y := \text{Mem}(r_y) \times p!][r_x := \text{Mem}(r_x - 1] \end{aligned}$$

Using theorem 4.2, we can compute $\text{lfp}(F) = \sqcup(\{F^n(\bot) \mid n \geq 0\})$ from $F^0(\bot) = \bot : \text{Mem} \mapsto \text{undef}$. By definition, we have:

$$F^1(\bot) = F(\bot) : \text{Mem} \mapsto \begin{cases} \text{undef if } \text{Mem}(r_x) \neq 1 \\ \text{Mem} \;\; \text{if } \text{Mem}(r_x) = 1 \end{cases}$$

Similarly:

$$F^2(\bot) = F(F^1(\bot)) : \text{Mem} \mapsto \begin{cases} F^1(\bot)([\![c3]\!]_{\text{Env}}(\text{Mem})) & \text{if } \text{Mem}(r_x) \neq 1 \\ \text{Mem} & \text{if } \text{Mem}(r_x) = 1 \end{cases}$$

and since $[\![c3]\!]_{\text{Env}}(\text{Mem})(r_x) = \text{Mem}(r_x) - 1$, we have:

$$F^2(\bot) : \text{Mem} \mapsto \begin{cases} \text{undef} & \text{if } \text{Mem}(r_x) \neq 1 \text{ and } \text{Mem}(r_x) \neq 2 \\ [\![c3]\!]_{\text{Env}}(\text{Mem}) & \text{if } \text{Mem}(r_x) = 2 \\ \text{Mem} & \text{if } \text{Mem}(r_x) = 1 \end{cases}$$

With a further iteration, we obtain:

$$F^3(\bot) = F(F^2(\bot)) : \text{Mem} \mapsto \begin{cases} F^2(\bot)([\![c3]\!]_{\text{Env}}(\text{Mem})) & \text{if } \text{Mem}(r_x) \neq 1 \\ \text{Mem} & \text{if } \text{Mem}(r_x) = 1 \end{cases}$$

hence:

$$F^3(\bot) : \text{Mem} \mapsto \begin{cases} \text{undef} & \text{if } \text{Mem}(r_x) \neq 1 \text{ and } \text{Mem}(r_x) \neq 2 \\ & \text{and } \text{Mem}(r_x) \neq 3 \\ [\![c3]\!]^2_{\text{Env}}(\text{Mem}) & \text{if } \text{Mem}(r_x) = 3 \\ [\![c3]\!]_{\text{Env}}(\text{Mem}) & \text{if } \text{Mem}(r_x) = 2 \\ \text{Mem} & \text{if } \text{Mem}(r_x) = 1 \end{cases}$$

By induction, we obtain:

$$F^n(\bot) : \text{Mem} \mapsto \begin{cases} \text{undef} & \text{if } \text{Mem}(r_x) \neq 1 \text{ and } \text{Mem}(r_x) \neq 2 \\ & \text{and } \cdots \text{ and } \text{Mem}(r_x) \neq n \\ [\![c3]\!]^{n-1}_{\text{Env}}(\text{Mem}) & \text{if } \text{Mem}(r_x) = n \\ [\![c3]\!]^{n-2}_{\text{Env}}(\text{Mem}) & \text{if } \text{Mem}(r_x) = n-1 \\ \cdots & \\ [\![c3]\!]^1_{\text{Env}}(\text{Mem}) & \text{if } \text{Mem}(r_x) = 2 \\ [\![c3]\!]^0_{\text{Env}}(\text{Mem}) & \text{if } \text{Mem}(r_x) = 1 \end{cases}$$

which can be written in a simpler form:

$$F^n(\bot) : \text{Mem} \mapsto \begin{cases} \text{undef} & \text{if } \text{Mem}(r_x) < 1 \text{ or } \text{Mem}(r_x) > n \\ [\![c_3]\!]^{p-1}_{\text{Env}}(\text{Mem}) & \text{if } \text{Mem}(r_x) = p \text{ and } 1 \leq p \leq n \end{cases}$$

and the function $[\![c_2]\!]_{\text{Env}} = \text{lfp}(F) = \sqcup_{n \geq 0}\{F^n(\bot)\}$ is thus defined by:

$$[\![c_2]\!]_{\text{Env}} : \text{Mem} \mapsto \begin{cases} \text{undef} & \text{if } \text{Mem}(r_x) < 1 \\ [\![c_3]\!]^{p-1}_{\text{Env}}(\text{Mem}) & \text{if } \text{Mem}(r_x) = p \text{ and } p \geq 1 \end{cases}$$

Using the expression of $[\![c_3]\!]^{p-1}_{\text{Env}}$ established earlier, we finally obtain:

$$[\![c_2]\!]_{\text{Env}} : \text{Mem} \mapsto \begin{cases} \text{undef if } \text{Mem}(r_x) < 1 \\ \text{Mem}[r_y := \text{Mem}(r_y) \times p!][r_x := \text{Mem}(r_x) - 1] \\ \text{if } \text{Mem}(r_x) = p \text{ and } p \geq 1 \end{cases}$$

It is now possible to compute $[\![c]\!]_{\text{Env}_0}(\text{Mem}_0)$ as follows:

$$\begin{aligned} [\![c]\!]_{\text{Env}_0}(\text{Mem}_0) &= [\![c_2]\!]_{\text{Env}_0}([\![c_1]\!]_{\text{Env}_0}(\text{Mem}_0)) = [\![c_2]\!]_{\text{Env}_0}(\text{Mem}_0[r_y := 1]) \\ &= \text{Mem}_0[r_y := 5!][r_x := 1] \end{aligned}$$

At the end of execution of the program c started in the state $(\text{Env}_0, \text{Mem}_0)$, the value contained in the cell designated by y is thus $\text{Mem}_0(\text{Env}_0(x))! = 5! = 120$, and the value contained in the cell designated by x is 1.

4.7.2. *Axiomatic semantics, Hoare logic*

The operational and denotational semantics of the language **Lang**$_3$ presented in this chapter characterize the transformations of the state along the execution of a program. The state descriptions provided by these semantics are "complete" in the sense that they determine the values contained in every memory cell. Axiomatic semantics describes these states in terms of the properties they verify. These properties, which are often called *assertions*, may concern the contents of any location, but it is also possible to consider properties defined only for a subset of the locations. Moreover, this axiomatic semantics is generally used to prove properties concerning the expected results of programs, rather than to describe their execution. It also provides a formalism for establishing the correctness of a program with regard to desired properties, i.e. with regard to a program specification. This formalism, introduced in [HOA 83], is based on a specification language used to create formal descriptions of tasks assigned to programs, and to express the relation between program input and output. The correctness property is expressed using a Hoare triple:

$$\{P_1\}c\{P_2\}$$

where P_1 and P_2 are expressions given in some language of specifications and c is a program. Intuitively, this triple expresses the notion that if the memory state satisfies the assertion P_1 (the precondition), then the execution of program c will lead to a memory state satisfying assertion P_2 (the postcondition).

Axiomatic semantics characterizes Hoare triples that are "true" (for example the triple $\{\downarrow x = 2\}x :=\downarrow x + 1\{\downarrow x = 3\}$ is "true", while the triple $\{\downarrow x = 2\}x :=\downarrow x + 1\{\downarrow x = 5\}$ is "false"). The logical formalism used to express the assertions may correspond to a fragment of first-order logic (with deduction methods), or to an extensional approach where an assertion P is considered to be a predicate over states such that $P((\text{Env}, \text{Mem}))$ is satisfied if and only if $[\![P]\!]_{\text{Env}}^{\text{Mem}} = 1$. In this case, assertions are assimilated to the boolean expressions used in the language. We shall take this approach here.

4.7.2.1. *Partial correctness*

4.7.2.1.1. Provable Hoare triples

The inference system in Box 4.8 contains rules that characterize a set of provable Hoare triples expressing the partial correctness of c for precondition P_1 and postcondition P_2.

$$(\text{Skp})\frac{}{\{P\}\mathsf{skip}\{P\}} \qquad (\text{Aff})\frac{}{\{P[\downarrow x \leftarrow e]\}x := e\{P\}}$$

$$(\text{Seq})\frac{\{P_1\}c_1\{P_2\} \qquad \{P_2\}c_2\{P_3\}}{\{P_1\}c_1 \,;\, c_2\{P_3\}}$$

$$(\text{Cond})\frac{\{P_1 \text{ and } e\}c_1\{P_2\} \quad \{P_1 \text{ and not } e\}c_2\{P_2\}}{\{P_1\}\mathsf{if}\ e\ \mathsf{then}\ c_1\ \mathsf{else}\ c_2\{P_2\}}$$

$$(\text{W}_\text{P})\frac{\{P_I \text{ and } e\}c\{P_I\}}{\{P_I\}\mathsf{while}\ e\ \mathsf{do}\ c\{P_I \text{ and not } e\}}$$

$$(\text{Cons})\frac{\{P_1'\}c\{P_2'\}}{\{P_1\}c\{P_2\}}\begin{pmatrix} P_1 \Rightarrow P_1' \\ P_2' \Rightarrow P_2 \end{pmatrix}$$

Box 4.8. *Axiomatic semantics: partial correctness*

Since the instruction skip does not modify the state of the memory, rule (Skp) simply indicates that the states before and after the execution of skip verify the same properties. Rule (Aff) states that for a property to be verified for the contents of a memory cell designated by x after execution of $x := e$, it must be verified for e prior to the execution of the assignment. For example, we have:

$$(\text{Aff})\frac{}{\{\downarrow x + 1 = 3\}x :=\downarrow x + 1\{\downarrow x = 3\}}$$

Note here that the precondition $\downarrow x+1 = 3$ (equivalent to $\downarrow x = 2$) corresponds to the postcondition $\downarrow x = 3$ where $\downarrow x$ (where x is the left member of the assignment) has been replaced by $\downarrow x + 1$ (the right member of the assignment), noted $(\downarrow x + 3)$ $[\downarrow x \leftarrow \downarrow x+1]$. In this rule, $P[\downarrow x \leftarrow e]$ designates the assertion obtained by replacing $\downarrow x$ by e in P: the substitution mechanism depends on the specification language used to express the assertions.

Using the extensional approach described above, the assertion $P[\downarrow x \leftarrow e]$ satisfies:

$$[\![P[\downarrow x \leftarrow e]]\!]_{\text{Env}}^{\text{Mem}} = [\![P]\!]_{\text{Env}}^{\text{Mem}'} \text{ where } \text{Mem}' = \text{Mem}[\text{Env}(x) := [\![e]\!]_{\text{Env}}^{\text{Mem}}]$$

Rule (Seq) states that if the execution of c_1 in a state that satisfies P_1 leads to a state that satisfies P_2 and if the execution of c_2 in such a state leads to a state that satisfies P_3, then the execution of the sequence $c_1\,;\,c_2$ in a state that satisfies P_1 leads to a state that satisfies P_3. For example, we have:

$$(\text{Seq})\dfrac{(\text{Aff})\dfrac{}{\{0+x \geq 0\}a := 0\{\downarrow a + x \geq 0\}} \quad (\text{Aff})\dfrac{}{\{\downarrow a + x \geq 0\}b := x\{\downarrow a + \downarrow b \geq 0\}}}{\{0+x \geq 0\}a := 0\,;\,b := x\{\downarrow a + \downarrow b \geq 0\}}$$

In this example, we wish to write the precondition in a simpler form, replacing $0+x \geq 0$ with $x \geq 0$ and expressing the postcondition in the form $\downarrow a \geq -\downarrow b$. This transformation is possible using rule (Cons): if every state which satisfies $x \geq 0$ also satisfies $0 + x \geq 0$ and if the execution of the sequence $a := 0\,;\,b := x$ from such a state leads to a state which satisfies $\downarrow a + \downarrow b \geq 0$, which must verify $\downarrow a \geq -\downarrow b$, then $\{x \geq 0\}a := 0\,;\,b := x\{\downarrow a \geq -\downarrow b\}$ is a provable triple:

$$(\text{Cons})\dfrac{(\text{Seq})\dfrac{\vdots}{\{0+x \geq 0\}a := 0\,;\,b := x\{\downarrow a + \downarrow b \geq 0\}}}{\{x \geq 0\}a := 0\,;\,b := x\{\downarrow a \geq -\downarrow b\}}\begin{pmatrix} x \geq 0 \Rightarrow 0+x \geq 0 \\ \downarrow a + \downarrow b \geq 0 \Rightarrow \downarrow a \geq -\downarrow b \end{pmatrix}$$

The transformation of the assertions of a Hoare triple done by this rule is based on logical implications, which must be established. With the extensional approach, proving $P_1 \Rightarrow P_2$ requires us to show that for any state (Env, Mem), if $[\![P_1]\!]_{\text{Env}}^{\text{Mem}} = 1$, then $[\![P_2]\!]_{\text{Env}}^{\text{Mem}} = 1$. It is possible to only transform one of the two assertions in a Hoare triple, using one of the following two rules:

$$(\text{Cons}_\text{G})\dfrac{\{P_1'\}c\{P_2\}}{\{P_1\}c\{P_2\}}(P_1 \Rightarrow P_1') \qquad (\text{Cons}_\text{D})\dfrac{\{P_1\}c\{P_2'\}}{\{P_1\}c\{P_2\}}(P_2' \Rightarrow P_2)$$

These rules can be deduced directly from rule (Cons). Rule (Cond) expresses the postcondition of a conditional instruction as a function of the boolean expression found in this instruction and of the postconditions associated with the programs found in the two branches of the instruction. For example, we have:

$$\text{(Cond)}\dfrac{\text{(Cons}_\text{G})\dfrac{\text{(Aff)}\dfrac{}{\{y=0\}x:=y\{\downarrow x=0\}}}{\left\{\begin{array}{c}\mathit{true}\\ \text{and}\\ y=0\end{array}\right\} x:=y\{\downarrow x=0\}} \quad \text{(Cons}_\text{G})\dfrac{\text{(Aff)}\dfrac{}{\{\text{not}\, y=0\}x:=0\{\downarrow x=0\}}}{\left\{\begin{array}{c}\mathit{true}\\ \text{and}\\ \text{not}\, y=0\end{array}\right\} x:=0\{\downarrow x=0\}}}{\{\mathit{true}\}\mathsf{if}\ y=0\ \mathsf{then}\ x:=y\ \mathsf{else}\ x:=0\{\downarrow x=0\}}$$

Finally, considering a property P_I, the rule (W_P) establishes that:

– P_I is verified at the beginning of the execution of a loop;

– if both P_I and the condition to enter the loop are verified, P_I remains true after executing the body of the loop, then after the execution of the loop, P_I holds and the condition to enter the loop is no longer satisfied.

P_I is known as a *loop invariant* and is used to express properties of the computation done by the loop. For example:

$$\text{(Cons}_\text{D})\dfrac{\text{(W}_\text{P})\dfrac{\text{(Cons}_\text{G})\dfrac{\text{(Aff)}\dfrac{}{\{\underbrace{\downarrow x+1\le b}_{P_1'}\}x:=\downarrow x+1\{\downarrow x\le b\}}}{\{\underbrace{\downarrow x\le b\ \text{and}\ \downarrow x<b}_{P_1}\}x:=\downarrow x+1\{\downarrow x\le b\}}(P_1\Rightarrow P_1')}{\{\downarrow x\le b\}\mathsf{while}\downarrow x<b\ \mathsf{do}\ x:=\downarrow x+1\{\underbrace{\downarrow x\le b\ \text{and not}\ \downarrow x<b}_{P_2'}\}} \quad (P_2'\Rightarrow P_2)}{\{\downarrow x\le b\}\mathsf{while}\downarrow x<b\ \mathsf{do}\ x:=\downarrow x+1\{\underbrace{\downarrow x=b}_{P_2}\}}$$

4.7.2.1.2. Validity and completeness

A program c is partially correct with regard to precondition P_1 and postcondition P_2 if and only if for every state that satisfies P_1 and in which the execution of c terminates, the state obtained after execution satisfies P_2. Since the operational semantics presented in section 4.4.1 can be used to characterize terminating executions and the states resulting from these executions, we can establish that provable Hoare triples $\{P_1\}c\{P_2\}$ (with the extensional approach used here) correspond exactly to triples expressing the partial correctness of c in relation to P_1

and P_2. First, the inference system c can be shown to be valid (if a triple is provable, then the program is partially correct):

$$\{P_1\}c\{P_2\}$$
$$\Rightarrow \forall(\text{Env}, \text{Mem}) \in \mathbf{E} \times \mathbf{M} \left(\begin{array}{c} [\![P_1]\!]_{\text{Env}}^{\text{Mem}} = 1 \\ \text{and } \langle \text{Mem}, c \rangle \twoheadrightarrow_{\text{Env}} \text{Mem}' \end{array} \right) \Rightarrow [\![P_2]\!]_{\text{Env}}^{\text{Mem}'} = 1$$

Second, we can show that this system is complete (if a program is partially correct, then the corresponding Hoare triple is provable):

$$\left(\forall(\text{Env}, \text{Mem}) \in \mathbf{E} \times \mathbf{M} \left(\begin{array}{c} [\![P_1]\!]_{\text{Env}}^{\text{Mem}} = 1 \\ \text{and } \langle \text{Mem}, c \rangle \twoheadrightarrow_{\text{Env}} \text{Mem}' \end{array} \right) \Rightarrow [\![P_2]\!]_{\text{Env}}^{\text{Mem}'} = 1 \right)$$
$$\Rightarrow \{P_1\}c\{P_2\}$$

Proofs of these two properties can be found in [NIE 07].

4.7.2.2. *Total correctness*

Partial correctness of a program relates to properties of the states obtained when the execution of the program terminates. For non-terminating executions, the properties that are obtained are not correct, even if the inference trees of the associated Hoare triples are finite. A typical example of this is the program while $true$ do skip, for which it is possible to prove the two triples $\{true\}$while $true$ do skip$\{true\}$ and $\{true\}$while $true$ do skip$\{false\}$:

$$(\text{Cons}_D)\dfrac{(\text{W}_P)\dfrac{(\text{Cons}_G)\dfrac{(\text{Skp})\dfrac{}{\{true\}\text{skip}\{true\}}}{\{true \text{ and } true\}\text{skip}\{true\}}((true \text{ and } true) \Rightarrow true)}{\{true\}\text{while } true \text{ do skip}\{true \text{ and not } true\}}}{\{true\}\text{while } true \text{ do skip}\{true/false\}}(I)$$

since the expression $true$ and not $true$ is never verified, so I can correspond to each of the implications $(true$ and not $true) \Rightarrow true$ and $(true$ and not $true) \Rightarrow false$.

A program c is totally correct with respect to a precondition P_1 and postcondition P_2 if and only if for any state satisfying P_1, the execution of c in this state terminates in a state that satisfies P_2. To characterize provable triples expressing the total correctness of a program, we simply modify rule (W_P) (as non-termination results from the construction while e do c alone). The new rule

$$(\text{W}_\text{T})\dfrac{\{P_I \text{ and } e \text{ and } V = n\}c\{P_I \text{ and } V < n\}}{\{P_I\}\text{while } e \text{ do } c\{P_I \text{ and not } e\}}(P_I \Rightarrow V \geq 0)$$

introduces a *loop variant* V that must be positive when the invariant P_I is verified, and must strictly decrease at each iteration of the loop. This guarantees that the loop will terminate, as no strictly decreasing infinite sequence exists over the set of natural numbers $\mathbb{N}$. For example, we can prove the total correctness expressed by the following triple:

$$\{\downarrow x \leq b\}\textsf{while} \downarrow x < b \textsf{ do } x :=\downarrow x + 1\{\downarrow x = b\}$$

by considering the variant $V = b- \downarrow x$. The precondition guarantees that $V \geq 0$ and V strictly decreases at each iteration of the loop (as $\downarrow x$ is strictly increasing).

$$\text{(Cons}_\text{D})\dfrac{\text{(W}_\text{T})\dfrac{\text{(Cons}_\text{G})\dfrac{\text{(Aff)}\dfrac{}{\left\{\underbrace{\begin{array}{c}\downarrow x+1 \leq b \text{ and}\\ b- \downarrow x - 1 < n\end{array}}_{A_6}\right\} x :=\downarrow x + 1\{A_5\}}}{\left\{\underbrace{\begin{array}{c}\downarrow x \leq b\\ \text{and}\\ \downarrow x < b\\ \text{and}\\ b- \downarrow x = n\end{array}}_{A_4}\right\} x :=\downarrow x+1 \left\{\underbrace{\begin{array}{c}\downarrow x \leq b\\ \text{and}\\ b- \downarrow x < n\end{array}}_{A_5}\right\}}(I_3)}{\{A_1\}\textsf{while} \downarrow x < b \textsf{ do } x :=\downarrow x + 1\{\underbrace{\downarrow x \leq b \text{ and not} \downarrow x < b}_{A_3}\}}(I_2)}{\{\underbrace{\downarrow x \leq b}_{A_1}\}\textsf{while} \downarrow x < b \textsf{ do } x :=\downarrow x + 1\{\underbrace{\downarrow x = b}_{A_2}\}}(I_1)$$

The choice of rules (Cons_G) and (Cons_D) is correct, as I_1 and I_3 are implications $A_3 \Rightarrow A_2$ and $A_4 \Rightarrow A_6$, which are easy to verify. Rule (W_T) is also used correctly, as I_2 corresponds to the implication $\downarrow x \leq b \Rightarrow b- \downarrow x \geq 0$.

4.8. Exercises

Exercise 4.1

Prove that:

1) $c_1 \,;\, (c_2 \,;\, c_3) \equiv (c_1 \,;\, c_2) \,;\, c_3$;

2) while e do $c \equiv$ if e then $(c;$ while e do $c)$ else skip.

Exercise 4.2

The language $\mathbf{Lang}_3$ is extended with the construct: repeat c until e where $c \in \mathbf{Lang}_3$ and $e \in \mathbf{Exp}_3$. Its informal semantics is: "execute c until e takes the value $true$".

1) Give the inference rules expressing the big-step semantics of this construct.

2) Give the execution tree for the following program c:

$$\text{repeat } (z :=\downarrow z* \downarrow x\ ;\ x :=\downarrow x - 1) \text{ until } \downarrow x = 0$$

in a state $(\text{Env}, \text{Mem}) = ((z, r_z) \oplus (x, r_x) \oplus \text{Env}_0, \text{Mem}_0[r_x := 2][r_z := 3])$.

3) Prove that:

$$\text{repeat } c \text{ until } e \equiv c;\ \text{if } e \text{ then skip else repeat } c \text{ until } e$$

4) Prove that:

$$\begin{aligned} \text{while } e \text{ do } c &\equiv \text{if } e \text{ then repeat } c \text{ until not } e \text{ else skip} \\ \text{repeat } c \text{ until } e &\equiv c;\ \text{while not } e \text{ do } c \end{aligned}$$

Hint. With the notations:

$$\begin{array}{ll} c_1 : \text{while } e \text{ do } c & c_2 : \text{if } e \text{ then repeat } c \text{ until not } e \text{ else skip} \\ c_3 : \text{repeat } c \text{ until } e & c_4 : c;\ \text{while not } e \text{ do } c \end{array}$$

prove the following properties and then conclude:

$$\begin{array}{ll} \text{(P1) } \forall \text{Mem}, \text{Mem}' \in \mathbf{M} & \langle \text{Mem}, c_1 \rangle \twoheadrightarrow_{\text{Env}} \text{Mem}' \Rightarrow \langle \text{Mem}, c_2 \rangle \twoheadrightarrow_{\text{Env}} \text{Mem}' \\ \text{(P2) } \forall \text{Mem}, \text{Mem}' \in \mathbf{M} & \langle \text{Mem}, c_2 \rangle \twoheadrightarrow_{\text{Env}} \text{Mem}' \Rightarrow \langle \text{Mem}, c_1 \rangle \twoheadrightarrow_{\text{Env}} \text{Mem}' \\ \text{(P3) } \forall \text{Mem}, \text{Mem}' \in \mathbf{M} & \langle \text{Mem}, c_3 \rangle \twoheadrightarrow_{\text{Env}} \text{Mem}' \Rightarrow \langle \text{Mem}, c_4 \rangle \twoheadrightarrow_{\text{Env}} \text{Mem}' \\ \text{(P4) } \forall \text{Mem}, \text{Mem}' \in \mathbf{M} & \langle \text{Mem}, c_4 \rangle \twoheadrightarrow_{\text{Env}} \text{Mem}' \Rightarrow \langle \text{Mem}, c_3 \rangle \twoheadrightarrow_{\text{Env}} \text{Mem}' \end{array}$$

Proofs of (P1) and (P3) can be obtained by structural induction over an inference tree; implications (P2) and (P4) are very easy to prove using (P1) and (P3).

Exercise 4.3

This exercise relates to denotational and axiomatic semantics for Euclidean division.

Consider the program c_d that computes the quotient and the remainder of the Euclidean division of x by y, by using successive subtractions defined by:

$$(a := 0\,;\, b := x)\,;\ \textsf{while} \downarrow b \geq y \ \textsf{do}\ b := \downarrow b - y\,;\, a := \downarrow a + 1$$

1) Compute $[\![c_d]\!]_{\text{Env}}$ and deduce that $[\![c_d]\!]_{\text{Env}}(\text{Mem})(\text{Env}(a))$ is the quotient of the integer division of $\text{Env}(x)$ by $\text{Env}(y)$ and $[\![c_d]\!]_{\text{Env}}(\text{Mem})(\text{Env}(b))$ is the remainder of this division.

2) Prove the total correctness of the following Hoare triple:

$$\{x \geq 0 \text{ and } y > 0\}c_d\{x = (\downarrow a \times y) + \downarrow b \text{ and } \downarrow b < y\}$$

5

Types

The syntax of the language $\mathbf{Exp}_2$, i.e. the rules defining the correct form of the expressions that make up the language, was presented in Chapter 3. The execution semantics of this language, which explains how programs written in the language are executed, was described using a variety of approaches.

However, when introducing expressions of the language (in particular, the applications, see section 3.1.1), then the evaluation function (see section 3.2) and then the interpretation of operators (see section 3.2.3), we noted that the syntax does not prevent us from writing expressions that cannot be evaluated, such as $3 + true$ or $(1\ 2)$ or $1 = false$. In other words, the execution semantics chosen for this language cannot assign a value to these expressions. The existing syntax also allows us to write the expression $f = g$ where f and g are functions, raising substantive questions concerning the meaning of such an expression. In the following, the semantics of these syntactically correct expressions having no clear meaning is more deeply discussed.

In the expression $3 + true$, the semantics for the primitive $+$ is the addition over $\mathbb{Z}$. The semantics assigned to $true$ is not an element in $\mathbb{Z}$, and thus does not belong to the domain of definition of this addition. Hence, this expression cannot be evaluated. However, it would be possible to assign a semantics to $true$ in $\mathbb{Z}$ by associating it with an integer, such as 1. This might be unwise, as $true$ would then have two roles, denoting a truth value as well as the integer 1. Developers, compilers, interpreters and code reviewers would then need to determine the role of each occurrence of $true$ in a program to assign it the right semantics (creating a significant risk of confusion, and, thus, error). Coding $true$ in source code by an integer is thus error-prone. For this reason, we introduced a set $\mathbb{B}$ for assigning semantics to booleans.

A function application must have a functional value as its left hand operand. Once the formal parameter has been replaced by the right operand, i.e. the argument, the evaluation of the body of the functional value returns the value of this application. In the expression "(1 2)"', the constant 1 is not a function, and therefore has no body; this application cannot be evaluated.

In mathematics, an equality is a binary relation defined over elements of a same set. As the semantics of $\mathbf{Exp_2}$ is defined mathematically, $\mathbb{Z}$ and $\mathbb{B}$ are two disjoint sets, and their elements cannot be compared. For example, the semantics of the expression $1 = false$ is undefined. Some languages get rid of this link with mathematics. For example, in Python there is a defined equality relation between integers in booleans. For example, $1 = false$ is evaluated as $false$, while $1 = true$ is evaluated as $true$ but $42 = true$ is evaluated as $false$. This holds true for other relational operators, so, for example, $true < 2$ is evaluated as $true$ in Python.

There are several possible semantics for an equality between functions such as $f = g$. It is best to avoid testing equalities of this kind in programs, as the results are often unpredictable. If the domains or codomains of f and g are different, they clearly cannot be equal. Their types will be said to be "incompatible" (a notion which will be discussed later). However, equalities between two functions with the same domain and codomain are often used in mathematics, for example, in the following statement: "If two functions f and g over reals verify $f = g$, then $\int_a^b f(t)dt = \int_a^b g(t)dt$". To use this theorem, we first need to prove that $f = g$. The evaluation of this expression along program execution requires an algorithm, either written in the source code or imported from a library, which is able to prove that, for any value x of the domain of f and g, the equality $f(x) = g(x)$ is true. At the beginning of the 20th century, logicians proved that there can be no general algorithm for determining whether any two functions are "identical" in a finite time. The problem of equality of functions is said to be *undecidable*. If the domain is finite, then it is still possible to find an adapted algorithm, but the time required to execute it depends on the number of elements in the domain. In $\mathbf{Exp_2}$, no semantics is assigned to equalities between functions as the meaning of these equalities heavily depends on the domain of the functions. Proving the equality of two functions over the set of booleans is easy. On the other hand, proving the equality of two functions over the set of integers (restricted to integers representable in the memory) may take some time. Nevertheless, for a variety of reasons, the designers of certain programming languages have chosen to allow function equality tests. The semantics assigned to an expression $f = g$ are consequently rather *ad hoc*. One version involves the syntactic equality of definitions: $x \mapsto x + 2$ is equal to itself, but is not equal to $x \mapsto 1 + x + 1$, nor to $x \mapsto 2 + x$ or to $y \mapsto y + 2$. Another option focuses on the equality of the locations of the codes for f and g in the memory after the compilation.

We encourage readers who use a language that tests function equality to consult the relevant reference manual concerning the meaning assigned to $f = g$. However, we do not advise using expressions of this kind.

5.1. Type checking: when and how?

In the examples presented at the start of this chapter, we referred to the notion of a mathematical set while explaining why certain expressions should be rejected. For reasons issued from mathematical logical studies, this notion of "set" is replaced by the notion of *type* in programming languages studies.

In a typed language, all expressions have a type, as do the functions themselves. For example, the addition of integers is a function, of which the parameters are two integers and the result is an integer. The addition of floating point numbers (called floats) is a function, of which the parameters and result are floats. However, the two additions cannot use the same algorithm as the internal representations of integers and floating points in memory are totally different: the structure of their encodings is different and the bits do not have the same meaning. To evaluate the expression $3 + 4.5$, for example, we must first determine which algorithm to use for the addition. Some strongly typed languages reject this expression as it is "ill-typed". Some others will accept it but 3 will be considered as a floating point value. Such conversions often remain implicit; at best, they can lead to programs, which are difficult to understand, and at worst, they can lead to execution errors, which can be hard to detect. These implicit conversions are hard to take into account in typing rules.

Typechecking a program consists of verifying that the construction of all its expressions respect the rules of well-formedness regarding types, which are called the *typing rules*. If the typing rules are well designed, a well-typed program can be shown to be *consistent*, that is, any execution of this program, if it does not loop, provides a result of the type given by the typechecking. In this case, the language in question is said to be *strongly typed*. If a type system does not guarantee this property, it is said to be *weakly typed*. This is the case, for example, when a system permits explicit conversions between types: programmers can force the compiler to consider that the type of an expression is arbitrarily different from the one declared or computed by the compiler itself.

Typechecking significantly reduces the risk of writing and executing incorrect programs. The type language and the typing rules of the expressions of a language define the *typing semantics* of this language, which is formally studied in this chapter.

5.1.1. *When to verify types?*

There are two possible typechecking mechanisms. The first consists of verifying types during the execution of the program and determining whether execution can safely go on or whether a runtime error will occur. This semantics, known as *dynamic typing*, interacts with the dynamic execution semantics of the language. When an expression is recognized as being correctly typed, its evaluation returns a value that cannot be a typing error; otherwise, an error will be raised (this mechanism was first presented in Chapter 3 in relation to the Err_T error). In this case, the semantic rules must include typing rules, and must account for errors resulting from type incompatibilities. The second mechanism consists of carrying out typechecking prior to execution: this is known as *static typing*. A program can only be executed if its typechecking did not detect type incompatibilities. Typing error detection therefore occurs earlier than in the case of dynamic typing, and the execution is guaranteed to be free of typing errors. The execution semantics only needs to deal with well-typed programs, making it easier to handle.

5.1.2. *How to verify types?*

Once a static or dynamic typing mechanism has been selected, there are two options for verification. Some programming languages ask programmers to declare the types of their variables explicitly. In C, for example, the function to compute $n!$, presented in example 3.3, is written as:

```
C
int fact (int n) {
  if (n <= 1) return 1 ;
  else return n * fact (n - 1) ;
}
```

where the types (`int`) of the parameter n and of the result are given.

If the code of this function uses a temporary variable, then the type of this variable should be declared:

```
C
int fact (int n) {
  if (n <= 1) return 1 ;
  else {
    int rec_result = fact (n - 1) ;
    return n * rec_result ;
  }
}
```

The typecheker uses this explicit type information when applying typing rules to guarantee that this program is well-typed. Some other languages do not require typing annotations in the source code and the type of any expression is elaborated automatically. This mechanism is called *type inference* or *type synthesis*. It is also based on typing rules, but requires additional analysis. The types obtained thus are said to be *synthesized* or *inferred*.

The design of $\mathbf{Exp}_2$ is done with a type inference mechanism. Section 5.3 studies the formal type verification rules used to define the type inference algorithm described in section 5.4. A more detailed study of typing can be found in [PIE 02].

5.2. Informal typing of a program Exp_2

5.2.1. *A first example*

Let us begin by computing, intuitively, the type of the expression [3.3] presented in section 3.1.4:

$$\text{let } x_{(1)} = 25 \text{ in let } f = \text{fun } x_{(2)} \mapsto x_{(3)} + 3 \text{ in } (f\ 7) + x_{(4)}$$

– let $x_{(1)} = 25$ in ...

This first let binds $x_{(1)}$ to the constant value 25. As this literal constant represents an integer, we deduce that $x_{(1)}$ must have an integer type, denoted as `int`.

– let $f = \text{fun } x_{(2)} \mapsto x_{(3)} + 3$ in ...

The second let binds f to a functional expression. Thus, its type must be that of a function taking an argument of a certain type and returning a value of another type. Now, we have to determine these two types. For the moment, the type of the argument $x_{(2)}$ is unknown as no type annotation is provided for it. We introduce an unknown α (also known as a *type variable*) to denote this unknown type. Let us examine the body of f to deduce the type of the result. This body contains an application of the $+$ operator, the semantics of which is the addition of two integers. This means that the result of f must be of type `int`; furthermore, the two subexpressions that are arguments of $+$ must also be of type `int`. This is the case for 3 as it is a literal constant representing an integer. The subexpression on the left, $x_{(3)}$, also needs to be of type `int`. However, $x_{(3)}$ is an occurrence of the identifier $x_{(2)}$, bound by this function f. This function is well-typed only if $x_{(3)}$ and $x_{(2)}$ (to which we assigned the unknown type α) have the same type. Hence $\alpha =$ `int`. Finally, the type of f is that of a function for which the argument is an integer and which returns an integer. This type is denoted as `int` $\rightarrow$ `int`, where the arrow represents the type of a function, the type of its parameter is on the left of the arrow and the type of the return value on the right.

– $(f\ 7) + x_{(4)}$

This expression is an application of the $+$ operator, which takes two arguments of the type `int`. On the left, the function f is clearly applied to an integer (7 is of type `int`) and returns an integer. On the right, $x_{(4)}$ is an occurrence of the identifier $x_{(1)}$ bound by let $x_{(1)} = 25$ in ..., and is also of type `int`. Thus, the application of $+$ is well-typed and returns a result of type `int`, which is the type of the whole expression.

In this informal description of the typechecking mechanism, we had to retrieve the types of the identifiers. Just as expressions are evaluated in an execution environment, a typing environment is required for typechecking. As a first approach, we shall consider that this environment binds identifiers to their type; this point will be discussed in more detail later.

5.2.2. *Typing a conditional expression*

Staying with our informal approach, we shall now compute the type of the expression fun $x \mapsto$ if x then $true$ else 0. This is a function with a parameter x. As in the previous example, we have no type annotation, so the type of x is an unknown α. Analyzing the body of the function, we see that x acts as a condition, implying that it must be a boolean, denoted as `bool`. We must therefore have $\alpha =$ `bool`. The type of the result of this expression is the one of the conditional expression. The type of the then branch is `bool`, and the type of the else branch is `int`. Which type to choose? Let us look at the question from another angle. When applying this function to an actual value, the type of the result will depend on the value of the argument, a point that prevents the further use of this result in source code. For this reason, a well-typed conditional expression must apply a test to a boolean value and return a value of the same type in both possible cases. In conclusion, the expression we are studying here is ill-typed.

5.2.3. *Typing without type constraints*

In this final part of our presentation of typechecking, we consider the following expression:

$$\text{let } f_{(1)} = \text{fun } x_{(1)} \mapsto x_{(2)} \text{ in if } (f_{(2)}\, true) \text{ then } 8 \text{ else } (f_{(3)}\, 7)$$

– let $f_{(1)} =$ fun $x_{(1)} \mapsto x_{(2)}$ in ...

As in our first example, let binds $f_{(1)}$ to a functional expression. We thus need to determine the types of its argument and of its body. Once again, as there is no type annotation for the occurrence $x_{(1)}$, its type is unknown and will be denoted by α. The body of f simply returns the argument; this does not provide any further information about the type of x. Thus, there are no constraints on this type, meaning that α can be replaced by any type. Indeed, this function may be applied to any value, since it simply returns this value as a result. Hence, the type of $f_{(1)}$ is $\alpha \to \alpha$.

– if $(f_{(2)}\, true)$, then 8 else $(f_{(3)}\, 7)$

In a conditional expression, the condition must be of the type `bool`. We must therefore ensure that $(f_{(2)}\, true)$ is of the type `bool`. The type of the function $f_{(1)}$ is $\alpha \to \alpha$, and, as it is applied to the constant $true$ of type `bool`, we must have $\alpha =$ `bool`. The type of the condition is `bool` since α is also the type of the result of $f_{(1)}$. Now, we must ensure that the expressions found in each of the two branches of this conditional have the same type. As the then branch is of type `int`, the else branch must also have type `int`. This last branch is an application of $f_{(3)}$ to the argument 7 of type `int`, so the result is also of type `int`. Both branches of the conditional are therefore of the same type, which is also the type of the full expression. However, there is a subtle consistency issue in our reasoning, to be explained in section 5.2.4.

5.2.4. *Polymorphism*

In the last example in section 5.2.3, in applying $f_{(2)}$ to *true*, we used $\alpha = \texttt{bool}$. Then, when applying $f_{(3)}$ to 7, we implicitly used $\alpha = \texttt{int}$. By transitivity of the equality, we get $\texttt{bool} = \texttt{int}$, which is certainly wrong. However, as f is the identity function, both uses of f appear to be licit. While there are no constraints concerning the type of $f_{(1)}$, the type expression $\alpha \to \alpha$ cannot serve directly to establish equality constraints between α and several different types which may be incompatible. For each occurrence of f, we must use a "copy" of its type expression and replace the type variable α with a new type variable (known as a *fresh variable*). This prevents interferences between type constraints introduced by different occurrences of f. Thus, the typing environment binds identifiers to the "form" of their types, rather than to the types themselves; this form is called a *type scheme*. This scheme may contain special type variables called *generalized* type variables. When searching for the type of an identifier in a typing environment, a "fresh" copy of the type scheme is produced, in which the generalized type variables are replaced by new variables. This copy is known as an *instance* of the scheme, and these new variables are not generalized, as they denote some yet unknown types. A type scheme, noted σ, is thus made up of the list of generalized type variables (to be copied) and the shape of the type written with these variables. The type scheme of the function f described above is thus denoted as $\forall\alpha.\alpha \to \alpha$.

An expression for which the type includes type variables at the end of the typechecking process is said to be *polymorphic*. As we shall see in section 5.6, at the end of the typechecking process, the types of some expressions may contain unknown variables that cannot be generalized.

5.3. Typing rules in $\mathbf{Exp}_2$

We turn now to the formal definitions of the rules of typechecking. These rules do not yet allow us to synthesize types in a program, but they make it possible to check that a program is well-typed. These rules will be used in section 5.4 to develop an inference algorithm for automatic type computation.

5.3.1. *Types, type schemes and typing environments*

First, the language used in expressions denoting types, known as the *type algebra*, must be defined. Each type τ corresponds to a set of values. $\mathbf{Exp}_2$ includes integer, boolean and functional values. As we have seen, type variables are also required. Types are defined inductively using the grammar shown in Table 5.1.

In what follows, $\alpha_1 \to \alpha_2 \to \cdots \to \alpha_{n+1}$ will be used to denote the type:

$$\alpha_1 \to (\alpha_2 \to (\cdots(\alpha_n \to \alpha_{n+1})))$$

$\tau :=$	α	Variable
|	int	Integer
|	bool	Boolean
|	$\tau_1 \rightarrow \tau_2$	Function

Table 5.1. *Type algebra for* $\mathbf{Exp}_2$

This notation reflects the fact that a function $f : \alpha_1 \times \alpha_2 \times \cdots \times \alpha_n \rightarrow \alpha_{n+1}$ with n arguments can always be defined as a single-argument function, which returns a function with $n-1$ arguments.

```
OCaml
type var_name_t = string
type type_t = T_var of var_name_t | T_int | T_bool | T_fun of (type_t * type_t)

let new_ty_var =
  let cpt = ref 0 in
  function () -> cpt := !cpt + 1 ; T_var ("t" ^ (string_of_int !cpt))
val new_ty_var : unit -> type_t
```

The type algebra is implemented in OCaml using a sum type. Variables are differentiated using names (character strings). We also define a function `new_ty_var` used to create a new type variable whose name differs from all already-created ones. This unicity in terms of names is obtained using a counter, which is local to the function; this counter is incremented each time a variable is created, and its value supplies the name of the variable. As in the case of previous implementations, classes are used in Python to encode the sum types available in OCaml.

```
Python
class T_int:
    def __init__ (self):
        pass

class T_bool:
    def __init__ (self):
        pass

class T_fun:
    def __init__ (self, t1, t2):
        self.t1 = t1
        self.t2 = t2

class T_var:
    def __init__ (self, vname):
        self.vname = vname

var_cpt = 0
def new_ty_var ():
    global var_cpt
    var_cpt = var_cpt + 1
    return T_var ("t" + str (var_cpt))
```

Each class is defined to represent a constructor of the type algebra. While OCaml allows the creation of a persistent reference local to a function, in Python the counter used to assign a unique name to each new variable must be defined as a global variable. Obviously, only the function `new_ty_var` will use this counter. For example, the first application of this function will return the string `"t0"`, the second will return `"t1"`, etc.

Instead of giving typing rules using natural language, we introduce a mathematical definition of the typing relation $\text{Env} \vdash e : \tau$, which expresses that, in a typing environment Env, the expression e is of type τ. To type an expression, we need a typing environment that associates a type scheme to each identifier, as we saw, in section 5.2.4, that a single expression may have several types issued from the same type scheme. In this case, each type is derived from a "skeleton" *type scheme*: this is the definition of *polymorphism*.

DEFINITION 5.1.– *Type scheme: the syntax of a type scheme is defined by the following grammar:* $\sigma ::= \forall \alpha_1 \ldots \alpha_n . \tau$ *where* τ *is the* body *of the scheme and* $\alpha_1, \ldots, \alpha_n$ *are the generalized variables.*

The order in which universally quantified variables α_i appear in a scheme is not important. It is assumed that their names are all different. For example, $\forall \alpha_1\, \alpha_2 . \alpha_1 \to \alpha_2$ and $\forall \alpha_2\, \alpha_1 . \alpha_1 \to \alpha_2$ denote the same type scheme. Furthermore, two type schemes differing only by quantified variable names are considered to be identical (these schemes may also be said to be equivalent modulo α-conversion). For example, $\forall \alpha_1\, \alpha_2 . \alpha_1 \to \alpha_2$ and $\forall \alpha_3\, \alpha_4 . \alpha_3 \to \alpha_4$ denote the same scheme.

A *typing environment* is represented by a list of bindings:

$$\text{Env} = [(x_1, \sigma_1), \cdots , (x_n, \sigma_n)]$$

where each identifier x_i is bound to its type scheme σ_i.

OCaml
```
type scheme_t = (var_name_t list * type_t)
```

A type scheme is encoded in OCaml as a pair, the first component being the list of generalized variables and the second component being the type representing the body of the scheme. Python does not permit type definitions, so an encoding in this language cannot be given. However, we use the same data structure in Python and OCaml (i.e. a pair in which the first component is a list). The representation of typing environments is similar to the one used for execution environments. The values bound to identifiers are simply replaced by type schemes.

OCaml
```
type ty_env_t = (string * scheme_t) list
exception Unbound_var of string

let rec find_id_ty idname = function
  | [] -> raise (Unbound_var idname)
  | h :: q -> if fst h = idname then snd h else find_id_ty idname q
val find_id_ty : string -> (string * 'a) list -> 'a
```

Python
```
class Unbound_var (Exception):
    def __init__ (self, id):
        self.id = id

def find_id_ty (idname, env) :
    for (n, ty) in env :
        if n == idname : return ty
    raise Unbound_var (idname)
```

The typing environment is defined as an association list between identifiers and their type schemes. The function `find_id_ty` searches the first occurrence of its first argument (an identifier) in its second argument (the list encoding the current environment). It starts from the head of the list and raises a specific error if this argument is not found when reaching the end of this list. This representation is similar to the one used in execution environments; the values bound to the identifiers are simply replaced by type schemes.

5.3.2. *Generalization, substitution and instantiation*

To construct an environment, identifiers must be bound to type schemes, and so types must be transformed into schemes. Two steps are needed for this transformation. However, there are a few cases where no generalization should be done, so the type is converted to a so-called *trivial type scheme*.

DEFINITION 5.2.– *Trivial type scheme: a* trivial scheme *is a scheme with the form* $\forall\emptyset.\,\tau$.

OCaml
```
let trivial_scheme ty = (([], ty) : scheme_t)
val trivial_scheme : type_t -> scheme_t
```

Without the explicit type annotation `scheme_t` put on `trivial_scheme`, OCaml would have associated the "more general" type, `'a -> ('b list * 'a)`, with `trivial_scheme`. The annotation leads to a more precise type, a point that can help the developer.

Python
```
def trivial_scheme (ty):
    return ([], ty)
```

Conversely, generalizing variables appearing in a type expression needs a *type generalization* operation, which returns a polymorphic type scheme. However, only certain variables should be generalized:

– variables that are not found anywhere in the environment;

– variables that are generalized in the type schemes present in the environment; these variables can be renamed without changing the denotation of the scheme and are said to be *bound variables*.

In other words, variables that are present in the environment but not bound by a $\forall$ are not generalized: these variables are called the *free variables* of the environment in question. A free variable denotes a yet unknown type and all the free occurrences of this variable in the environment denote the same type.

DEFINITION 5.3.– *Free variables of a type scheme: the set $\mathcal{F}(\sigma)$ of* free variables *of the type scheme σ contains the type variables of σ, that is, those which are not generalized. This set is defined by:*

$$\mathcal{F}(\forall\alpha_1 \dots \alpha_n.\tau) = \mathcal{F}(\tau) \setminus \{\alpha_1, \dots, \alpha_n\}$$

where $\mathcal{F}(\tau)$ is defined by:

$$\mathcal{F}(\texttt{int}) = \mathcal{F}(\texttt{bool}) = \emptyset$$

$$\mathcal{F}(\alpha) = \{\alpha\}$$

$$\mathcal{F}(\tau_1 \to \tau_2) = \mathcal{F}(\tau_1) \cup \mathcal{F}(\tau_2)$$

This definition can be extended to typing environments:

$$\mathcal{F}(\text{Env}) = \bigcup_{(x,\sigma)\in\text{Env}} \mathcal{F}(\sigma)$$

DEFINITION 5.4.– *Generalization: the generalization of a type τ in an environment* Env *is the type scheme defined by:*

$$\texttt{Gen}\,(\tau, \text{Env}) = \forall\alpha_1 \dots \alpha_n.\tau \;\; \textit{with} \;\; \{\alpha_1 \dots \alpha_n\} = \mathcal{F}(\tau) \setminus \mathcal{F}(\text{Env})$$

EXAMPLE 5.1.– With the environment:

$$\text{Env} = [(x, \forall\alpha_1.\alpha_1 \to \alpha_1), (y, \forall\alpha_1.\alpha_1 \to \alpha_2), (z, \forall\emptyset.\alpha_3), (w, \forall\emptyset.\texttt{int})]$$

we have $\mathcal{F}(\text{Env}) = \{\alpha_2, \alpha_3\}$, and:

$$\texttt{Gen}\,((\alpha_1 \to \alpha_2) \to (\alpha_3 \to \alpha_4), \text{Env}) = \forall\alpha_1\,\alpha_4.\,(\alpha_1 \to \alpha_2) \to (\alpha_3 \to \alpha_4)$$

OCaml
```
let find_vars_in_type in_ty =
  let rec rec_find accu = function
    | T_int | T_bool -> accu
    | T_fun (t1, t2) ->
        let new_accu = rec_find accu t1 in rec_find new_accu t2
    | T_var v_name ->
        if not (List.mem v_name accu) then v_name :: accu else accu in
  rec_find [] in_ty
val find_vars : type_t -> var_name_t list
```

Python
```
def find_vars_in_type (in_ty):
    def rec_find (accu, ty):
        if isinstance (in_ty, T_int) or isinstance (in_ty, T_bool):
            return []
        elif isinstance (in_ty, T_fun):
            return find_vars_in_type (in_ty.t1) + find_vars_in_type (in_ty.t2)
        elif isinstance (in_ty, T_var):
            if in_ty.vname in accu: return accu
            else: return [in_ty.vname] + accu
        else: raise ValueError
    return rec_find ([], in_ty)
```

The function used to find free variables within a type operates recursively over the type structure. In the case of a type variable, we use the `List.mem` function from the standard library to verify whether an element belongs to a list; this prevents multiple insertions of the variable into the accumulator.

OCaml
```
let rec find_vars_in_env = function
  | [] -> []
  | (_, (gen_vars, body)) :: rem ->
      let vars_of_body = find_vars_in_type body in
      let sch_free_vars =
        List.filter (fun var -> not (List.mem var gen_vars)) vars_of_body in
      sch_free_vars @ (find_vars_in_env rem)
val find_vars_in_env :
  ('a * (var_name_t list * type_t)) list -> var_name_t list
```

Python
```
def find_vars_in_env (env):
    accu = []
    for (_, (gen_vars, body)) in env:
        vars_of_body = find_vars_in_type (body)
        sch_free_vars = \
            list (filter (lambda var: not (var in gen_vars), vars_of_body))
        accu = accu + sch_free_vars
    return accu
```

OCaml
```
let generalize env ty =
  let vars_of_ty = find_vars_in_type ty in
  let vars_of_env = find_vars_in_env env in
  let gen_vars =
    List.filter (fun var -> not (List.mem var vars_of_env)) vars_of_ty in
  (gen_vars, ty)
val generalize :
  ('a * ('b * type_t)) list -> type_t -> (var_name_t list * type_t)
```

Python
```
def generalize (env, ty) :
  vars_of_ty = find_vars_in_type (ty)
  vars_of_env = find_vars_in_env (env)
  gen_vars = list (filter (lambda var: not (var in vars_of_env), vars_of_ty))
  return (gen_vars, ty)
```

The generalization function proceeds in three stages. First, all variables present in the type to be generalized are fetched. Then the set of free variables in the environment is computed. The final step computes the list of variables present in the type, excluding those found in the environment. In OCaml, this filtering operation is carried out using the `List.filter` function from the standard library, which only retains those elements of a list that satisfy a given predicate. In Python, filtering is carried out by defining this predicate using the construct `lambda`.

Conversely, to type an expression which is an identifier, we need to generate a type from the type scheme bound to this identifier in the environment. This is done by the *instantiation* operation, which returns an *instance* of a scheme by "specializing" it. The formal definition of an instance uses the notion of *substitution*, presented as follows.

DEFINITION 5.5.– *Substitution: a substitution is a partial function from the set of type variables to the type algebra, which has a finite domain.* $[\alpha_1 \leftarrow \tau_1; \dots; \alpha_n \leftarrow \tau_n]$ *denotes the substitution* θ. *The domain of* θ, *denoted as* $\mathrm{dom}(\theta)$, *is the set* $\{\alpha_1, \dots, \alpha_n\}$. *The codomain of* θ, *denoted as* $\mathrm{codom}(\theta)$, *is the set* $\{\tau_1, \dots, \tau_n\}$. θ *associates the type* τ_i *to the variable* α_i.

For the sake of simplicity, in what follows, we suppose that the variables of $\mathrm{dom}(\theta)$ *are all different and that:*

$$\mathrm{dom}(\theta) \cap \bigcup_{\tau \in \mathrm{codom}(\theta)} \mathcal{F}(\tau) = \emptyset$$

Substitutions naturally extend to a partial function from types to types, still with a finite domain, in the following manner:

$$\overline{\theta}(\mathtt{int}) = \mathtt{int} \quad \overline{\theta}(\mathtt{bool}) = \mathtt{bool}$$
$$\overline{\theta}(\alpha) = \begin{cases} \theta(\alpha) & \text{if } \alpha \in \mathrm{dom}(\theta) \\ \alpha & \text{else} \end{cases}$$
$$\overline{\theta}(\tau_1 \rightarrow \tau_2) = \overline{\theta}(\tau_1) \rightarrow \overline{\theta}(\tau_2)$$

From now on, we shall consider the substitution θ and its extension $\overline{\theta}$ to be one and the same, and both shall be denoted as θ. The hypotheses used in defining a substitution guarantee that, for any type τ, $\overline{\theta}(\overline{\theta}(\tau)) = \overline{\theta}(\tau)$ (in this case, θ is said to be *idempotent*).

OCaml
```
type subst_t =  (var_name_t * type_t) list
let empty_subst = ([] : subst_t)
val empty_subst : subst_t
let singleton var_name ty = ([(var_name, ty)] : subst_t)
val singleton : var_name_t -> type_t -> subst_t
```

```
Python
empty_subst = []
def singleton (var_name, ty) : return [(var_name, ty)]
```

A substitution is represented by an association list, i.e. a list of pairs in which the first component represents the name of the type variable and the second component the type to replace the variable. If the association list contains only one element, then the substitution is said to be *elementary*.

```
OCaml
let rec apply_one transf = function
  | T_fun (t1, t2) -> T_fun (apply_one transf t1, apply_one transf t2)
  | T_var var_name ->
      if (fst transf) = var_name then snd transf else T_var var_name
  | any -> any
val apply_one : (var_name_t * type_t) -> type_t -> type_t
```

```
Python
def apply_one (transf, ty) :
    if isinstance (ty, T_int) or isinstance (ty, T_bool) : return ty
    elif isinstance (ty, T_fun) :
        return T_fun (apply_one (transf, ty.t1), apply_one (transf, ty.t2))
    elif isinstance (ty, T_var) :
        (fst_transf, snd_transf) = transf
        if fst_transf == ty.vname  : return snd_transf
        else : return ty
    else : raise ValueError
```

The first argument of the function `apply_one` is a list encoding an elementary substitution, and the second argument is a type. This function returns an update of this type, where the variable involved in the elementary substitution is replaced by its associated type.

```
OCaml
let rec apply ty = function
  | [] -> ty | h :: q -> apply (apply_one h ty) q
val apply : type_t -> (var_name_t * type_t) list -> type_t
```

```
Python
def apply (ty, subst)  :
    ty_res = ty
    for transf in subst : ty_res = apply_one (transf, ty_res)
    return ty_res
```

The `apply` function applies a substitution to a type. In other terms, the update of this type is done by iteratively performing the elementary substitutions contained in the list representing the substitution. This way of doing is correct by the hypotheses established in definition 5.5.

The notion of substitution is used to define instances of a type scheme.

DEFINITION 5.6.– *Instance: τ' is said to be an instance of the type scheme $\sigma = \forall \alpha_1 \ldots \alpha_k.\, \tau$ and we note it $\sigma \leq \tau'$ if there exists a substitution θ such that* $\mathrm{dom}(\theta) = \{\alpha_1, \ldots, \alpha_k\}$ *and $\tau' = \theta(\tau)$. In this case, the type τ' is said to be* compatible *with τ.*

5.3.3. *Typing rules and typing trees*

We can now give the rules used to ensure that an expression is well-typed. These rules are shown in Box 5.1, with the notation already used for the description of the big-step operational semantics of $\mathbf{Exp_2}$ in section 3.3. We give a formal definition of the relation $\mathrm{Env} \vdash e : \tau$, which means that in a typing environment Env, the expression e is of type τ.

$$(T_{\mathrm{true}})\frac{}{\mathrm{Env} \vdash \mathit{true} : \mathtt{bool}} \quad (T_{\mathrm{false}})\frac{}{\mathrm{Env} \vdash \mathit{false} : \mathtt{bool}} \quad (T_{\mathrm{int}})\frac{}{\mathrm{Env} \vdash i : \mathtt{int}}$$

$$(T_{\mathrm{var}})\frac{\mathrm{Env}(x) \leq \tau}{\mathrm{Env} \vdash x : \tau} \qquad (T_{\mathrm{unop}})\frac{\mathrm{Env} \vdash \mathrm{op} : \tau_1 \to \tau_2 \quad \mathrm{Env} \vdash e : \tau_1}{\mathrm{Env} \vdash \mathrm{op}\ e : \tau_2}$$

$$(T_{\mathrm{binop}})\frac{\mathrm{Env} \vdash \mathrm{op} : \tau_1 \to \tau_2 \to \tau_3 \quad \mathrm{Env} \vdash e_1 : \tau_1 \quad \mathrm{Env} \vdash e_2 : \tau_2}{\mathrm{Env} \vdash e_1\, \mathrm{op}\, e_2 : \tau_3}$$

$$(T_{\mathrm{ite}})\frac{\mathrm{Env} \vdash e_1 : \mathtt{bool} \quad \mathrm{Env} \vdash e_2 : \tau \quad \mathrm{Env} \vdash e_3 : \tau}{\mathrm{Env} \vdash \textsf{if } e_1 \textsf{ then } e_2 \textsf{ else } e_3 : \tau}$$

$$(T_{\mathrm{fun}})\frac{(x, \forall\emptyset.\, \tau_1) \oplus \mathrm{Env} \vdash e : \tau_2}{\mathrm{Env} \vdash \textsf{fun } x \mapsto e : \tau_1 \to \tau_2}$$

$$(T_{\mathrm{let}})\frac{\mathrm{Env} \vdash e_1 : \tau_1 \quad \sigma = \mathtt{Gen}\,(\tau_1, \mathrm{Env}) \quad (x, \sigma) \oplus \mathrm{Env} \vdash e_2 : \tau_2}{\mathrm{Env} \vdash \textsf{let } x = e_1 \textsf{ in } e_2 : \tau_1 \to \tau_2}$$

$$(T_{\mathrm{rec}})\frac{(f, \forall\,\emptyset.\, \tau_1 \to \tau_2) \oplus (x, \forall\,\emptyset.\, \tau_1) \oplus \mathrm{Env} \vdash e : \tau_2}{\mathrm{Env} \vdash \textsf{rec } f\ x = e : \tau_1 \to \tau_2}$$

$$(T_{\mathrm{app}})\frac{\mathrm{Env} \vdash e_1 : \tau_2 \to \tau_1 \quad \mathrm{Env} \vdash e_2 : \tau_2}{\mathrm{Env} \vdash (e_1\, e_2) : \tau_1}$$

Box 5.1. *Typing rules in* $\mathbf{Exp_2}$

Rules T_{true}, T_{false} and T_{int} indicate that a boolean constant has the type `bool` and an integer constant has the type `int`. According to rule T_{var}, the type of an identifier is an instance of the scheme to which it is bound in the typing environment and $\mathrm{Env}(x)$ denotes the type scheme bound to x in Env. Rule T_{unop} states that the type of the application of a unary operator to an argument is that returned by the operator, as long as the argument is of the type required by this operator. Similarly, rule T_{binop} indicates that the type of the application of a binary operator to two arguments is that returned by the operator, as long as the two arguments are of the types required by this operator. Rule T_{ite} indicates that the type of a conditional expression is that shared by the two subexpressions making up its then and else branches, as long as the condition subexpression is of type `bool`.

According to rule T_{fun}, a function of a variable x of which the body is an expression e is of type $\tau_1 \to \tau_2$ if the type of e is τ_2 when the type of x is taken to be τ_1. Note that the scheme bound to the argument x is trivial: the type variables that may appear in τ_1 are not generalized. If these variables were generalized, it would be possible to type incorrect programs. For example, consider the expression fun $x \mapsto$ if x then $x + 1$ else $x + 2$. Typing this expression must fail, as x is used both as a boolean (rule T_{ite}) and as an integer (rule T_{binop}). If the type of x was generalized to a scheme used to typecheck the body of the function, each occurrence of x would be typed using a new instance of the scheme: the first instance would be compatible with booleans, while the second one would be compatible with integers. For this reason, type τ_1 is not generalized, meaning that the same instance must be used for all occurrences of x in the body of the function.

Rule T_{let} indicates that the type of an expression let $x = e_1$ in e_2 is that of the subexpression e_2, under the hypothesis that x is bound to the scheme that generalizes the type of e_1. According to rule T_{rec}, a recursive function f of a variable x, of which the body is an expression e, is of type $\tau_1 \to \tau_2$ where τ_2 is the type of e under the hypothesis that the type of x is τ_1 and the hypothesis that the type scheme of f is compatible with $\tau_1 \to \tau_2$. Finally, rule T_{app} states that, in order to apply an expression e_1 to an argument e_2, e_1 must be of the functional type $\tau_1 \to \tau_2$ and e_2 must be of the type τ_1. The type of the overall expression is that of the result of the function, i.e. τ_2.

Note that rule T_{rec} introduces a binding between the recursively defined function and a trivial type scheme into the environment. Thus, the expression

rec $f\ x =$ let $y = (f\ 7)$ in $(f\ true)$

is rejected by the typechecking process. Indeed, f is used in the expression with both the type `int` $\to \alpha$ and the type `bool` $\to \alpha$.

Note that generalizing the type of f, i.e. assigning type $\beta \to \alpha$ to f, would still have been correct in terms of the execution semantics used in evaluating expressions. The scheme of f must be limited to a trivial scheme in order to guarantee the decidability of type inference (i.e. the possibility of designing a type computation algorithm which is guaranteed to terminate).

NOTE.– This restriction is also applicable to languages that permit the definition of mutually recursive functions, although if the mutually defined functions do not perform actual recursive calls. Thus, the following program – which is evidently correct – will be rejected by the typing process:

```
OCaml
let rec map f l = match l with [] -> [] | h :: q -> (f h) :: (map f q)
and map_succ l = map (fun x -> x + 1) l
and map_not l = map (fun x -> not x) l ;;
Error: This expression has type int but an expression was
       expected of type bool
```

This program redefines the function `map` that iterates a function `f`, given as an argument over a list `l`. Within this mutually recursive definition (introduced by let rec and and), it also defines a function that adds 1 to each element of a list of integers, and a function that complements each element of a list of booleans. This joint definition of the three functions is rejected by the OCaml compiler, as `map` is applied to a list of integers and to a list of booleans in the same definition. Typechecking of the subexpression `map` does not generalize its type. Generalization would have been done if `map` had been recursively defined independently of its two uses.

```
OCaml
let rec map f l = match l with [] -> [] | h :: q -> (f h) :: (map f q) ;;
val map : ('a -> 'b) -> 'a list -> 'b list = <fun>
```

```
OCaml
let map_succ l = map (fun x -> x + 1) l
and map_not l = map (fun x -> not x) l ;;
val map_succ : int list -> int list = <fun>
val map_not : bool list -> bool list = <fun>
```

To avoid such difficulties, OCaml provides the keyword `rec` to annotate the functions recognized as recursive by their author. Using this annotation avoids implicitly considering all functions as recursive ones (as is done in many languages without type inference), which would forbid the generalization of their types, hence cancelling any possibility of polymorphism.

Note that there are other advantages to distinguishing between recursive and non-recursive functions during compilation, including the capacity to generate more efficient code.

The different stages involved in computing the type of an expression can be described in the form of a *typing tree*, similar to the trees used to represent the evaluation of expressions in Chapter 3.

EXAMPLE 5.2.– The typing tree of the expression $\mathsf{let}\ f = \mathsf{fun}\ x \mapsto x\ \mathsf{in}\ (f\,5)$ is:

$$
(T_{\text{let}})\dfrac{
(T_{\text{fun}})\dfrac{(T_{\text{var}})\dfrac{\alpha \leq \alpha}{(x, \forall\emptyset.\,\alpha) \oplus \mathtt{Env} \vdash x : \alpha}}{\mathtt{Env} \vdash \mathsf{fun}\ x \mapsto x : \alpha \to \alpha}
\qquad
(T_{\text{app}})\dfrac{(T_{\text{var}})\dfrac{\beta \to \beta \leq \mathtt{int} \to \mathtt{int}}{\mathtt{Env}' \vdash f : \mathtt{int} \to \mathtt{int}} \quad \mathtt{Env}' \vdash 5 : \mathtt{int}}{\underbrace{(f, \forall\beta.\,\beta \to \beta) \oplus \mathtt{Env}}_{\mathtt{Env}'} \vdash (f\,5) : \mathtt{int}}
}{\mathtt{Env} \vdash \mathsf{let}\ f = \mathsf{fun}\ x \mapsto x\ \mathsf{in}\ (f\,5) : \mathtt{int}}
$$

5.4. Type inference algorithm in Exp_2

The rules presented in section 5.3 are used to verify the type of a program including type annotations. They also define the typing system of the language and, as we shall see in section 5.5, guarantee certain safety properties of well-typed programs. In this section, we describe an *algorithm* that generates types from expressions in a program. By extension, this *inference algorithm* can also be used to check whether a program is well-typed. Obviously, the type inference algorithm must be correct with regard to the type system and the associated rules. The advantage of type inference lies in the fact that programmers do not have to provide explicit type annotation within programs.

5.4.1. *Principal type*

The type inference algorithm is described formally by a set of rules. These rules establish the fact that, in a typing environment Env, an expression e is of type τ when a set of constraints described by a substitution θ over the types can be satisfied (for example, this set of constraints must not require a type variable α to be equal to both `int` and `bool`). This is noted $\mathtt{Env} \vdash e \rhd (\tau, \theta)$.

Furthermore, in case of polymorphism, the type τ computed by this algorithm must be as general as possible: it is called the *principal type* of e. For example, the type of $\mathsf{fun}\ x \mapsto \mathsf{fun}\ y \mapsto (x\,y)$ is $\tau_0 = (\alpha \to \beta) \to \alpha \to \beta$, but it could also be:

$$
\begin{aligned}
&\tau_1 : (\mathtt{int} \to \mathtt{int}) \to \mathtt{int} \to \mathtt{int}\\
&\tau_2 : (\mathtt{int} \to \mathtt{bool}) \to \mathtt{int} \to \mathtt{bool}\\
&\tau_3 : (\mathtt{int} \to \mathtt{bool} \to \mathtt{int}) \to \mathtt{int} \to \mathtt{bool} \to \mathtt{int}
\end{aligned}
$$

However, type τ_0 is the most general, as it allows this expression to be used in the highest number of contexts. The other possible types all result from a substitution applied to τ_0. For example, types τ_1, τ_2 and τ_3 may be obtained in the following manner:

$$
\begin{aligned}
&\tau_1 = \tau_0[\alpha \leftarrow \mathtt{int}, \beta \leftarrow \mathtt{int}]\\
&\tau_2 = \tau_0[\alpha \leftarrow \mathtt{int}, \beta \leftarrow \mathtt{bool}]\\
&\tau_3 = \tau_0[\alpha \leftarrow \mathtt{int}, \beta \leftarrow (\mathtt{bool} \to \mathtt{int})]
\end{aligned}
$$

Considering the principal type of an expression prevents some type conflicts. If the type inference algorithm does not use the principal type of an expression, it may reject expressions which are indeed well-typed. For this reason, any expression must have a principal type belonging to the type system (this is the case in the $\mathbf{Exp}_2$ language), otherwise inference will be impossible or users will be forced to provide type annotation for certain parts of their programs.

5.4.2. *Sets of constraints and unification*

The type inference algorithm establishes that an expression e is of type τ when a set of constraints can be satisfied, i.e. when a solution exists. A constraint introduces an equality between two types. For example, from the expression if $(f\,4)$ then 5 else $(g\,true)$, we deduce the following set of constraints to be satisfied:

– the type τ_f of f is functional: $\tau_f = \alpha \rightarrow \beta$;

– the argument of f is an integer: $\alpha = \texttt{int}$;

– as the result of f being used as a condition, it is a boolean: $\beta = \texttt{bool}$;

– the type τ_g of g is functional: $\tau_g = \gamma \rightarrow \delta$;

– the argument of g is a boolean: $\gamma = \texttt{bool}$;

– the result of g is of the same type as the expression 5, meaning that it is an integer: $\delta = \texttt{int}$.

If all these equality constraints can be satisfied together, then the expression is well-typed. This is true here if we take $\alpha = \texttt{int}$, $\beta = \texttt{bool}$, $\gamma = \texttt{bool}$ and $\delta = \texttt{int}$. The non-constrained variables are then generalizable (with certain conditions), and contribute to establishing the polymorphism of the type.

Proving that a constraint $\tau_1 = \tau_2$ is satisfied requires the ability to replace certain variables in the left and/or right of the equals sign with types, which may themselves contain their own variables. In other words, a substitution θ must exist such that $\theta(\tau_1) = \theta(\tau_2)$. Determining whether such a substitution exists is known as a *type unification problem*. If this substitution exists, then τ_1 and τ_2 are said to be *unifiable*, and substitution θ is said to be an *unifier* of terms τ_1 and τ_2. Thus, two terms are unifiable if all or part of their variables can be instantiated by the same substitution in a way which makes them structurally equal.

As several types may be assigned to an expression, several substitutions may exist to identify two terms (this is the reason for type plurality). Consider the two terms $\tau_1 = \alpha \rightarrow \beta$ and $\tau_2 = \alpha \rightarrow \texttt{bool}$. The two substitutions $\theta_1 = [\alpha \leftarrow \texttt{int}; \beta \leftarrow \texttt{bool}]$ and $\theta_2 = [\beta \leftarrow \texttt{bool}]$ are unifiers of τ_1 and τ_2 since $\theta_1(\tau_1) = \theta_1(\tau_2) = \texttt{int} \rightarrow \texttt{bool}$ and $\theta_2(\tau_1) = \theta_2(\tau_2) = \alpha \rightarrow \texttt{bool}$.

A *principal unifier*, also often called a *most general unifier*, of τ_1 and τ_2 is a substitution μ such that, for any other unifier θ, $\theta(\tau_1)$ is an instance of $\mu(\tau_1)$, i.e. $\theta(\tau_1)$ can be obtained by applying a substitution θ' to $\mu(\tau_1)$. This is essentially the same as saying that the principal unifier μ is a minimal substitution, and thus all other unifiers θ include this substitution and can be written as $\theta = \theta' \circ \mu$. In the previous example, θ_2 is a principal unifier, whereas θ_1 is not. In this case, we have $\theta_1 = [\alpha \leftarrow \texttt{int}] \circ \theta_2$.

The principal unifier of two types τ_1 and τ_2 is computed by solving the system of equations $\{\tau_1 = \tau_2\}$. The unification algorithm Mgu presented in section 5.4.2 can be used to solve systems of this kind. It takes a set Q of equalities between types, called *constraints*, as an argument. The solution of a system Q, when it exists, is a substitution θ such that, for every equation $\tau_1 = \tau_2$ in Q, we have $\theta(\tau_1) = \theta(\tau_2)$. This algorithm is recursive and computes the most general unifier of system Q. The system $Q[\alpha \leftarrow \tau]$ denotes the system Q in which α is replaced by τ in every equation. If none of the listed cases apply, the operation fails, triggering the failure of the type inference process. Notably, if one member of an equality is a variable, the other cannot be a type containing this variable (unless it is itself a variable). This constraint is needed to avoid infinite looping over problems of the form $\texttt{Mgu}\,(\{\alpha = \texttt{int} \to \alpha\})$ for which there is no type of finite size which solves this equation. If $\alpha = \texttt{int} \to \alpha$, we have $\alpha = \texttt{int} \to \texttt{int} \to \alpha$, $\alpha = \texttt{int} \to \texttt{int} \to \texttt{int} \to \alpha$, and so on.

$$
\begin{array}{c}
\texttt{Mgu}\,(\emptyset) = id \\
\texttt{Mgu}\,(\{\alpha = \alpha\} \cup Q) = \texttt{Mgu}\,(Q) \\
\texttt{Mgu}\,(\{\alpha = \tau\} \cup Q) = \texttt{Mgu}\,(Q[\alpha \leftarrow \tau]) \circ [\alpha \leftarrow \tau] \text{ if } \alpha \notin \mathcal{F}(\tau) \\
\texttt{Mgu}\,(\{\tau = \alpha\} \cup Q) = \texttt{Mgu}\,(Q[\alpha \leftarrow \tau]) \circ [\alpha \leftarrow \tau] \text{ if } \alpha \notin \mathcal{F}(\tau) \\
\texttt{Mgu}\,(\{\texttt{int} = \texttt{int}\} \cup Q) = \texttt{Mgu}\,(Q) \\
\texttt{Mgu}\,(\{\texttt{bool} = \texttt{bool}\} \cup Q) = \texttt{Mgu}\,(Q) \\
\texttt{Mgu}\,(\{\tau_1 \to \tau_1' = \tau_2 \to \tau_2'\} \cup Q) = \texttt{Mgu}\,(\{\tau_1 = \tau_2; \tau_1' = \tau_2'\} \cup Q)
\end{array}
$$

Box 5.2. *Type unification algorithm for* $\mathbf{Exp}_2$

EXAMPLE 5.3.– Let us compute the principal unifier of $\tau_1 = \alpha_1 \to (\alpha_2 \to \texttt{int})$ and $\tau_2 = \texttt{int} \to (\alpha_3 \to \alpha_1)$.

$$
\begin{array}{l}
\texttt{Mgu}\,(\{\alpha_1 \to (\alpha_2 \to \texttt{int}) = \texttt{int} \to (\alpha_3 \to \alpha_1)\}) \\
= \texttt{Mgu}\,(\{\alpha_1 = \texttt{int}; \alpha_2 \to \texttt{int} = \alpha_3 \to \alpha_1\}) \\
= \texttt{Mgu}\,(\{\alpha_2 \to \texttt{int} = \alpha_3 \to \texttt{int}\}) \circ [\alpha_1 \leftarrow \texttt{int}] \\
= \texttt{Mgu}\,(\{\alpha_2 = \alpha_3; \texttt{int} = \texttt{int}\}) \circ [\alpha_1 \leftarrow \texttt{int}] \\
= \texttt{Mgu}\,(\{\texttt{int} = \texttt{int}\}) \circ [\alpha_2 \leftarrow \alpha_3] \circ [\alpha_1 \leftarrow \texttt{int}] \\
= \texttt{Mgu}\,(\emptyset) \circ [\alpha_2 \leftarrow \alpha_3] \circ [\alpha_1 \leftarrow \texttt{int}] \\
= id \circ [\alpha_2 \leftarrow \alpha_3] \circ [\alpha_1 \leftarrow \texttt{int}] \\
= [\alpha_2 \leftarrow \alpha_3; \alpha_1 \leftarrow \texttt{int}] = \theta
\end{array}
$$

We can verify that $\theta(\tau_1) = \theta(\tau_2) = \texttt{int} \to (\alpha_2 \to \texttt{int})$.

OCaml
```
exception Cycle
exception Incompatible_tys of (type_t * type_t)

let rec occur_check v_name = function
  | T_int | T_bool -> false
  | T_fun (t1, t2) -> (occur_check v_name t1) || (occur_check v_name t2)
  | T_var n -> v_name = n
val occur_check : var_name_t -> type_t -> bool
```

```
let rec unify t1 t2 =
  match (t1, t2) with
  | (T_int, T_int) | (T_bool, T_bool) -> empty_subst
  | (T_fun (a, b), T_fun (c, d)) ->
      let subst_left = unify a c in
      let b_substed = apply b subst_left in
      let d_substed = apply d subst_left in
      compose (unify b_substed d_substed) subst_left
  | (T_var v, other) | (other, T_var v) ->
      if occur_check v other then raise Cycle ;
      singleton v other
  | _ -> raise (Incompatible_tys (t1, t2))
val unify : type_t -> type_t -> subst_t
```

Python
```
class Cycle (Exception) :
    pass
class Incompatible_tys (Exception) :
    def __init__ (self, ty1, ty2) :
        self.ty1 = ty1
        self.ty2 = ty2

def occur_check (v_name, ty)  :
    if isinstance (ty, T_int) or isinstance (ty, T_bool) : return False
    elif isinstance (ty, T_fun) :
        return occur_check (v_name, ty.t1) or occur_check (v_name, ty.t2)
    elif isinstance (ty, T_var) : return ty.vname == v_name
    else : raise ValueError
```

Python
```
def unify (ty1, ty2) :
    if (isinstance (ty1, T_int) and isinstance (ty2, T_int)) or\
       (isinstance (ty1, T_bool) and isinstance (ty2, T_bool)) :
        return empty_subst
    elif isinstance (ty1, T_fun) and isinstance (ty2, T_fun) :
        a, b, c, d = ty1.t1, ty1.t2, ty2.t1, ty2.t2
        subst_left = unify (a, c)
        b_substed = apply (b, subst_left)
        d_substed = apply (d, subst_left)
        return compose (unify (b_substed, d_substed), subst_left)
```

```
    elif isinstance (ty1, T_var) or isinstance (ty2, T_var) :
        (other, v) = (ty1, ty2.vname) if isinstance (ty2, T_var) \
            else (ty2, ty1.vname)
        if occur_check (v, other) : raise Cycle
        return singleton (v, other)
    else : raise Incompatible_tys (ty1, ty2)
```

The verification "if $\alpha \notin \mathcal{F}(\tau)$" is carried out using the function `occur_check`, which checks whether a variable appears in a type. As we saw from the rules, this function is used in the two cases when a variable must be unified with a type. The function `unify` works through the structure of both types, verifying equalities in the case of basic types, carrying out recursive calls for functional types and extending the substitution result in cases where one of the two types is a variable. The exception `Incompatible_tys` has two parameters: the two types that raise an unification conflict. Such a possibility is useful for creating error messages, providing users with more details concerning the nature of the problem. For reasons of efficiency, the implementation given here does not operate over a set of constraints, but only over the two types to unify.

The unification function `unify` can now be tested using example 5.3, defining the two types of this example using Python syntax.

```
Python
type1 = T_fun ((T_var ("t1")), (T_fun ((T_var ("t2")), (T_int ()))))
type2 = T_fun ((T_int ()), (T_fun ((T_var ("t3")), (T_var ("t1")))))
mgu = unify (type1, type2)
print_subst (mgu)
[('t1', <type.T_int object at 0x10cb29290>),
 ('t3', <type.T_var object at 0x10cb29150>)]
(t1 <- T_int); (t3 <- T_var (t2))
```

Using the general function `print` to display a substitution do not reveal the full structure of the substitution itself. Using a function to display types and substitutions (`print_subst`, not included in this implementation), we see that the result returned by the implementation corresponds to the result that we computed theoretically. Note, however, that the implementation returns a substitution in which α_3 is replaced by α_2; we computed the reverse in our example, although both are correct. This difference is due to the fact that the implementation handles in a single case the two disjunct cases $\texttt{Mgu}\,(\{\alpha = \tau\} \cup Q)$ and $\texttt{Mgu}\,(\{\tau = \alpha\} \cup Q)$ (causing an inversion of the domain and codomain variables in certain cases).

Using the type unification algorithm, we can now present the inference algorithm used to compute the type of any expression.

5.4.3. *Type inference algorithm*

5.4.3.1. *Instantiation of type schemes*

DEFINITION 5.7.– *A fresh instance of a type scheme. We define a function* Inst (σ) *which returns an instance of the type scheme σ, in which the generalized-type variables α_i are replaced by fresh variables α'_i, i.e. variables having no free occurrence in the current typing environment.*

$$\texttt{Inst}\,(\forall\alpha_1,\ldots,\alpha_n.\tau) = \tau[\alpha_1 \leftarrow \alpha'_1,\ldots,\alpha_n \leftarrow \alpha'_n]$$

Giving a formal meaning to the notion of fresh variables implies including, in the whole formalization, the existence of an infinite countable set of variables and keeping trace of the set of variables used in the hypothesis and the conclusion of each rule within the inference algorithm presented in section 5.4.3. For the sake of simplicity, we shall omit this technical detail from our notation.

OCaml
```
let specialize (vars, body) =
  let fresh_assoc =
    List.map
      (fun var -> let fresh_var = new_ty_var () in (var, fresh_var)) vars in
  let rec copy ty =
    match ty with
    | T_int | T_bool -> ty
    | T_fun (t1, t2) ->
        let t1_copied = copy t1 in
        let t2_copied = copy t2 in
        T_fun (t1_copied, t2_copied)
    | T_var v_name ->
        (try List.assoc v_name fresh_assoc with Not_found -> ty) in
  copy body
val specialize : (var_name_t list * type_t) -> type_t
```

Python
```
class Not_found (Exception) :
    pass

def list_assoc (key, l) :
    for (k, v) in l :
        if k == key : return v
    raise Not_found

def specialize (sch) :
    (vars, body) = sch
    fresh_assoc = list (map (lambda var : (var, new_ty_var ()), vars))
    def copy (ty) :
        if isinstance (ty, T_int) or isinstance (ty, T_bool) : return ty
        elif isinstance (ty, T_fun) :
            t1_copied = copy (ty.t1)
            t2_copied = copy (ty.t2)
            return T_fun (t1_copied, t2_copied)
```

```
        elif isinstance (ty, T_var) :
            try :
                return list_assoc (ty.vname, fresh_assoc)
            except Not_found : return ty
        else : raise ValueError
    return copy (body)
```

The function `specialize`, used to obtain an instance of a type scheme, creates an association list that binds each generalized variable of the scheme to a fresh variable. It then follows a recursive path, copying the structure of the body of the scheme and replacing each generalized variable with its corresponding fresh version (otherwise, the non-generalized variable is retained). The implementation in OCaml uses the function `List.assoc` from the standard library, which searches through an association list for the value associated with a key, and, if none is found, the exception `Not_found` is raised. This function does not exist in Python, so we define it under the name `list_assoc`.

As the environment contains type schemes, we also need to be able to apply a substitution to a scheme and to an environment.

DEFINITION 5.8.– *Substitution applied to a type scheme: take a type scheme* $\sigma = \forall \alpha_1 \ldots \alpha_n . \tau$ *and a substitution* θ *such that:*

$$(\mathrm{dom}(\theta) \cup \mathcal{F}(\theta(\tau))) \cap \{\alpha_1, \ldots, \alpha_n\} = \emptyset$$

we define $\theta(\sigma) = \forall \alpha_1 \ldots \alpha_n . \theta(\tau)$.

This definition notably prohibits substitution of the generalized variables of a type scheme. In the following implementation, we do not verify that the variables from the substitution domain and codomain are not present in the scheme. This choice has no effect on the correctness of the inference algorithm, which only establishes equality constraints between types. The types here never include generalized variables, as they are produced by the instantiation operation that replaces generalized variables with fresh variables. Thus, the constructed substitutions will never contain generalized variables.

NOTE.– A more general definition of the application of a substitution θ to a type scheme σ, with no hypotheses concerning σ and θ, is possible. This is obtained by renaming the generalized variables $\alpha_1 \ldots \alpha_n$ of σ with variables found neither in the domain of θ nor in $\theta(\tau)$.

Applying a substitution to an environment comes down to creating an environment in which the bodies of the schemes involved in the bindings have been transformed by the substitution.

OCaml
```
let subst_env subst env =
  List.map
    (fun (name, (vars, body)) ->
      let body' = apply body subst in
      let sch' = (vars, body') in
      (name, sch'))
    env
val subst_env :
  subst_t -> ('a * ('b * type_t)) list -> ('a * ('b * type_t)) list
```

Python
```
def subst_env (subst, env) :
    def do_binding (binding) :
        (name, sch) = binding
        (vars, body) = sch
        body_new = apply (body, subst)
        sch_new = (vars, body_new)
        return (name, sch_new)
```

5.4.3.2. *Composition of substitutions*

The inference algorithm works through expressions recursively, determining equality constraints. At each step, it constructs a substitution and a type. The type is that of the current expression, and the substitution memorizes the equality constraints to be satisfied to obtain this type. The substitutions that are constructed for each subexpression must be combined. The composition of two substitutions is defined as follows.

Composing two substitutions θ_1 and θ_2, represented by the lists

$$\theta_1 = [(\alpha_1, \tau_1), \cdots, (\alpha_n, \tau_n)] \qquad \theta_2 = [(\beta_1, \tau_1'), \cdots, (\beta_k, \tau_k')]$$

comes down to computing the substitution $\theta_1 \circ \theta_2$, such that $\theta_1 \circ \theta_2(x) = \theta_1(\theta_2(x))$. We therefore apply θ_1 to all the types contained in θ_2; the only elements of substitution θ_1 that are retained are the bindings for variables, which do not occur in the domain of θ_2. The substitution $\theta_1 \circ \theta_2$ is thus represented by the list:

$$[(\beta_1, \theta_1(\tau_1')), \cdots, (\beta_k, \theta_1(\tau_k')), (\alpha_1, \tau_1), \cdots, (\alpha_n, \tau_n)]$$

from which we have eliminated all pairs $(\beta_i, \theta_1(\tau_i'))$ such that $\beta_i = \theta_1(\tau_i')$ (since, in this case, β_i no longer belongs to the domain of $\theta_1 \circ \theta_2$) and all pairs (α_i, τ_i) such that $\alpha_i \in \{\beta_1, \cdots, \beta_k\}$ (these pairs are no longer useful, as the substitution θ_2 is applied before the substitution θ_1).

EXAMPLE 5.4.– Given two substitutions:

$$\theta_1 = [(\alpha_1, \tau_1), (\alpha_2, \alpha_4)] \quad \text{and} \quad \theta_2 = [(\alpha_3, \alpha_1 \rightarrow \tau_2), (\alpha_1, \tau_3), (\alpha_4, \alpha_2)]$$

the composition $\theta_1 \circ \theta_2$ is the substitution:

$$\theta_1 \circ \theta_2 = [(\alpha_1, \tau_3), (\alpha_3, \tau_1 \to \tau_2), (\alpha_2, \alpha_4)]$$

In this case, the binding (α_1, τ_1) from θ_1 is not retained since α_1 belongs to the domain of θ_2 and the binding $(\alpha_4, \theta_1(\alpha_2))$ is not retained since $\alpha_4 = \theta_1(\alpha_2)$.

OCaml
```
let compose subst1 subst2 =
  let subt_1_2 =
    (List.map (fun (name, ty) -> (name, apply ty subst1)) subst2) @
    (List.filter (fun (name, _) -> not (List.mem_assoc name subst2)) subst1) in
  List.filter
    (fun (name, ty) ->
      match ty with Types.T_var name' -> name <> name' | _ -> true)
    subt_1_2
val compose : subst_t -> subst_t -> subst_t
```

Python
```
def list_mem_assoc (key, l) :
    for (k, v) in l :
        if k == key : return True
    return False

def compose (subst1, subst2) :
    def subst1_on_binding (binding) :
        (name, ty) = binding
        return (name, apply (ty, subst1))
    def filter_not_in_subst2 (binding) :
        (name, _) = binding
        return not (utils.list_mem_assoc (name, subst2))
    def filter_not_identity (binding) :
        (name, ty) = binding
        if isinstance (ty, T_var) : return name != ty.vname
        return True
    part1 = list (map (subst1_on_binding, subst2))
    part2 = list (filter (filter_not_in_subst2, subst1))
    return list (filter (filter_not_identity, part1 + part2))
```

The implementation uses the function `List.mem_assoc` from the standard OCaml library. The arguments of this function are a key and an association list, and it returns the boolean value stating whether or not the key occurs in the domain of the association list. This function does not exist in Python, so we define it under the name `list_mem_assoc`.

5.4.3.3. *Type inference*

In order to limit the number of inference rules, we have chosen to consider applications of unary and binary operators as specific cases of the general application of a function. Usually (and with no loss of generality), the type of an operator is found in the initial environment of any program. The name of an operator plays exactly the same role as that of an identifier declared by the programmer. In the rules presented in Box 5.3, id denotes the identity substitution.

EXAMPLE 5.5.– The tree used to infer the type of expression e defined by:

$$\text{let } f = \underbrace{\text{fun } x \mapsto \underbrace{\text{fun } y \mapsto \underbrace{\text{fun } z \mapsto \underbrace{\text{if } x \text{ then } y \text{ else } z}_{e_4}}_{e_3}}_{e_2}}_{e_1} \text{ in } \underbrace{(\underbrace{(\underbrace{(f\ true)}_{e_5}\ 2)}_{e_6}\ 3)}_{e_7}$$

is:

$$(I_{let})\ \dfrac{\underset{1}{\nabla} \quad \underset{2}{\nabla}}{\text{Env} \vdash e \rhd (\texttt{int}, \theta)}$$

where ∇_1 is the typing tree $\text{Env} \vdash e_1 \rhd (\texttt{bool} \to (\alpha_3 \to (\alpha_3 \to \alpha_3)), \theta_1)$ shown below, and obtained considering:

$$\begin{array}{lll}
\mu_1 & = \textbf{Mgu}\,(\{\alpha_1 = \texttt{bool}\}) = \textbf{Mgu}\,(\emptyset) \circ [\alpha_1 \leftarrow \texttt{bool}] = id \circ [\alpha_1 \leftarrow \texttt{bool}] \\
 & = [\alpha_1 \leftarrow \texttt{bool}] \\
\text{Env}_3' & = \mu_1(id(\text{Env}_3)) = [(z, \forall\emptyset.\,\alpha_3), (y, \forall\emptyset.\,\alpha_2), (x, \forall\emptyset.\,\texttt{bool})] \\
\mu_2 & = \textbf{Mgu}\,(\{id(\alpha_2) = \alpha_3\}) = \textbf{Mgu}\,(\emptyset) \circ [\alpha_2 \leftarrow \alpha_3] = [\alpha_2 \leftarrow \alpha_3] \\
\theta_1 & = \mu_2 \circ id \circ id \circ \mu_1 \circ id = \mu_2 \circ \mu_1 = [\alpha_2 \leftarrow \alpha_3; \alpha_1 \leftarrow \texttt{bool}]
\end{array}$$

$$(I_{true})\ \dfrac{}{\text{Env} \vdash true \rhd (\texttt{bool}, id)} \qquad (I_{false})\ \dfrac{}{\text{Env} \vdash false \rhd (\texttt{bool}, id)}$$

$$(I_{int})\ \dfrac{}{\text{Env} \vdash v \rhd (\texttt{int}, id)} \qquad (I_{var})\ \dfrac{x \in \text{dom}(\text{Env})}{\text{Env} \vdash x \rhd (\textbf{Inst}\,(\text{Env}(x)), id)}$$

$$(I_{ite})\ \dfrac{\begin{array}{c}\text{Env} \vdash e_1 \rhd (\tau_1, \theta_1) \quad \mu_1 = \textbf{Mgu}\,(\{\tau_1 = \texttt{bool}\}) \quad \mu_1(\theta_1(\text{Env})) \vdash e_2 \rhd (\tau_2, \theta_2) \\ \theta_2(\mu_1(\theta_1(\text{Env}))) \vdash e_3 \rhd (\tau_3, \theta_3) \quad \mu_2 = \textbf{Mgu}\,(\{\theta_3(\tau_2) = \tau_3\})\end{array}}{\text{Env} \vdash \text{if } e_1 \text{ then } e_2 \text{ else } e_3 \rhd (\mu_2(\tau_3), \mu_2 \circ \theta_3 \circ \theta_2 \circ \mu_1 \circ \theta_1)}$$

$$(I_{app})\ \dfrac{\begin{array}{c}\text{Env} \vdash e_1 \rhd (\tau_1, \theta_1) \quad \theta_1(\text{Env}) \vdash e_2 \rhd (\tau_2, \theta_2) \\ \alpha \text{ fresh} \quad \mu = \textbf{Mgu}\,(\{\theta_2(\tau_1) = \tau_2 \to \alpha\})\end{array}}{\text{Env} \vdash (e_1\ e_2) \rhd (\mu(\alpha), \mu \circ \theta_2 \circ \theta_1)}$$

$$(I_{fun})\ \dfrac{\alpha \text{ fresh} \quad (x, \forall\emptyset.\,\alpha) \oplus \text{Env} \vdash e \rhd (\tau_1, \theta_1)}{\text{Env} \vdash \text{fun } f \mapsto e \rhd (\theta_1(\alpha) \to \tau_1, \theta_1)}$$

$$(I_{let})\ \dfrac{\text{Env} \vdash e_1 \rhd (\tau_1, \theta_1) \quad (x, \textbf{Gen}\,(\tau_1, \theta_1(\text{Env}))) \oplus \theta_1(\text{Env}) \vdash e_2 \rhd (\tau_2, \theta_2)}{\text{Env} \vdash \text{let } x = e_1 \text{ in } e_2 \rhd (\tau_2, \theta_2 \circ \theta_1)}$$

$$(I_{rec})\ \dfrac{\begin{array}{c}\alpha, \beta \text{ fresh} \quad (f, \forall\emptyset.\,\alpha \to \beta) \oplus (x, \forall\emptyset.\,\alpha) \oplus \text{Env} \vdash e \rhd (\tau_1, \theta_1) \\ \mu = \textbf{Mgu}\,(\{\theta_1(\beta) = \tau_1\})\end{array}}{\text{Env} \vdash \text{rec } f\ x = e \rhd (\mu(\theta_1(\alpha \to \beta)), \mu \circ \theta_1)}$$

Box 5.3. *Type inference algorithm for* $\mathbf{Exp_2}$

The typing tree ∇_1 is thus:

$$
(I_{ite})\dfrac{(I_{var})\dfrac{}{\text{Env}_3 \vdash x \rhd (\alpha_1, id)} \quad (I_{var})\dfrac{}{\text{Env}'_3 \vdash y \rhd (\alpha_2, id)} \quad (I_{var})\dfrac{}{\text{Env}'_3 \vdash z \rhd (\alpha_3, id)}}{\underbrace{(z, \forall\emptyset.\, \alpha_3) \oplus \text{Env}_2}_{\text{Env}_3} \vdash e_4 \rhd (\alpha_3, \theta_1)}
$$

$$
(I_{fun})\dfrac{\qquad\qquad\qquad\qquad}{\underbrace{(y, \forall\emptyset.\, \alpha_2) \oplus \text{Env}_1}_{\text{Env}_2} \vdash e_3 \rhd (\alpha_3 \to \alpha_3, \theta_1)}
$$

$$
(I_{fun})\dfrac{\qquad\qquad\qquad\qquad}{\underbrace{(x, \forall\emptyset.\, \alpha_1) \oplus \text{Env}}_{\text{Env}_1} \vdash e_2 \rhd (\alpha_3 \to (\alpha_3 \to \alpha_3), \theta_1)}
$$

$$
(I_{fun})\dfrac{\qquad\qquad\qquad\qquad}{\text{Env} \vdash e_1 \rhd (\texttt{bool} \to (\alpha_3 \to (\alpha_3 \to \alpha_3)), \theta_1)}
$$

and where ∇_2 is the typing tree:

$$
(I_{app})\dfrac{(I_{app})\dfrac{(I_{app})\dfrac{(I_{var})\dfrac{}{\text{Env}_4 \vdash f \rhd (\texttt{bool} \to (\alpha_4 \to (\alpha_4 \to \alpha_4)), id)} \quad \nabla_5}{\text{Env}_4 \vdash e_5 \rhd (\alpha_4 \to (\alpha_4 \to \alpha_4), \mu_3 \circ id)} \quad \nabla_4}{\text{Env}_4 \vdash e_6 \rhd (\texttt{int} \to \texttt{int}, \mu_4 \circ id \circ \mu_3)} \quad \nabla_3}{\text{Env}_4 \vdash e_7 \rhd (\texttt{int}, \mu_5 \circ id \circ \mu_4 \circ id \circ \mu_3)}
$$

where $\text{Env}_4 = (x, \forall \alpha_3.\, (\texttt{bool} \to (\alpha_3 \to (\alpha_3 \to \alpha_3)))) \oplus \text{Env}$ and the typing of e_5 is obtained considering:

$$
\begin{aligned}
\mu_3 &= \textbf{Mgu}\,(\{id(\texttt{bool} \to (\alpha_4 \to (\alpha_4 \to \alpha_4))) = \texttt{bool} \to \alpha_5\}) \\
&= \textbf{Mgu}\,(\{\texttt{bool} = \texttt{bool}; \alpha_4 \to (\alpha_4 \to \alpha_4) = \alpha_5\}) \\
&= \textbf{Mgu}\,(\{\alpha_4 \to (\alpha_4 \to \alpha_4) = \alpha_5\}) \quad = \textbf{Mgu}\,(\{\emptyset\}) \circ [\alpha_5 \leftarrow \alpha_4 \to (\alpha_4 \to \alpha_4)] \\
&= [\alpha_5 \leftarrow \alpha_4 \to (\alpha_4 \to \alpha_4)] \\
\mu_3(\alpha_5) &= \alpha_4 \to (\alpha_4 \to \alpha_4)
\end{aligned}
$$

to type e_6:

$$
\begin{aligned}
\mu_3(\text{Env}_4) &= \text{Env}_4 \\
\mu_4 &= \textbf{Mgu}\,(\{id(\alpha_4 \to (\alpha_4 \to \alpha_4)) = \texttt{int} \to \alpha_6\}) \\
&= \textbf{Mgu}\,(\{\alpha_4 = \texttt{int}; \alpha_4 \to \alpha_4 = \alpha_6\}) \\
&= \textbf{Mgu}\,(\{\texttt{int} \to \texttt{int} = \alpha_6\}) \circ [\alpha_4 \leftarrow \texttt{int}] \\
&= \textbf{Mgu}\,(\emptyset) \circ [\alpha_6 \leftarrow \texttt{int} \to \texttt{int}] \circ [\alpha_4 \leftarrow \texttt{int}] \\
&= [\alpha_6 \leftarrow \texttt{int} \to \texttt{int}; \alpha_4 \leftarrow \texttt{int}] \\
\mu_4(\alpha_6) &= \texttt{int} \to \texttt{int}
\end{aligned}
$$

to type e_7:

$$
\begin{aligned}
\mu_4(\text{Env}_4) &= \text{Env}_4 \\
\mu_5 &= \textbf{Mgu}\,(\{id(\texttt{int} \to \texttt{int}) = \texttt{int} \to \alpha_7\}) = \textbf{Mgu}\,(\{\texttt{int} = \texttt{int}; \texttt{int} = \alpha_7\}) \\
&= \textbf{Mgu}\,(\{\texttt{int} = \alpha_7\}) = \textbf{Mgu}\,(\emptyset) \circ [\alpha_7 \leftarrow \texttt{int}] = [\alpha_7 \leftarrow \texttt{int}] \\
\mu_5(\alpha_7) &= \texttt{int}
\end{aligned}
$$

$$\theta = \mu_5 \circ id \circ \mu_4 \circ id \circ \mu_3 = \mu_5 \circ \mu_4 \circ \mu_3$$
$$= [\alpha_7 \leftarrow \texttt{int}; \alpha_6 \leftarrow \texttt{int} \rightarrow \texttt{int}; \alpha_4 \leftarrow \texttt{int}; \alpha 5 \leftarrow \texttt{int} \rightarrow (\texttt{int} \rightarrow \texttt{int})]$$

∇_3, ∇_4 and ∇_5 are, respectively, the following trees:

$$(I_{int})\frac{}{\text{Env}_4 \vdash 3 \rhd (\texttt{int}, id)} \qquad (I_{int})\frac{}{\text{Env}_4 \vdash 2 \rhd (\texttt{int}, id)}$$

$$(I_{true})\frac{}{\text{Env}_4 \vdash true \rhd (\texttt{bool}, id)}$$

OCaml
```
let rec type_expr env = function
  | Cste2 (CInt2 _) -> (T_int, empty_subst)
  | Cste2 (CBool2 _) -> (T_bool, empty_subst)
  | Var2 v -> (specialize (find_id_ty v env), empty_subst)
  | Fun2 (arg_name, body) ->
      let arg_ty = new_ty_var () in
      let (body_ty, body_subst) =
         type_expr
            ((arg_name, (trivial_scheme arg_ty)) :: env) body in
      (T_fun ((apply arg_ty body_subst, body_ty)),
       body_subst)
  | Letin2 (v_name, e1, e2) ->
      let (e1_ty, e1_subst) = type_expr env e1 in
      let new_env = (v_name, (generalize env e1_ty)) :: env in
      let new_env2 = subst_env e1_subst new_env in
      let (e2_ty, e2_subst) = type_expr new_env2 e2 in
      (e2_ty, compose e2_subst e1_subst)
  | App2 (e1, e2) ->
      let (e1_ty, e1_subst) = type_expr env e1 in
      let (e2_ty, e2_subst) =
         type_expr (subst_env e1_subst env) e2 in
      let fun_res_ty = new_ty_var () in
      let e1_ty_substed = (apply e1_ty e2_subst) in
      let mu = unify (T_fun (e2_ty, fun_res_ty)) e1_ty_substed in
      ((apply fun_res_ty mu),
       (compose mu (compose e2_subst e1_subst)))
  | If2 (cond_e, then_e, else_e) -> (
      let (cond_ty, cond_subst) = type_expr env cond_e in
      let mu1 = unify cond_ty T_bool in
      let env_then = subst_env mu1 (subst_env cond_subst env) in
      let (then_ty, then_subst) = type_expr env_then then_e in
      let env_else = subst_env then_subst env_then in
      let (else_ty, else_subst) = type_expr env_else else_e in
      let mu2 = unify (apply then_ty else_subst) else_ty in
      ((apply else_ty mu2),
       (compose mu2
          (compose else_subst
             (compose then_subst
                (compose mu1 cond_subst)))))
    )
```

```
  | Rec2 (f_name, arg_name, body) ->
      let arg_ty = new_ty_var () in
      let fun_res_ty = new_ty_var () in
      let fun_ty = T_fun (arg_ty, fun_res_ty) in
      let new_env =
        (f_name, (trivial_scheme fun_ty)) ::
        (arg_name, (trivial_scheme arg_ty)) :: env in
      let (body_ty, body_subst) = type_expr new_env body in
      let mu = unify (apply fun_res_ty body_subst) body_ty in
      ((apply (apply fun_ty body_subst) mu),
       (compose mu body_subst))
val type_expr : (string * scheme_t) list -> exp2 -> (type_t * subst_t)
```

Python
```
def type_expr (env, exp):
    if isinstance (exp, Cste2) :
        if isinstance (exp.cst, CInt2) : return (T_int (), empty)
        elif isinstance (exp.cst, CBool2) : return (T_bool (), empty)
        else : raise AstValError
    elif isinstance (exp, Var2) :
        return (specialize (find_id_ty (exp.name, env)), empty)
    elif isinstance (exp, Fun2) :
        arg_ty = new_ty_var ()
        arg_name = exp.id_arg
        body = exp.body
        (body_ty, body_subst) = \
            type_expr ([(arg_name, trivial_scheme (arg_ty))] + env, \
                       body)
        return (T_fun (apply (arg_ty, body_subst), body_ty), \
                body_subst)
    elif isinstance (exp, Letin2) :
        (e1_ty, e1_subst) = type_expr (env, exp.exp1)
        new_env = [(exp.name, generalize (env, e1_ty))] + env
        new_env2 = subst_env (e1_subst, new_env)
        (e2_ty, e2_subst) = type_expr (new_env2, exp.exp2)
        return (e2_ty, compose (e2_subst, e1_subst))
    elif isinstance (exp, App2) :
        (e1_ty, e1_subst) = type_expr (env, exp.exp1)
        (e2_ty, e2_subst) = \
            type_expr (subst_env (e1_subst, env), exp.exp2)
        fun_res_ty = new_ty_var ()
        e1_ty_substed = apply (e1_ty, e2_subst)
        mu = unify.unify (T_fun (e2_ty, fun_res_ty), e1_ty_substed)
        return \
            (apply (fun_res_ty, mu),\
             compose (mu, compose (e2_subst, e1_subst)))
    elif isinstance (exp, If2) :
        (cond_ty, cond_subst) = type_expr (env, exp.cond_e)
        mu1 = unify.unify (cond_ty, T_bool ())
        env_then = subst_env (mu1, subst_env (cond_subst, env))
        (then_ty, then_subst) = type_expr (env_then, exp.then_e)
        env_else = subst_env (then_subst, env_then)
        (else_ty, else_subst) = type_expr (env_else, exp.else_e)
```

```
        mu2 = unify.unify (apply (then_ty, else_subst), else_ty)
        return \
            (apply (else_ty, mu2), \
             compose \
               (mu2,\
                compose \
                  (else_subst,\
                   compose \
                     (then_subst, compose (mu1, cond_subst)))))
    elif isinstance (exp, Rec2) :
        arg_ty = new_ty_var ()
        fun_res_ty = new_ty_var ()
        fun_ty = T_fun (arg_ty, fun_res_ty)
        new_env = \
            [(f_name, trivial_scheme (fun_ty))] + \
            [(arg_name, trivial_scheme (arg_ty))] + env
        (body_ty, body_subst) = type_expr (new_env, body)
        mu = unify.unify (apply (fun_res_ty, body_subst), body_ty)
        return \
            (apply (apply (fun_ty, body_subst), mu),\
             compose (mu, body_subst))
     else: raise AstValError
```

The function `type_expr` is defined by case over the structure of the expressions. It returns a pair in which the first component is the type of the typed expression and the second component is the substitution resulting from the typing. As we explained earlier, the cases of unary and binary operators will not be addressed separately, as these are simply specific cases of the general application `App2`.

5.5. Properties

5.5.1. *Properties of typechecking*

We can establish several properties of the typing system set out in Box 5.1 which guarantee its correctness with regard to the semantics of the language. The most important of these properties is set out here, although we shall not provide a demonstration; the relevant proof can be found in [DAM 82].

PROPOSITION 5.1.– Strong safety: let e be an expression, τ a type, Env_t a typing environment and Env an evaluation environment such that $\forall x \in \mathbf{X}$, $\mathrm{Env}_t \vdash \mathrm{Env}(x) : \mathrm{Env}_t(x)$. If $\mathrm{Env}_t \vdash e : \tau$ and $\mathrm{Env} \vdash e \rightsquigarrow v$, then v is a value of type τ.

5.5.2. *Properties of the inference algorithm*

Just as typing rules need to satisfy a certain number of properties, the algorithm itself must meet four requirements: termination, correctness, completeness and principality. Termination guarantees that the type of an expression will be inferred in a finite period of time (the compiler will not enter into an infinite loop). The algorithm is correct if, for a given expression e and a typing environment Env, it returns a type which is correct with respect to the typing rules set out in Box 5.1. This correctness means that the inference algorithm will not produce *wrong* results. Completeness means that if a type exists for a given expression e and a typing environment Env, then it will be found by the algorithm: in other words, the algorithm will not "miss" a solution. Principality guarantees that the algorithm will always compute the most general type, in the sense presented in section 5.4.

PROPOSITION 5.2.– Correctness of the inference algorithm: if $\text{Env} \vdash e \rhd (\tau, \theta)$, then $\theta(\text{Env}) \vdash e : \tau$.

PROPOSITION 5.3.– Completeness and principality of the inference: let Env be an environment and e an expression. If there exists a substitution θ' and a type τ' such that $\theta'(\text{Env}) \vdash e : \tau'$, then $\text{Env} \vdash e \rhd (\tau, \theta)$ and there exists a substitution θ'' such that $\tau' = \theta''(\tau)$ and $\theta' = \theta'' \circ \theta$.

5.6. Typechecking of imperative constructs

To introduce typechecking of imperative languages, we shall define rules for the language $\mathbf{Lang_3}$ introduced in Chapter 4, extended using the block structure (see section 4.6.1). The type system given here is not an extension of the rules shown in Box 5.1, as we are not considering functional features, only mutable variables. We thus need to type the language of definitions (given in Table 4.1), the language of expressions (given in Table 4.2) and the instructions (described in Table 4.3). For reasons of simplicity, the same symbol will be used to denote all three typing relations.

5.6.1. *Type algebra*

As references belong to the set of values, they must be assigned a type, which is called `ref`. To handle, in a consistent manner, the value stored at the address denoted by a reference, the type `ref` must also indicate the type of the referenced value. The kernel of $\mathbf{Lang_3}$ contains no function nor polymorphism and only integer and boolean constants; the type algebra is thus very simple. This algebra, presented in Table 5.2, contains an integer type `int`, a boolean type `bool` and a type of references on a value of type τ, denoted as τ `ref`.

As polymorphism is not considered, the typing environment does not bind identifiers to type schemes, but simply to types.

$\tau ::=$ int	Integer
| bool	Boolean
| τ ref	Reference

Table 5.2. *Type algebra of* **Lang**$_3$

5.6.2. *Typing rules*

Let us begin by setting out the typing rules for expressions in the language. As in the case of the rules presented in Box 5.1, the type relation $\text{Env} \vdash e : \tau$ establishes that, in a typing environment Env, an expression e is of type τ.

$$(T_{\text{int}})\ \frac{k \in \mathbb{Z}}{\text{Env} \vdash k : \texttt{int}} \qquad (T_{\text{bool}})\ \frac{k \in \mathbb{B}}{\text{Env} \vdash k : \texttt{bool}} \qquad (T_{\text{var}})\ \frac{\text{Env}(x) = \tau}{\text{Env} \vdash x : \tau}$$

$$(T_{\text{unop}})\ \frac{\vdash \text{op} : \tau_1 \to \tau_2 \qquad \text{Env} \vdash e : \tau_1}{\text{Env} \vdash \text{op}\ e : \tau_2}$$

$$(T_{\text{binop}})\ \frac{\vdash \text{op} : \tau_1 \to \tau_2 \to \tau_3 \qquad \text{Env} \vdash e_1 : \tau_1 \qquad \text{Env} \vdash e_2 : \tau_2}{\text{Env} \vdash e_1\, \text{op}\, e_2 : \tau_3}$$

$$(T_{\text{ref}})\ \frac{\text{Env} \vdash e : \tau}{\text{Env} \vdash \uparrow e : \tau\ \texttt{ref}} \qquad (T_{\text{deref}})\ \frac{\text{Env} \vdash e : \tau\ \texttt{ref}}{\text{Env} \vdash \downarrow e : \tau}$$

Box 5.4. *Typing rules for expressions in* **Lang**$_3$

Now, let us consider the typing of variable declarations and blocks. A variable declaration d may be used to extend the typing environment by binding the declared identifier to the type of its defining expression. Thus, the relation $\text{Env} \vdash d : \text{Env}'$ expresses the fact that, in an environment Env, typing the declaration d will create, from Env, a new environment noted Env$'$. We recall that a block is written begin **d** c end, where **d** is a (possibly empty) sequence of declarations (see section 4.6.1) and c an instruction. We thus need rules for typing a sequence of declarations. Let d **d** denote a sequence beginning with a declaration d, followed by the sequence **d**. The empty sequence is noted $\emptyset$.

We have chosen to consider that an instruction c has no type as it has no value. Nevertheless, given a typing environment Env, the typing algorithm must check that the expressions involved in an instruction are correctly typed. This typing relation is noted $\text{Env} \vdash c : WT$ (WT denotes *well-typed*). Note that in certain languages, such as OCaml or certain extensions of C (such as those available with the GCC compiler), instructions are expressions, meaning that they have values and, consequently, types. The nature of these types depends on choices inherent to the language.

$$(T_{\text{let}})\ \frac{\text{Env} \vdash e : \tau}{\text{Env} \vdash \textsf{let}\ x = e; : (x, \tau) \oplus \text{Env}}$$

$$(T_{\text{edseq}})\ \frac{}{\text{Env} \vdash \emptyset : \text{Env}} \qquad (T_{\text{dseq}})\ \frac{\text{Env} \vdash d : \text{Env}_1 \qquad \text{Env}_1 \vdash \mathbf{d} : \text{Env}_2}{\text{Env} \vdash d\,\mathbf{d} : \text{Env}_2}$$

Box 5.5. *Typing rules for declarations in* $\mathbf{Lang}_3$

$$(T_{\text{assign}})\ \frac{\text{Env}(x) = \tau\ \texttt{ref} \qquad \text{Env} \vdash e : \tau}{\text{Env} \vdash x := e : WT} \qquad (T_{\text{skip}})\ \frac{}{\text{Env} \vdash \textsf{skip} : WT}$$

$$(T_{\text{seq}})\ \frac{\text{Env} \vdash c_1 : WT \quad \text{Env} \vdash c_2 : WT}{\text{Env} \vdash c_1\ ;\ c_2 : WT} \qquad (T_{\text{loop}})\ \frac{\text{Env} \vdash e : \texttt{bool} \quad \text{Env} \vdash c : WT}{\text{Env} \vdash \textsf{while}\ e\ \textsf{do}\ c : WT}$$

$$(T_{\text{cond}})\ \frac{\text{Env} \vdash e : \texttt{bool} \qquad \text{Env} \vdash c_1 : WT \qquad \text{Env} \vdash c_2 : WT}{\text{Env} \vdash \textsf{if}\ e\ \textsf{then}\ c_1\ \textsf{else}\ c_2 : WT}$$

$$(T_{\text{block}})\ \frac{\text{Env} \vdash \mathbf{d} : \text{Env}_1 \qquad \text{Env}_1 \vdash c : WT}{\text{Env} \vdash \textsf{begin}\ \mathbf{d}\ c\ \textsf{end} : WT}$$

Box 5.6. *Typing rules for instructions in* $\mathbf{Lang}_3$

EXAMPLE 5.6.– Consider the GCD computation program presented in section 4.1 and defined by:

$$\begin{array}{l}\textsf{begin let}\ x = 45;\textsf{let}\ y = 5;\\ \textsf{while}\ \underbrace{\textsf{not}(\downarrow x = \downarrow y)}_{e_1}\ \textsf{do}\ \underbrace{\textsf{if}\ \underbrace{\downarrow x \leq \downarrow y}_{e_2}\ \textsf{then}\ \underbrace{y := \downarrow y - \downarrow x}_{i_2}\ \textsf{else}\ \underbrace{x := \downarrow x - \downarrow y}_{i_3}}_{i_1}\\ \textsf{end}\end{array}$$

Its typechecking tree is:

$$(T_{\text{block}})\ \frac{(T_{\text{deseq}})\ \dfrac{(T_{\text{let}})\ \dfrac{(T_{\text{int}})\ \dfrac{}{\text{Env} \vdash 45 : \texttt{int}}}{\text{Env} \vdash \textsf{let}\ x = 45; : \text{Env}_1} \quad (T_{\text{let}})\ \dfrac{(T_{\text{int}})\ \dfrac{}{\text{Env}_1 \vdash 5 : \texttt{int}}}{\text{Env}_1 \vdash \textsf{let}\ y = 5; : \text{Env}_2}}{\text{Env} \vdash \textsf{let}\ x = 45;\textsf{let}\ y = 5; : \text{Env}_2} \qquad \underset{1}{\nabla}}{\text{Env} \vdash \textsf{begin let}\ x = 45;\textsf{let}\ y = 5;\ \textsf{while}\ e_1\ \textsf{do}\ i_1\ \textsf{end} : WT}$$

where $\text{Env}_1 = (x, \texttt{int ref}) \oplus \text{Env}$, $\text{Env}_2 = (y, \texttt{int ref}) \oplus \text{Env}_1$ and ∇_1 is the typing tree while e_1 do i_1, written as:

$$(T_{\text{loop}})\dfrac{(T_{\text{unop}})\dfrac{\text{Env}_2 \vdash \text{not} : \texttt{bool} \rightarrow \texttt{bool} \quad (T_{\text{binop}})\dfrac{\text{Env}_2 \vdash = : \texttt{int} \rightarrow \texttt{int} \rightarrow \texttt{bool} \quad \nabla_2 \quad \nabla_3}{\text{Env}_2 \vdash \downarrow x =\downarrow y : \texttt{bool}}}{\text{Env}_2 \vdash \text{not}(\downarrow x =\downarrow y) : \texttt{bool}} \quad \nabla_4}{\text{Env}_2 \vdash \text{while } e_1 \text{ do } i_1 : WT}$$

and ∇_2 and ∇_3 are the typing trees for $\downarrow x$ and $\downarrow y$:

$$(T_{\text{deref}})\dfrac{(T_{\text{var}})\dfrac{\text{Env}_2(x) = \texttt{int ref}}{\text{Env}_2 \vdash x : \texttt{int ref}}}{\text{Env}_2 \vdash \downarrow x : \texttt{int}} \qquad (T_{\text{deref}})\dfrac{(T_{\text{var}})\dfrac{\text{Env}_2(y) = \texttt{int ref}}{\text{Env}_2 \vdash y : \texttt{int ref}}}{\text{Env}_2 \vdash \downarrow y : \texttt{int}}$$

and ∇_4 is the typing tree for if e_2 then i_2 else i_3:

$$(T_{\text{cond}})\dfrac{(T_{\text{binop}})\dfrac{\text{Env}_2 \vdash \leq : \texttt{int} \rightarrow \texttt{int} \rightarrow \texttt{bool} \quad \nabla_2 \quad \nabla_3}{\text{Env}_2 \vdash e_2 : \texttt{bool}} \quad \nabla_5 \quad \nabla_6}{\text{Env}_2 \vdash \text{if } e_2 \text{ then } i_2 \text{ else } i_3 : WT}$$

and ∇_5 and ∇_6 are the typing trees for instructions i_2 and i_3:

$$(T_{\text{assign}})\dfrac{\text{Env}_2(y) = \texttt{int ref} \quad (T_{\text{binop}})\dfrac{\text{Env}_2 \vdash - : \texttt{int} \rightarrow \texttt{int} \rightarrow \texttt{int} \quad \nabla_3 \quad \nabla_2}{\text{Env}_2 \vdash \downarrow y - \downarrow x : \texttt{int}}}{\text{Env}_2 \vdash i_2 : WT}$$

$$(T_{\text{assign}})\dfrac{\text{Env}_2(x) = \texttt{int ref} \quad (T_{\text{binop}})\dfrac{\text{Env}_2 \vdash - : \texttt{int} \rightarrow \texttt{int} \rightarrow \texttt{int} \quad \nabla_2 \quad \nabla_3}{\text{Env}_2 \vdash \downarrow x - \downarrow y : \texttt{int}}}{\text{Env}_2 \vdash i_3 : WT}$$

5.6.3. *Typing polymorphic definitions*

In a language that accepts polymorphism, the introduction of references creates an additional difficulty. Consider the definition let $x =\uparrow []$; where $[]$ is the value of an empty list. Suppose that the type of this list is polymorphic, and thus of the form α `list`, where α is the type of the elements. If we assign the type scheme $\forall \alpha. (\alpha\ \texttt{list})\ \texttt{ref}$ to x, then the rules of polymorphism can be used, for example, in:

```
x := [1] ;
x := [true] ;
1 + head (↓ x) ;
```

where the function head (supposed to exist) returns the first element of a (non-empty) list. In this program, as the type of x is polymorphic, it is possible to assign x with any given list. The final instruction in the program is well-typed: x is of type (α list) ref, so $\downarrow x$ is of type α list, and so head returns a value of type α, the latter being instantiated in int in the addition expression. However, in practice, x contains a boolean value, so execution will necessarily raise an error, as addition is only defined for integers.

Conversely, consider the following program:

```
let f = fun x -> ↑ x ;
```

which creates a reference upon an input value. The type of the function f is $\alpha \rightarrow \alpha$ ref. The generalization of α is not problematic. Function f can be applied to a value of any type t, and will return a reference of type t ref, which is not polymorphic.

One simple solution to this problem is to only generalize certain variables. The types of expressions that correspond to constants, variables or functions are generalized, and those of applications of functions or operators are not generalized. In the example above, in the declaration let $x =\uparrow []$; the expression $\uparrow []$ is an application of the operator $\uparrow$, hence the type α list ref should not be generalized. This solution relies on a syntactic criterion alone. Other methods have been developed (e.g. [LER 91, PIE 02]) to relax this restriction and increase the number of safe generalizations, but these lie outside the scope of this book.

5.7. Subtyping and overloading

Many languages offer extensions to the notion of type, designed to make the type algorithm accept expressions which it would otherwise reject. This increases flexibility when writing code, but makes it harder to review: in addition to typing rules, users must also master the rules of subtyping and overloading.

Subtyping makes it possible to relax the constraint concerning the type t_1 of the argument of a function, which may then be applied to an expression of type t_2, as long as the values of type t_2 share at least the characteristics of the values of type t_1.

Overloading enables the same name f to be used to denote different values (usually functions), leaving the typing algorithm to determine which binding to use for each occurrence of f in an expression. The aim, in this case, is to relax the constraint brought by the typing discipline: an identifier must have one, and only one, type.

In this section, we provide a brief presentation of subtyping and overloading. These features will be discussed in greater detail in Volume 2, in Chapters 1 and 3, with examples provided in Chapters 2 and 4.

5.7.1. *Subtyping*

If the typing algorithm used by a language satisfies the partial correctness property, any well-typed program will be evaluated as a value of the expected type or will loop. Certain programs rejected by the typing process also respect this property of partial correctness, but the typechecking is unable to determine this. For this reason, the type language and the typing algorithm may be enriched in order to reduce program rejection, while maintaining the partial correctness property.

Subtyping is a way of refining the typing process using a comparison relation on types noted $t_s \preceq t_b$ [1] and a typing algorithm that permits any expression e of type t_s to be accepted as an expression of type t_b. The comparison relation must be reflexive and transitive: in other words, this relation $\preceq$ is a partial preorder (there is no need to compare any two types using $\preceq$). We have already encountered such a preorder in section 5.4. The relation " τ' is an instance of the type scheme $\sigma = \forall \alpha_1 \ldots \alpha_k . \tau$ ", denoted as $\sigma \leq \tau'$, is indeed a partial preorder over polymorphic types. For example, type schemes $\forall \emptyset.\,\texttt{int}$ and $\forall \emptyset.\,\texttt{bool}$ are not comparable, and $\forall \alpha.\, \alpha \rightarrow \alpha$ is smaller than $\texttt{int} \rightarrow \texttt{int}$.

A preorder on types can be defined in the following manner. We first choose the family F of types, which we already know how to compare. Next, we extend (when necessary: the preorder does not need to be total) the preorder chosen for F to other types. Since type languages are often constructed by induction, extension is also usually carried out by induction. This gives some guarantees on the consistency between inductive construction and comparison. For example, consider $t_s \preceq t_b$ and $s_s \preceq s_b$. If the type language includes product types, it seems consistent to extend this order by taking $t_s \times s_s \preceq t_b \times s_b$.

Suppose now that the order on types is the order of inclusion of their sets of values: $t_s \preceq t_b$ if the set of values of type t_s is included in the set of values of type t_b. For example, this would be the case if values of type t_b represent individual people, and values of type t_s represent individual employees. Suppose that $t_s \preceq t_b$ and $s_s \preceq s_b$. consider a function $f : t_b \rightarrow s_s$. In this case, f allows us to define a function $g : t_s \rightarrow s_b$ by simply restricting f to the subset of values of type t_s. Furthermore, as g takes values in s_s, it also takes values in s_b, as shown in Figure 5.1.

Function f is defined over a domain that contains the domain of g and the codomain of f is included in that of g. Thus, anything which is treated by g is also treated by f, and any value returned by f is included in the values returned by g. From this intuitive interpretation, it is possible to prove that the subtyping relation defined over the type family F can be extended by taking the following rule: if $t_s \preceq t_b$ and $s_s \preceq s_b$, then $(t_b \rightarrow s_s) \preceq (t_s \rightarrow s_b)$.

1 Read t_{small} for t_s and t_{big} for t_b.

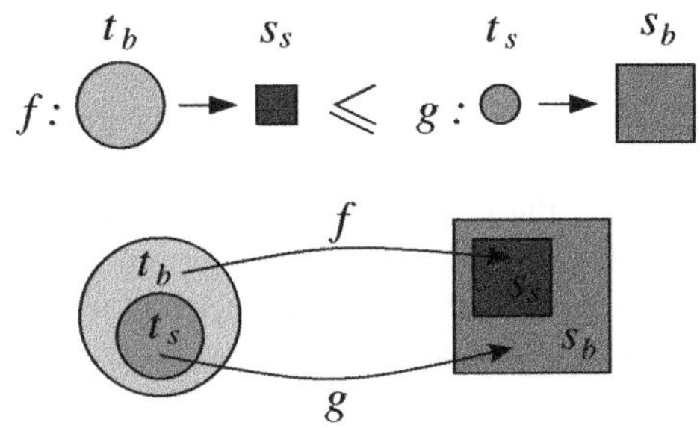

Figure 5.1. *Covariance and contravariance. For a color version of this figure, see www.iste.co.uk/hardin/programming1.zip*

A preorder over types that verify this rule is said to be *contravariant* over the type of the function arguments and *covariant* over the type of the results. It can be proved that a typing algorithm using such a preorder can be constructed from the two rules shown in Box 5.7 (and, potentially, from similar rules for certain other type constructors).

$$(T_{\preceq})\frac{\text{Env} \vdash e : t_s \quad t_s \preceq t_b}{\text{Env} \vdash e : t_b} \qquad (T_{app\,\preceq})\frac{\text{Env} \vdash e_1 : t_b \to t \quad \text{Env} \vdash e_1 : t_s \quad t_s \preceq t_b}{\text{Env} \vdash (e_1\, e_2) : t}$$

Box 5.7. *Subtyping rules*

Rule $T_{\preceq}$ defines the semantics of the order relation over types: any expression e of a type t_s that is smaller than a type t_b is also of type t_b. Rule $T_{app\,\preceq}$ is a consequence of this semantics: any function that requires an expression of type t_b as its argument will accept an argument with a type t_s, which is a subtype of t_b.

This order is frequently used in the case described below using our example. Consider a function f, taking individuals as its parameter and returning, for example, their birth year (f is therefore of the type $t_b \to \texttt{int}$). f may be restricted to data of type t_s, i.e. for individual employees. Noting this restriction g, the type of g is thus $t_s \to \texttt{int}$. We have $(t_b \to \texttt{int}) \preceq (t_s \to \texttt{int})$, ensuring the consistency of this restriction in terms of typing.

Conventionally, the name f is retained for the restriction (instead of introducing a new name g). Programming language manuals often specify that any function of type $t_1 \to t_2$ can be applied to any value of type t, if t is a subtype of t_1, and the type of the result t_2 remains unchanged. This approach does not exploit the full potential of functional type comparison. We shall return to this point in Volume 2, Chapter 1.

These rules extending the typing relation cannot be generalized indiscriminately to any operation; however, this does not decrease the utility of subtyping, as only a partial preorder is required. For example, the subtyping relation cannot be extended to assignment. Consider $t_s \preceq t_b$: in our example, t_s is the type of individual employees and gives their name and salary, and t_b is the type of all individuals, giving only their name. Now, suppose that the values of type t_s are stored over a greater number of bytes than the values of type t_b. This is the case in our example. The assignment of $v : t_b$ to $x : t_s$ `ref` will leave a number of "free" bytes in the zone pointed by x; evidently, this is not acceptable, unless the compiler identifies the issue and handles it in a suitable manner. Now, let us examine how this "common sense" constraint is obtained through a detailed examination of extension to assignment. Consider an assignment operator := of type t_b `ref` $\times$ $t_b \rightarrow$ `unit`. If we extend subtyping to this operator by induction, then := will also be of the type t_s `ref` $\times$ $t_b \rightarrow$ `unit`. This means that a reference to a value of type t_s will become a reference to a value of type t_b. This raises the question of the effective meaning of the word "reference" (see Chapter 6).

We do not intend to go into detail concerning subtyping and its properties here; our presentation is limited to the most commonly encountered notion. Other approaches exist, for example, using different orders induced over functional types, which may be covariant on the argument types. In all cases, complex theoretical studies are required to prove the partial correctness properties of these notions of subtyping, and, finally, to create effective and efficient typing algorithms. A number of examples of subtyping will be studied in Volume 2, notably in Chapters 2 and 4.

5.7.2. *Overloading*

Identifier overloading is a syntactic facility proposed by certain programming languages. In this case, the same identifier can be used for functions or operators with different definitions, and, consequently, different types and values. Overloading is often used to give the same name to operations, which have very similar specifications, but are applied to different data types. For example, the expressions `3 + 4`, `3.5 + 4.6` and `"aa" + "bb"` all use +, but the code and type of the operations denoted by this identifier are very different. The compiler is responsible for choosing the correct version of + to use in evaluating each of these expressions. This task is known as *overloading resolution*.

In order to take account of overloading, the environment model must be adapted. Without overloading, adding a new binding for an identifier f, which is already bound in an environment E, will mask the pre-existing bindings of f in E. In cases where a licit overloading is present, several bindings of f may be accessible in E. The typing algorithm must determine which binding of f to use when typing the application $f(e)$, according to the type of e. Overloading resolution may be carried out during compilation or execution of the program. The overloading resolution algorithm used is highly dependent on choices made by the language. Different forms, such as operator overloading, function and procedure overloading, constant overloading and method overloading, may or may not be included, independently of one another. The overloading resolution method may be described, informally, as follows. Let exp be an expression containing an overloaded identifier f. To type this expression and thus determine which value of f to use, we must:

1) associate a type variable with each occurrence of f, denoting one of the possible types for this identifier;

2) establish the system of equations which these variables must satisfy, using the typing rules of the language. Additional information may also be used, such as the names of formal parameters if these are used in passing the actual parameters;

3) at the end of this step, each type variable is associated with at least a type;

4) verify that there is only one solution to the system of equations. To do this, we substitute each type variable, one by one, with one of its associated types. If there is no solution, then exp is ill-typed. If several solutions exist, then we have an ambiguity and exp is rejected by the typing process.

Consider the following example, written in a pseudo-language that permits overloading for functions and for constants defining enumerated types, i.e. types defined by a set of explicitly named and non-mutable values:

```
type color is (blue, green, red)
type light is (green, red)
function F(c:color):light = ...
function F(x:light):color = ...
```

Let `F_a` be the first declaration of `F` and let `F_b` be the second declaration. Let `green_C` and `red_C` be the overloaded values of the type `color` and `green_F` and `red_F` the overloaded values of the type `light`. The expression `F(blue)` is clearly typable, taking `F_a` for `F` since `blue` is clearly and unambiguously of type `color`. The expressions `F(green)` and `F(F(green))`, on the other hand, are evidently ambiguous and will therefore be rejected.

This ambiguity is removed in `F(F(green)) = F(blue)`: typing this expression leads us to resolve the overloading of `F` and `green`. To do this, we write the expression in the form `F_1(F_2(green)) = F_3(blue)`. Let $\tau_1 = \tau_{11} \to \tau_{12}$ be the type of `F_1`, $\tau_2 = \tau_{21} \to \tau_{22}$ the type of `F_2` and $\tau_3 = \tau_{31} \to \tau_{32}$ that of `F_3`, where t_{ij} denotes the type variables. The possible values of τ_1 are `light` $\to$ `color` and `color` $\to$ `light`. Since `F_1` is applied to the result of `F_2`, we deduce the equation $\tau_{11} = \tau_{22}$. The results of `F_1` and `F_3` must be of the same type, giving us the equation $\tau_{12} = \tau_{32}$. Since `F_3` is applied to `blue`, we have $\tau_{31} =$ `color`, hence

τ_{32} = `light`, hence τ_{12} = `light` and $\tau_{11} = \tau_{22}$ = `color`, and thus the type of `green` is `light`. The expression `F(F(green)) = F(blue)` is then analyzed by the typing algorithm, following resolution, as `F_a(F_b(green)) = F_a(blue)`.

The rules used in overloading resolution may be more complex in cases where functions may have a variable number of arguments, in the presence of optional arguments, where the language includes object features that permit overloading for methods interacting with inheritance and subclassing. A brief presentation of overloading is given in Chapters 1 and 3 of Volume 2, and the subject is also discussed in relation to each of the languages presented in Chapters 2 and 4 of Volume 2. Note that overloading is not permitted in C.

It is important to note that, although overloading makes it easier to write programs, it can make them difficult to review; users must have an in-depth knowledge of the associated semantics, which may differ from one language to the next. This point will be addressed further in Volume 2, Chapters 3 and 4.

6

Data Types

6.1. Basic types

The languages $\mathbf{Exp}_2, \ldots, \mathbf{Lang}_3$ that we used to describe the notions of semantics and typing only consider two data types: integers and booleans. However, most programming languages include many more basic types. In this section, we shall present the most widespread data types, along with the ways in which they differ between languages.

6.1.1. *Booleans*

For certain languages, such as OCaml, Java, Ada and Pascal, the boolean type `bool` is a constructor of the type algebra and has two possible values: $true$ and $false$. Other languages, such as C and C++, encode boolean values as integers, with a value of 0 conventionally used to mean "false" and any other value corresponding to "true". Some of these languages, such as C++, include a type `bool` and predefined identifiers `false` and `true` but these values are fully compatible with integer values. Any integer expression may be used as the condition for a loop or an `if-then-else`. Any boolean expression may be used as an index in an array, or as an operand in an arithmetic operation. This makes the source text harder to understand and may lead to errors, which are difficult to detect.

Python behaves somewhat differently. It supplies two values, `True` and `False`, but allows integers and booleans to be mixed. It is thus possible to test whether or not an integer value is equal or different to a boolean value, or to use an integer value as the test for a condition. These practices are not recommended, see the examples given in the following. Remember that an integer, such as the value 2 in example 6.1, may be the result of evaluating a more complex expression. If using this value 2 as a boolean leads to an execution error, it may be difficult to retrieve its origin in the source code.

EXAMPLE 6.1.–

```
Python
$ python3.7
>>> 0 == True
False
>>> 0 == False
True
>>> 1 == True
True
>>> 1 == False
False
>>> 2 == True
False
>>> 2 == False
False
```

Example 6.1 shows that version 3.7 of Python interprets `False` as the value 0 and `True` as the value 1. It is interesting to note that the value 2 (or any other value which is not 0 or 1) is neither `True` nor `False`. Thus, the two expressions in example 6.2 – which have identical intuitive semantics – give different results.

EXAMPLE 6.2.–

```
Python
>>> if 2 == True :
...     print ("2 == True")
... else :
...     print ("2 != True")
...
2 != True
>>> if 2 == False :
...     print ("2 != True")
... else :
...     print ("2 == True")
...
2 == True
```

As the tests `2 == True` and `2 == False` are both always false, these two expressions evaluate the `else` branch, displaying contradictory results. However, if we use the integer value 2 directly as a condition, the `then` branch will be evaluated, as if 2 was equivalent to `True`.

EXAMPLE 6.3.–

```
Python
>>> if 2 :
...     print ("Then branch")
... else :
...     print ("Else branch")
...
Then branch
```

6.1.2. *Integers*

Given the finite size of a register, it is only possible to represent a finite number of binary configurations. As every integer is encoded in binary (base 2), a register can only encode a finite number of integers.

For positive integers, the bit of rank n in a binary configuration represents the value 2^n. The value of the integer encoded by a binary configuration is the sum of its powers. For example, the configuration 01100101 over 8 bits corresponds to $2^6+2^5+2^2+2^0 = 101$.

Positive and negative integers are differentiated by the use of a *sign bit*, which is the most significant bit (MSB) in the binary representation of the integer. Furthermore, negative integers are encoded using *two's complement*. To obtain the opposite of a positive number, we invert the bits of the binary representation of the absolute value, and then add 1 to the result. Thus, the value 4 is represented over 4 bits as 0100, and the value -4 is represented as 1100. This encoding guarantees the uniqueness of the value 0 and provides an addition instruction, which is not dependent on the signs of the operands.

6.1.2.1. *Size of integers*

Up to this point, we have only used one integer type, `int`, representing relative integers. Some languages, such as C, C++ and Java, include several integer types: `char`, `short`, `long`, `long long` and `int`. These differ in terms of the number of bits used to encode values. In Java, the number of bits for each integer type is clearly specified; in C, however, successive versions of the standard simply indicate a *minimum* size for each integer type. The values of `int` in C/C++ are generally coded over 32 bits and those of `long int` are coded over 64 bits, at least in current architectures.

Java	Bits	C/C++	Bits
`byte`	8	`char`	≥ 8
`short`	16	`short (int)`	≥ 16
`int`	32	`int`	≥ 16
`long`	64	`long (int)`	≥ 32
		`long long (int)`	≥ 64

Table 6.1. *Integer size by language*

In other languages, such as OCaml or Python 3 (Python 2 permitted the specification of `long`), integer size cannot be specified, and is set by the standard of the language. In any case, users should keep integer size in mind in order to avoid certain errors.

6.1.2.2. *Signed and unsigned integers*

Working in C and C++, programmers may choose to work with signed integers (used to represent relative integers) or unsigned integers (used to represent natural numbers). This choice defines the way the MSB is used. For signed integers, this bit is used for the sign, using two's complement encoding. For unsigned integers, this bit is used in the same way as the others to encode values. This choice is not available in OCaml, Java and Python, where integers are always signed.

The range of data that may be represented by an integer type depends on the presence of a sign and on the number of bits assigned to the type. This is presented in Table 6.2.

Bits	Sign	Min	Max
8	No	0	255 ($2^8 - 1$)
8	Yes	-128 (-2^7)	127 ($2^7 - 1$)
16	No	0	65535 ($2^{16} - 1$)
16	Yes	-32768 (-2^{15})	32767 ($2^{15} - 1$)
32	No	0	4294967294 ($2^{32} - 1$)
32	Yes	-2147483648 (-2^{31})	2147483647 ($2^{31} - 1$)
64	No	0	18446744073709551615 ($2^{64} - 1$)
64	Yes	-9223372036854775808 (-2^{63})	9223372036854775807 ($2^{63} - 1$)

Table 6.2. *Data range by type*

6.1.2.3. *Overflow*

As integer size is predetermined, *overflow* may occur in cases where an integer is "too big" or "too small" to be represented over the number n of bits assigned to its type. Integer operations and their properties do not, therefore, belong to $\mathbb{Z}$ but to arithmetic *modulo* $\mathbb{Z}/n\mathbb{Z}$.

Consider the following example. The decimal value 9 is represented in $\mathbb{Z}$ in binary by 1001; the value 7 is represented by 0111 and the value of their sum is represented by 10000. Take a representation using unsigned integers, coded over 4 bits. The representations of 9 and 7 are those found in $\mathbb{Z}$, as they only require 4 bits. The representation of 16, however, becomes 0000: the MSB has been lost. The actual computation is $(9 + 7)\, mod\, 2^4$ and not $9 + 7$. Symmetrically, consider $-8 - 1$: assuming that these are signed values, do we obtain -9? The decimal value -8 is represented in binary by 1000, and the value 1 is represented by 0001. Their difference is thus 11111, which now requires 5 bits. Again, the MSB has been lost, and we obtain a result of 1111, equal to -1 in decimal.

Similarly, if the location of a given integer variable is fixed to n bits and if this variable is assigned with a value requiring m bits with $m > n$, then overflow will occur: $(m - n)$ bits are lost.

6.1.2.4. *Mixing integers*

Assigning an unsigned variable with the signed result of a computation, and vice-versa, may also result in overflow. The value of the computation must therefore be converted to a value with the same type as the variable. This "signed $\leftrightarrow$ unsigned" conversion is somewhat dangerous.

Let us consider the conversion of the signed number -4 into an unsigned number using integers encoded over 4 bits. Using two's complement, the encoding for -4 is 1100. If an unsigned variable is assigned this bit sequence (without two's complement), then its value is decoded as an unsigned value and represents the (necessarily) positive value 12. Now, consider the conversion of the unsigned number 15 into a signed number. The encoding for 15 is the sequence 1111, which is interpreted as a signed number using two's complement as the value -1.

Similarly, mixing signed and unsigned integers in computations can lead to overflow. In such cases, the value of one of the operands must be converted in order to acquire the type of the other operand. Example 6.4 shows a case of mixing and its effects.

EXAMPLE 6.4.–

```
C
#include <stdio.h>

int main () {
  int i = -10 ;
  unsigned int u = 1 ;
  if (i + u < 0) printf ("%d + %u < 0\n", i, u) ;
  else printf ("%d + %u >= 0\n", i, u) ;
  return (0) ;
}
```

```
$ ./a.out
-10 + 1 >= 0
```

If `i + u` is negative, the computation will have been carried out using `int`; otherwise, `unsigned int` will have been used. The C standard stipulates that if both signed and unsigned operands are present, the signed operand must be converted into an unsigned operand. The value of `i` is thus interpreted as unsigned, leading to overflow, as the sign bit is considered as a "normal" value bit and `i` no longer represents -10, but 4294967286. Therefore, the result of the operation is positive. If the computation had been carried out using `int`, however, the value of `i` would have remained negative, and the result of the computation would have been negative.

The (needed) use of *modulo* n arithmetic, where n is the size of the integers, creates particularly tricky overflow problems, often occurring at the bounds of the representable integer domains. Thus, a program may operate correctly for a time, for many different inputs, before suddenly and silently producing an inconsistent result (without raising any error) for a case that falls near to the limits. In the absence of an error message, this inconsistent result may propagate throughout the execution, far from the point at which it appeared; this makes the source of the error particularly difficult to identify.

6.1.2.5. *Integer arithmetic*

We shall conclude our study of integers in computer science according to their characteristics in most programming languages by considering some properties of integer arithmetic in a mathematical context. Addition is a total operation over $\mathbb{N}$, extended to a total operation over $\mathbb{Z}$. Subtraction is not total over $\mathbb{N}$ ($5 - 25$ cannot be computed in $\mathbb{N}$) but becomes total in $\mathbb{Z}$. The division encountered in $\mathbb{N}$ is the Euclidean division (with an integer quotient and remainder). The exact division operation is defined in $\mathbb{Q}$.

In practice, the passage from $\mathbb{N}$ to $\mathbb{Z}$ is generally implicit: "the temperature is $10°$, it falls by $25°$, it is now $-15°$". This implicit passage from $\mathbb{N}$ to $\mathbb{Z}$ or from $\mathbb{Z}$ to $\mathbb{R}$ corresponds to a type change in computing. In $\mathbb{N}$, the expression $1\,/\,4$ denotes a Euclidean division with a result of 0, while in $\mathbb{R}$, $1\,/\,4$ is equal to 0.25. Ignoring these shifts between number sets can produce some surprising results: for example, $(1\,/\,4) * 4$ gives a result of 0, whereas $(1 * 4)\,/\,4$ gives a result of 1. These results are perfectly correct, since in integer arithmetic, / is used to denote Euclidean division. However, / is often interpreted implicitly as representing exact division, hence the surprise.

When writing arithmetic expressions using integers, the order of evaluation of certain operations (such as multiplications and divisions) is essential, and should be specified using parentheses. Otherwise, in the expression `1 / 4 * 4`, the compiler will choose to compute either the division or the multiplication first, depending on the implementation. Switching to another compiler may therefore lead to a change in the result and, possibly, an error. This error does not necessarily become apparent immediately: it may remain latent, giving the impression that the program is correct.

Note that Python 3, which is dynamically typed, may change the type of an arithmetic expression imperceptibly, moving from an integer expression to a floating expression, as we see in example 6.5. The division operator automatically returns a value of the floating (real) type.

EXAMPLE 6.5.–

```
Python
>>> x = 5
>>> type (x)
<class 'int'>
>>> type (x + x)
<class 'int'>
```

```
>>> type (x / x)
<class 'float'>
>>> x / x
1.0
```

The variable x, initialized with the value 5, is taken to be of type integer. The result of `x + x` is also of type integer. The result of `x / x`, on the other hand, is automatically converted to a float, as we see from the type of `x / x` and the displayed value, which includes a decimal element.

6.1.2.6. *Implicit conversion of integers in C*

The C standard stipulates that intermediate computations in an arithmetic expression are to be applied to `int` by default. This semantic feature of C explains the result obtained in example 6.6, which otherwise appears surprising.

EXAMPLE 6.6.–

```
C
#include <stdio.h>

int main (int argc, char *argv[]) {
  signed char i = -128 ;
  i = (i - 1) / 2 ;
  printf ("%d\n", i) ;
  return (0) ;
}
```

```
$ ./a.out
-64
```

The variable `i`, which is signed over 8 bits (`signed char`), is initialized with the smallest possible value. Therefore, the evaluation of `(i – 1)` using modulo arithmetic results in an overflow, generating a value of 0; dividing this result by 2, we obtain 0. However, as the arithmetic expression is implicitly executed using `int`, i.e. 32-bits integers, the computation of `i – 1` does not generate an overflow and gives a result of – 129. This result is then divided by 2 using integer division, producing a final result of – 64. This semantic feature prevented the expected overflow from occurring, but it can also be a cause of unexpected results, as we see from example 6.7.

EXAMPLE 6.7.–

```
C
#include <stdio.h>

int main (int argc, char *argv[]) {
  int a = 100000, b = 100000 ;
  long int c = a * b ;
  printf ("%ld\n", c) ;
  c = (long int) (a * b) ;
  printf ("%ld\n", c) ;
  c = (long int) a * b ;
```

```
  printf ("%ld\n", c) ;
  return (0) ;
}
```

```
$ ./a.out
1410065408
1410065408
10000000000
```

This program attempts to compute $10^5 \times 10^5$, storing the result in a `long int`, which is sufficiently large to hold the result of 10^{10}. The result of the first computation is not 10^{10}. The multiplication was carried out using two `int`s, resulting in overflow. The result was then stored in a `long int c`. Imposing a type for the result of the multiplication does not solve the problem, as we see from the second computation. The conversion occurs too late in the process: the overflow has already occurred. To obtain a correct result, the type of one of the operands, for example `a`, has to be converted to `long`, as shown in the third computation.

6.1.3. *Characters*

Prior to the introduction of the Unicode standard and UTF-8, UTF-16 and UTF-32 encodings, characters were represented by integers over 8 bits (one byte). Each of the first 128 values was associated with a "letter", following the ASCII standard. The integer value used to represent each character was referred to as its *ASCII code*. The ASCII standard was extended, with different variants in which the 128 remaining values were used to represent other “letters” (including accented characters).

Unicode extended character encoding by allowing “letters” to take up several bytes, vastly broadening the field of possibilities. Certain languages have dedicated types for handling Unicode encoding; for example, Java uses the `char` type (an integer over 16 bits) in addition to the `byte` type to handle ASCII encoding. Others, such as OCaml and C/C++ only include a `char` type, used for integers over 8 bits. In this case, the programmer is left to encode characters manually using a different type if extended ASCII is not sufficient.

In Python 3, no difference is made between character strings and individual characters. At syntax level, anything that looks like a character (shown between apostrophes in many languages) produces a value of the same type used for character strings (usually shown between speech marks). Example 6.8 provides an illustration of this point.

EXAMPLE 6.8.–

```
Python
>>> u = 'c'
>>> u
'c'
>>> v = "c"
>>> v
'c'
>>> type (u)
<class 'str'>
>>> type (v)
<class 'str'>
```

Similarly, the extraction of a character located at a given position in a string (e.g. the second character in example 6.9) returns a string made up of a single character.

EXAMPLE 6.9.–

```
Python
>>> str = "Aloha"
>>> str[1]
'l'
```

Python 3 includes a further type, `byte`, which is – as the name suggests – used to represent bytes. This type is mainly used to represent data byte by byte, instead of encoding single characters. Byte values less than 128 correspond to characters in ASCII standard. Any higher values are not interpreted as characters, as we see in example 6.10.

EXAMPLE 6.10.–

```
Python
>>> b'\x41'
b'A'
>>> b'\xC3'
b'\xc3'
```

6.1.4. *Floating point numbers*

Floating point numbers are encoded in an entirely different way to integers. The encoding of floating point numbers is governed by the IEEE 754 standard, which specifies two different lengths: `floats`, over 32 bits, and `doubles` over 64 bits. The standard IEC60559 (which is compatible with IEEE 754 for 32- and 64-bits floats) introduced an additional size, `long double`, which is not numerically fixed. These `long double` integers are intended to enable greater precision than `double`, with at least a range as large as that of `doubles`.

According to these standards, a floating point number is represented by three groups of bits with different meanings. The MSB is used to represents the sign. The next group of bits represents the exponent. Finally, the *significand* (also known as a *mantissa*) is coded over the remaining bits. As a first approximation – which is technically erroneous, but sufficient to establish an intuitive understanding – we may consider that the value of a floating point number is $sign \times mantissa \times 2^{exponent}$.

As in the case of integers, certain languages, such as C/C++, allow users to specify the size of floating point numbers; others, such as Java, OCaml and Python, have a predefined default size.

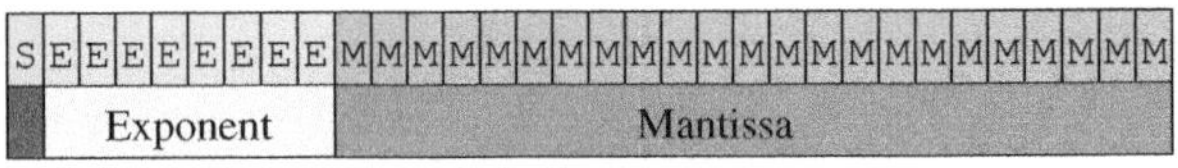

Figure 6.1. *Encoding a floating point number over 32 bits. For a color version of this figure, see www.iste.co.uk/hardin/programming1.zip*

Floating point numbers are encoded over a finite number of bits, meaning that the set of these numbers is finite. As floating point numbers are indeed rational, any rational (or real) number falling between two consecutive floating point numbers cannot be represented exactly. In other terms, there are "holes" between two consecutive rational numbers exactly represented using floating point numbers, as shown in example 6.11.

EXAMPLE 6.11.–

```
C
#include <stdio.h>

int main () {
  float fl1 = 48431.1231 ;
  float fl2 = 48431.1239 ;
  float fl3 = 48431.1250 ;
  printf ("fl1: %f  fl2: %f  fl3: %f\n", fl1, fl2, fl3) ;
  printf ("fl1 == fl2 ? %d\n", (fl1 == fl2)) ;
  printf ("fl2 == fl3 ? %d\n", (fl2 == fl3)) ;
  printf ("fl1 == fl3 ? %d\n", (fl1 == fl3)) ;
  fl1 = fl1 + 0.0001 ;
  printf ("fl1 + 0.0001: %f\n", fl1) ;
  return (0) ;
}
```

```
$ ./a.out
fl1: 48431.125000  fl2: 48431.125000  fl3: 48431.125000
fl1 == fl2 ? 1
fl2 == fl3 ? 1
fl1 == fl3 ? 1
fl1 + 0.0001: 48431.125000
```

Three variables of type `float` are declared and initialized with syntactically different constants. The display seems to indicate that they are equal, but the accuracy and precision of this display may be questionable. For this reason, we compare these values with each other and consider the results of the equality tests (booleans). This verification indicates that the three variables are, in fact, equal: two of the initialization values were lost in a "hole". The same holds true for the result obtained by adding 0.0001 to the first variable, which does not change: this, too, is lost in a "hole". Floating point numbers are not the ideal rational numbers of mathematics. As there is only a finite number of `float` values, approximation and rounding cannot be avoided in evaluating declarations and arithmetic operations. Example 6.12 shows the verification of an expected result for a simple addition of `doubles`.

EXAMPLE 6.12.–

```
C
#include <stdio.h>

int main () {
  double fl1 = 0.1 ;
  double fl2 = 0.2 ;
  double fl3 = 0.3 ;
  double fl4 = fl1 + fl2 ;
  printf ("fl1: %f  fl2: %f  fl3: %f  fl4: %f\n", fl1, fl2, fl3, fl4) ;
  printf ("fl3 = fl4 ? %d\n", (fl3 == fl4)) ;
  return (0) ;
}
```

```
$ ./a.out
fl1: 0.100000  fl2: 0.200000  fl3: 0.300000  fl4: 0.300000
fl3 = fl4 ? 0
```

Displaying the variables and the result of the computation seems to indicate that $0.1 + 0.2 = 0.3$. However, the equality test between the addition and the variable `fl3` assigned to 0.3 tells a different story: the computation has been affected by rounding. Finally, the displayed values have also been subject to rounding, and do not correspond to the actual values stored in the memory.

To avoid certain problems related to value rounding and computations using floating point numbers, the equality operators found in programming languages should never be used to test the equality of two floating point numbers v_1 and v_2. It is better to use a so-called ϵ *comparison* (also said a test $\pm\,\epsilon$), i.e. $-\,\epsilon \leq v_1 - v_2 \leq \epsilon$, or $\mid v_1 - v_2 \mid \leq \epsilon$, as we show in example 6.13.

EXAMPLE 6.13.–

```
C
#include <stdio.h>
#include <math.h>
#define EPSILON (1e-6)  /* Very small epsilon... */

int main () {
  double fl1 = 0.1 ;
  double fl2 = 0.2 ;
  double fl3 = 0.3 ;
  double fl4 = fl1 + fl2 ;
  printf ("fl1: %f  fl2: %f  fl3: %f  fl4: %f\n", fl1, fl2, fl3, fl4) ;
  printf ("fl3 ~ fl4 ? %d\n", (fabs (fl3 - fl4) < EPSILON)) ;
  return (0) ;
}
```

```
$ ./a.out
fl1: 0.100000  fl2: 0.200000  fl3: 0.300000  fl4: 0.300000
fl3 ~ fl4 ? 1
```

The `fabs` function found in the mathematical library of C is used to compute the absolute value of the difference between `fl3` and `fl4`. The chosen value of ϵ allows us to ignore rounding that took place during computation, meaning that we can consider the two values to be equal. Note that the value of ϵ depends on the order of magnitude of the values involved in the computation. In astronomy, for example, if the result of a computation states that two stars are within 100 m of one another, this very likely means that the stars have, in fact, collided. In the context of particle physics, on the other hand, if a computation states that two particles are 0.0001 m apart, they have clearly not collided.

The examples shown above all use relatively “simple” numbers, without long sequences of digits behind the decimal point. The problems that we have described may occur in any program using floating point numbers. Furthermore, as the encoding of floating point numbers is standardized, these problems apply equally to any programming language.

6.1.4.1. *Mixing integers and floating point numbers*

Certain languages, such as OCaml, prohibit arithmetic operations, which include both integers and floating point numbers. As we see from example 6.14, an addition operation using integers is denoted as + and is of type `int` $\rightarrow$ `int` $\rightarrow$ `int`; an addition operation using floating point numbers is denoted as `+.` and is of type `float` $\rightarrow$ `float` $\rightarrow$ `float`.

EXAMPLE 6.14.–

```
OCaml
# (+) ;;
- : int -> int -> int = <fun>
# (+.) ;;
- : float -> float -> float = <fun>
# 3 + 4.5 ;;
Error: This expression has type float but an expression was expected of type
        int
```

Any attempt to use a value of the wrong type will result in a typing error. Developers must explicitly specify the conversion of one of the operands using functions from the language's standard library, as shown in example 6.15.

EXAMPLE 6.15.–

```
OCaml
# (float_of_int 3) +. 4.5 ;;
- : float = 7.5
# 3 + (int_of_float 4.5) ;;
- : int = 7
```

This choice prevents any "silent" value conversions: the program does exactly what the developer specifies. The loss of flexibility resulting from this choice is counterbalanced by increased ease of understanding and more reliable maintainability. Other languages carry out implicit conversions between integers and floating point numbers according to their own specific rules. For example, the C standard (ISO/IEC 9899:201x) stipulates that if one of the operands of an arithmetic operator is a floating point number, then the other operand must be converted to a floating point number, and the result will also be of this type. In cases where the result is assigned to a variable, the type of the variable has no impact on the type of the arithmetic computation. Once obtained, the result will be converted to conform to the type of the variable.

6.2. Arrays

An *array* is a data structure that groups data elements of the same type into a contiguous sequence of fixed size. Any element in the array can be accessed in a constant time via an indexing mechanism. Given the location of the first element of an array in memory (the *base* address) and the size of a single element, an element with index i can be accessed at the address base $+ i \times$ size of an element. The indices of the elements in the array are therefore integer values (or, more generally in certain languages, values of a discrete and totally ordered type). As the size of an array is fixed, attempting to access an element outside the array is a programming error, and may not immediately, or indeed necessarily, trigger an execution error: the error is thus latent. Indeed, attempts to access a location outside of the bounds of an array may very well concern a memory location that belongs to the program being executed; in this case, there will be no violation of access rights, and the operating system will not react. The program will read – or, in cases of assignment, silently destroy – a zone corresponding to other variables, making the error very hard to detect. Once a change in a variable behavior has been detected – which may take some time – the user needs to work backwards to find the illegal access in the array which is responsible for corrupting the data. In the most favorable case, the zone which the program attempts to access does not belong to the program itself: this causes the program to be immediately aborted by the operating system.

Certain languages include a mechanism to detect attempts to access locations outside the bounds of an array. For example, during the execution, Python, Ada and Java check that indices remain within the bounds of the array. These verifications may be deactivated in OCaml, resulting in a mechanism similar to that found in C/C++, where no verification is carried out. However, the improvement in performance gained in this way (faster access due to the lack of verification) comes at the price of reduced program safety.

In some languages, such as OCaml, C/C++ and Java, indices run from 0 to the size of the array – 1. In other languages, such as Ada and Pascal, the index range can be defined using upper and lower bounds. For example, an array might be defined with indices varying from 3 to 15 inclusive. The compiler then translates these indices so that the minimum index corresponds to the base of the array, i.e. to a shift of 0.

Python takes a very different approach to arrays. This combines some features of arrays in other languages, as described above, with list mechanisms. In other terms, it is possible to access elements directly via indexing, but it is also possible to add new elements or suppress some of them at any "place" in the array, say at index i, resulting in a modification of indices of elements whose indices are greater than i. This flexibility means that the complexity of insertion operations varies, as one or more other elements may need to be moved. The internal structure of an array in Python cannot simply be described by the address of the base. Similarly, the size of arrays is not fixed, and it is adjusted at runtime as elements are added and deleted. Each time an array is full, Python reallocates a greater array by adding several cells at once to avoid too frequent extensions. When an array includes a certain proportion of unused cells, a smaller array is allocated. The precise enlargement/reduction policy used in Python lies outside of the scope of this book. Note that arrays behave in a more complex manner in Python than in other programming languages, as their internal structure is not so easy to grasp.

Python also permits negative indices, which correspond to the nth element counted back from the end of the array, thus traversing it from "right" to "left". It also allows users to manipulate *slices* of an array, taken as subarrays containing the elements of the array present between the two bounds defining the slice, with possible application of an incrementation step, as shown in example 6.16.

EXAMPLE 6.16.–

```
Python
>>> t = [0, 2, 4, 6, 8, 10, 12, 14, 16, 18, 20, 22, 24]
>>> t[2:10]
[4, 6, 8, 10, 12, 14, 16, 18]
```

The constructed slice begins with the element of index 2 and ends at index 10 excluded. Let us continue this example.

EXAMPLE 6.17.–

```
Python
>>> t[2:10:3]
[4, 10, 16]
```

With an incrementation step of 3, the slice still begins with the element of index 2 and runs to index 10 excluded, but it jumps two elements from the initial array each time (i.e. the index is incremented by 3 each time).

The general syntax of a slice is `array[start:end:increment]`: any of the three components may be absent, positive, or negative, and in the latter case, the array is processed in reverse order. The first slice from the array `t` in example 6.18 fetches all elements of the array in reverse order, starting with the fifth from the end and working back to the first element of the array. In other words, it fetches all elements in reverse order, with the exception of the final four values in the initial array. The second slice only takes one element out of two.

EXAMPLE 6.18.–

```
Python
>>> t[-5::-1]
[16, 14, 12, 10, 8, 6, 4, 2, 0]
>>> t[-5::-2]
[16, 12, 8, 4, 0]
```

This flexibility in expressing arrays comes at a cost, as the meaning of these syntactic variants of array operations are more involved. Purely for entertainment purposes, readers may wish to determine the result of `t[-6:2:-3]`, `t[6:-2:-1]` or `t[6:-2:1]`.

6.3. Strings

A character string (or *string*) is a contiguous sequence of characters. A string is, intrinsically, an array of characters, independent of the way an individual character is represented (see section 6.1.3). Many languages have a predefined type for strings, along with a range of associated operations (concatenation, access to characters, length computation, etc.). Java and C++ offer a dedicated class for strings; OCaml and Python use a predefined type and a module providing operations. In C, strings are encoded by the developer as a character array. A special character (known as the *sentinel*), with value 0 and noted `'\0'`, is used to mark the end of a string. This representation is used internally by C++, ensuring compatibility with C-style strings.

Depending on the chosen language, it may or may not be possible to modify some characters in a string. Python, Java and OCaml do not permit modifications, strings in these languages are not *mutable*. These languages provide alternative types, which implement sequences of characters using mutable bytes. However, a suitable function from the language's standard library may be used to convert these sequences into strings. In C (and C++ if the dedicated class is not used), although strings take the form of arrays, mutability is not entirely straightforward. A string may be defined in two ways, as shown in example 6.19.

EXAMPLE 6.19.–

```
C
char arr[] = "A string" ;
char *str = "A string" ;
```

In the first case, the compiler creates an array of characters that is stored in the memory zone of the program variables; characters in the string can be modified. In the second case, the compiler creates a non-mutable string of characters, which is stored in a read-only area of the memory. Any attempt to modify characters in the string will result in an execution error, as the operating system will detect an illegal attempt to write to a read-only zone.

All of the languages, which include a specific type for strings, also supply comparison primitives, which are used to test equality or order relations between strings. And C, which does not include a dedicated type, does not provide comparison operators for strings. In this case, a comparison operator applied to two strings will simply compare the addresses of the respective arrays, rather than inspecting the contents of the strings (see section 6.6). The language's standard library provides a string comparison function (`strcmp`).

6.4. Type definitions

We presented typechecking and inference mechanisms in section 5.4. In this case, the type algebra was fixed, and comprised basic types and the "function" type constructor. In section 6.1, we examined a number of basic types used in modern programming languages. In addition to these basic types added to the type algebra, all languages include constructs for defining new types. Type definition provides a response to three orthogonal needs. First, users may wish to create new values different to those of the basic types (such as `int` or `bool`). Second, it is sometimes necessary to combine several types in order to create a more complex data structure. Finally, it may be helpful to rename an existing type to make code easier to read and maintain. In all cases, the definition of a type involves adding a new name called *type constructor* to the type algebra.

The introduction of type definitions modifies the existing equality relation between types. The equality of terms of the type algebra, i.e. structural equality, was used implicitly in Chapter 5; two types were considered to be equal if their terms were identical. When we introduce new type names – either to rename existing types or to create structured types – there are two possible meanings of equality of types. In the case of *nominal* typing, we consider that two types are equal if and only if they have the same name. Otherwise, the equality is defined by verifying that the structures of the types bound to the names are equal: this is known as *structural typing*. The choice of one of these options is a semantical decision, and it does not depend on the forms of definition available in a language.

6.4.1. *Type abbreviations*

Type abbreviation consists of naming a type expression. In C and C++, this naming operation is carried out using the `typedef` construct. In OCaml, a new name is introduced by means of a type definition, which must relate to a simple type expression (i.e. one comprising only type constructors).

Leaving aside the confinement mechanisms presented in Volume 2, Chapter 1. the creation of abbreviations is a means of shortening type expressions. Consider example 6.20, written in C. Instead of `const unsigned long long int`, we may wish to use a more concise alias, introducing the constructor `cullint_t` and using it in declarations.

EXAMPLE 6.20.–

```
C
typedef const unsigned long long int cullint_t ;
```

```
cullint_t t[10] ;

cullint_t add (cullint_t x, cullint_t x) {
 return (x + y) ;
}
```

Using type abbreviations also makes code easier to read and maintain. Consider a function for which the inputs are a pressure and a speed, and which returns a power order for a machine. The type of this function is `float` $\rightarrow$ `float` $\rightarrow$ `float`. With this type, it is not possible to differentiate between the `float` numbers representing these different kinds of values, and the type of the function provides no semantical information on its arguments. The introduction of three type abbreviations, as shown in example 6.21, allows a more informative function type.

EXAMPLE 6.21.–

```
OCaml
# type pressure_t = float ;;
type pressure_t = float
# type speed_t = float ;;
type speed_t = float
# type command_t = float ;;
type command_t = float
# let regulate (p : pressure_t) (s : speed_t) : command_t = ... ;;
val regulate : pressure_t -> speed_t -> command_t = <fun>
```

In many languages, such as C, C++ and OCaml, type abbreviations do not imply nominal typing. An alias remains equal to the body of its definition. This makes it possible to define a variable p of type `pressure_t` and a variable s of type `speed_t`, and to compare them or use the operator `+.` of type `float` $\rightarrow$ `float` $\rightarrow$ `float`, as shown in example 6.22.

EXAMPLE 6.22.–

```
OCaml
# let p : pressure_t = 2.0 ;;
val p : pressure_t = 2.
# let s : speed_t = 25.0 ;;
val s : speed_t = 25.
# p +. 2.5 = s ;;
- : bool = false
```

6.4.2. *Records*

An individual person may be characterized by data including their surname, first name and date of birth. Surnames and first names are strings, whereas a date of birth is composed of three integers. As these data elements are of different types, they cannot be stored in an array. The notion of a *record* (or *structure*) is used to aggregate data of different types within a single type.

Each data element is named and represents a *field* of the record. Each field is thus assigned the type corresponding to the data element, which it is intended to contain. The definition of a record type is given by the name of the type and the enumeration of the field names, along with their respective types.

EXAMPLE 6.23.–

```
OCaml
type person = {
  surname : string ; firstname : string ; month : int ; day : int ; year : int ;
}
```

EXAMPLE 6.24.–

```
C
struct person {
  char *surname ; char *firstname ; int month ; int day ; int year ;
};
```

As we saw in section 6.3, strings are not a base type in C, which explains the specific way in which they are declared.

If e is an expression of a record type that contains a field fld, the "dot notation" $e.fld$ denotes the field fld of the value of e.

Record type values are constructed by specifying the value associated with each field. In OCaml, as we see from example 6.25, a record value is defined by specifying the names of fields in order to enable type synthesis. One advantage of this seemingly verbose approach is that field values can be defined in any order. This also holds true for Ada despite the absence of a type inference mechanism.

EXAMPLE 6.25.–

```
OCaml
let p = { firstname = "John" ; month = 7 ; day = 5 ;
  year = 1900 ;  surname = "Doe" }
```

In C, field names are not shown, and so field values must be specified in the same order used when declaring the record type.

EXAMPLE 6.26.–

```
C
struct person p = { "Doe", "John", 7, 5, 1900 } ;
```

In many programming languages, the fields of a record cannot be defined in relation to one another, or in relation to the fields which precede them. Examples 6.27 and 6.28 illustrate this point.

EXAMPLE 6.27.–

```
OCaml
# type t = { a : int ; b : int } ;;
type t = { a : int; b : int; }
# let v = { a = 5 ; b = v.a } ;;
Error: Unbound value v
# let rec v = { a = 5 ; b = v.a } ;;
Error: This kind of expression is not allowed as right-hand side of 'let rec'
```

EXAMPLE 6.28.–

```
C
#include <stdio.h>

struct t { int a ; int b ; } ;

void f () {
  struct t v = { 15, v.a } ;
  printf ("%d\n", v.b) ;
}

int main () {
  f () ;
  return (0) ;
}
```

```
$ gcc foo.c
foo.c:7:22: warning: variable 'v' is uninitialized when used within its own
      initialization [-Wuninitialized]
  struct t v = { 15, v.a } ;
           ~         ^
1 warning generated.
$ ./a.out
0
```

In C, the compiler does not reject definitions of this type, but warns the user that the value of the record field used in defining the record itself is not initialized. Executing the program obtained in this way clearly shows that field `b` was not initialized properly.

6.4.2.1. *Record typing*

The definition of a record type introduces a new name (type constructor) into the typing environment. It is therefore important to consider how the type of a record value should be determined, and, by extension, how record types with different names but the same structure interact with one another. There are two main approaches to record typing.

The first method, which permits type inference, consists of guiding typing on the base of field names found in the expression. The value of a record expression must therefore include the names of fields in addition to their values. This is the case in OCaml. Thus, a type definition `type t = { a : int ; b : float }` introduces two new names, a and b, into the namespace for record field names in the environment, while recording the fact that these belong to the type t. When a record expression is analyzed, its fields must all belong to the same type, guiding its verification. Hence, two different record types cannot share the same field names (without significant effort on the part of the programmer). However, if type definitions are carried out in different compilation units or modules, ambiguity can be avoided by specifying the compilation unit containing these definitions. This point is illustrated in example 6.29.

EXAMPLE 6.29.–

```
OCaml
# type foo = { a : int ; b : float } ;;
type foo = { a : int; b : float; }
# type bar = { a : int ; b : float } ;;
type bar = { a : int; b : float; }
# let v = { a = 42 ; b = 3.14 } ;;
val v : bar = {a = 42; b = 3.14}
```

The expression bound to the value v is inferred as being of type bar, as it was responsible for the most recent binding of the field name a to a record type. Values of the foo type can no longer be created as the corresponding field names are masked. Note, however, that it is still possible to manipulate values of type foo by acting on their fields. Using a form of type annotation with the capacity to override inference, a function get_foo_a can be written to fetch the value of field a from a value of type foo.

EXAMPLE 6.30.–

```
OCaml
# let get_foo_a (x : foo) = x.a ;;
val get_foo_a : foo -> int = <fun>
```

The second method, used in C, among others, makes use of the type assigned to variables which take record type values. In this case, there is no need to specify field names in record values. Thus, the value { 42, 3.14 } does not have a type *per se*, except that of a two-field structure in which the first field is an integer and the second is a floating point number. However, at the point of assignment or initialization, the type given to the receiving variable at the time of its definition will be checked for compatibility with the fields of the expression. This means that two different record types may share the same field names, as shown in example 6.31.

EXAMPLE 6.31.–

```
C
struct foo { int a ; float b ; } ;
struct bar { int a ; float b ; } ;

void f () {
  struct foo v1 = { 42, 3.14 } ;
  struct bar v2 = { 43, 3.15 } ;
}
```

Note that, in C, record expressions can only be used to initialize variables. Thus, an expression of the form `f ({ 42, 3.14 })` is illegal.

In C, Java or OCaml, if two record types have the same structure but different names, they will not be compatible: the typing of records is nominal and not structural. Let us add an assignment of `v2` to `v1` in our previous program, written in C. In the example 6.32, the assignment of `v2` to `v1` is rejected, despite the fact that both of these variables are initialized by an expression of the same form.

EXAMPLE 6.32.–

```
C
...

void f () {
  struct foo v1 = { 42, 3.14 } ;
  struct bar v2 = { 43, 3.15 }  ;
  v1 = v2 ;
}
```

```
$ gcc -Wall foo.c
foo.c:14:6: error: assigning to 'struct foo' from incompatible type 'struct bar'
  v1 = v2 ;
     ^ ~~
1 error generated.
```

6.4.3. *Enumerated types*

The "values" of a traffic light are `red`, `yellow` and `green`. We wish to write a function `next`, which takes the current state of a traffic light and returns the following state.

The first solution is to choose a value from a basic type, such as `int`, for each value of the traffic light: for example, take `red` = 54, `yellow` = 12 and `green` = 36. However, there is nothing to prevent combining the "traffic light values" with any other integer value. Expressions such as `red`/2 or (`next` 23) are perfectly well-typed, although they are semantically incorrect. The function `next` is intended to operate on only three specific values – so what should it do if it receives an unexpected input value?

Imagine that we inadvertently define both `red` = 36 and `green` = 36. No error will be raised, despite the fact that these two traffic light values, which should be disjunct, are now equal. An expression used to test a traffic light value will no longer differentiate between a red light and a green light.

We explicitly chose the integer values to assign to the light values. There is no real reason for this choice, given that the values used to encode the colors have no influence on the algorithms acting upon them. Certain languages include a construct that allow users to perform tests enumerating all possible cases of a type's values and verifying exhaustiveness (no values are left out). In a case where light values are encoded by integers, we would need to write, at best, a general case that should never occur, or, at worst, all other possible cases that should never occur. The capacity to define a new type with two-by-two disjunct values, only compatible with other values of this type, eliminates these drawbacks. An *enumerated* type definition of this kind enumerates the names given to values, known as *value constructors*.

Languages such as Java, C and C++ allow these definitions to be created by means of the `enum` type construct. Once value constructors have been introduced, they can be used in the same ways as any other value. This is illustrated in example 6.33.

EXAMPLE 6.33.–

```
C
enum signal_t { Green, Yellow, Red } ;

enum signal_t next (enum signal_t state) {
  if (state == Green) return Yellow ;
  if (state == Orange) return Red ;
  return Green ;
}
```

The language compiler is in charge of encoding these values by integers internally. Value constructors may be more or less strongly typed depending on the language. In C, they remain integers in their own right: it is always possible to call the `next` function with any given integer. In C++ and Java, constructors are only values of the type which declaration introduced them and are not compatible with integers. For example, calling `next` using `0` as an argument in C++ results in a type error:

```
foo.cpp:14:3: error: no matching function for call to 'next'
  next (0) ;
  ^~~~
bar.cpp:4:15: note: candidate function not viable: no known conversion from
      'int' to 'enum signal_t' for 1st argument
enum signal_t next (enum signal_t state)
              ^
1 error generated.
```

6.4.4. *Sum types*

OCaml uses sum types that are more elaborate than simple enumerations of constructors. A sum type is used to group the values of several types into a single type. Take two types t_1 and t_2. Let us define a type t, in which all values are either the values of type t_1, marked by C_1, or the values of type t_2, marked by C_2. The construction of a sum type is similar to the disjoint sum operation on sets (hence the name *sum type*). C_1 defines the injection of v_1 of type t_1 that results in the value $C_1(v_1)$ of type t; similarly, v_2 of type t_2 is injected as $C_2(v_2)$ in t. C_1 and C_2 are called the *value constructors* of type t.

The OCaml code presented in this book has already used a lot of sum types to represent the abstract syntaxes of languages and the different algebras encountered. In example 6.34, playing cards are encoded by a sum type.

EXAMPLE 6.34.–

```
OCaml
type card_t = Jack | Queen | King | Number of int
```

A card may be a picture card, identified by its name alone. Alternatively, it may be a number card, and the number must be known in order to differentiate it from the other cards in the suit. The value constructor `Number` is applied to values of the `int` type. The value constructor `Jack` defines a unique value of the type `card_t`.

A sum type may be polymorphic. Some of its constructors may be applied to types written using type variables, introduced by the definition of the sum type. The predefined lists in the standard library of OCaml are given by a recursive and polymorphic type definition, similar to that shown in example 6.35.

EXAMPLE 6.35.–

```
OCaml
type 'a list = Nil | Cons of ('a * 'a list)
```

In C/C++, there is no syntactic construct for value constructors. Programmers must encode constructors themselves, using `union` and `struct` types. A union type is literally a disjointed sum of the different types mentioned in its definition. A value of the union is represented on a memory range whose size is the maximum of the sizes of the values of these different types. Thus, it differs from a record, where data is aggregated. Each field of the union allows access to the data of the union, which will be interpreted as having the type associated with the field. This point is illustrated in example 6.36.

EXAMPLE 6.36.–

```
C
#include <stdio.h>

union int_float_t { int as_int ; float as_float ; } ;

int main () {
  union int_float_t v ;
  v.as_float = 3.14 ;
  printf ("As the float %f as the integer %d\n", v.as_float, v.as_int) ;
  return (0) ;
}
```

```
$ ./a.out
As the float 3.140000 as the integer 1078523331
```

Any value of the union type `int_float_t` can receive two interpretations. The variable v is initialized by considering the union as a float. Two views of this value are displayed. The first one, as a float, allows the interpretation of the data contained in memory in accordance with the interpretation made during writing. As such, we find the value 3.14. The second display is performed by interpreting the value as an integer. Since the encoding of integers and floats are fundamentally different, we see a value that seems irrelevant.

Union types in C do not have value constructors like sum types in OCaml. Thus, it is necessary to add a marker to know, at runtime, which field of the union to use and how to interpret the value of this union. Example 6.36 shows that an incorrect interpretation gives semantically incorrect results. A suitable marker may be an enumerated type with at least as many values as the union has fields. This marker and the union should then be grouped together by means of a record. Let us see this with example 6.37.

EXAMPLE 6.37.–

```
C
#include <stdio.h>

enum tag_t { Val_int, Val_float } ;
union int_float_t { int as_int ; float as_float ; } ;
struct sum_t { enum tag_t tag ; union int_float_t val ; } ;

struct sum_t read_int () {
  struct sum_t v ;
  scanf ("%d", &v.val.as_int) ;
  v.tag = Val_int ;
  return (v) ;
}

struct sum_t read_float () {
  struct sum_t v ;
```

```
  scanf ("%f", &v.val.as_float) ;
  v.tag = Val_float ;
  return (v) ;
}

void print (struct sum_t v) {
  if (v.tag == Val_int) printf ("%d\n", v.val.as_int) ;
  else if (v.tag == Val_float) printf ("%f\n", v.val.as_float) ;
}

int main () {
  print (read_int ()) ;
  print (read_float ()) ;
  return (0) ;
}
```

```
$ ./a.out
42
42
45.7
45.700001
```

This program defines a sum type, which is equivalent to the OCaml definition `type sum_t = Val_int of int | Val_float of float`. Two functions, `read_int` and `read_float`, allow users to input an integer and a floating point number, and to generate a sum type value by initializing the union value with the appropriate marker and field. The `print` function selects an option according to the marker, and calls `printf` with the print format (`%f` or `%d`) and the field of the union, which corresponds to this marker. Note the presence of a rounding error: 45.7 is displayed as 45.700001.

Java does not feature `union` or sum type constructs. The union notion may be replaced by a record, a field being assigned to each type of the union and keeping the marker field in order to discriminate between the other fields. However, in this case, the size of the data element will be the sum of the sizes of all fields, rather than the size of the largest field, resulting in a memory waste. The usual solution to this problem is to create a class hierarchy, extending a base class, representing the sum type without its constructors (readers who are not familiar with the notions of object-oriented programming may wish to consult the Volume 2, Chapter 4, before returning to this explanation). Each value constructor is encoded by an inheriting class of the base class, its arguments being defined by the attributes of this class. Moreover, the class must possess methods returning the values of these attributes. To discriminate between the different kinds of values in the union, there are some solutions: either the base class provides a marker, or a dynamic test on the types of the union is done, or the algorithms must be adapted to the object-oriented style.

Python, too, does not propose union or sum type mechanisms. This is the reason why, in this book, the implementation of semantic notions with Python uses classes and the `isinstance` construct. This allows the mimicking of the OCaml pattern matching, in order to obtain rather similar codes. Note that this programming style in Python is rather "wordy" and does not always conform to best-practice rules for object-oriented programming.

An additional construct, known as *pattern matching*, used to discriminate between the values of sum or enumerated types on a case-by-case basis will be presented in section 6.5.

6.4.4.1. *Typing sum types*

In languages where enumerated types are reduced to integers (via the `enum` construct), such as C or Java, sum types expressions are typed in exactly the same way as integer type expressions.

In OCaml, the typing of sum types is directed by the names of the value constructors present in the expression, just as the typing of records is directed by field names. A definition `type t = A | B of int` introduces the names `t`, `A` and `B` into the typing environment; the fact that `A` and `B` are the constructors of `t`, the fact that `A` has no argument and the fact that `B` has an argument of type `int` are also recorded. Two different sum types cannot share the same constructor names; if this happens, the most recent value constructor will mask the same named constructor of the previous sum type. If definitions are done in different compilation units or modules, it is possible to distinguish between constructors by precising their compilation unit.

6.5. Generalized conditional

Most programming languages include a construct for discriminating between the different values of a type in order to avoid cascading "if... then... if" situations. This is a form of generalized conditional that only tests equalities. Such a control structure allows to describe a treatment according to the different possible values of a given expression.

6.5.1. *C style* `switch/case`

In its simplest form, the generalized conditional is an enumeration of some or all values of the type of the evaluated expression, where each case is bound to the code to execute. The `switch` construct in C allows the discrimination between values of scalar types (more precisely, types represented by integers: this includes `enum` but excludes floats). In Java, since version 7, strings can also be processed in this way. A construct of this kind in C is shown in example 6.38.

EXAMPLE 6.38.–

```
C
void f (int x) {
  switch (x) {
  case 0: printf ("x is 0\n") ; break ;
  case 1: printf ("x is 1\n") ; break ;
  default: printf ("x is something else\n") ; break ;
  }
}
```

The function `f` evaluates `x` and executes the code of the `case` corresponding to the value of `x`. The execution of `f (1)` will therefore display "x is 1". Each `case` is ended, hence separated from the next one, by a `break` instruction.

Since the integer type has a very large (while not infinite) number of values, a special case labeled `default` is used to consider all the values that do not appear in the explicit cases. It is important to note that the `default` case is not mandatory. If it is not supplied and if the value does not correspond to any of the listed cases, then no action will be applied (silently). For instance, if we were to remove the `default` case from our example before computing `f (2)`, the function would not display anything.

Using a `switch` without including a `default` case is not recommended. It is impossible to know, when reading code, if the case was forgotten or if it was simply not written, because in "all other cases" the "empty" behavior is suitable. In order to produce maintainable code, it is good practice to explicitly include an empty `default` case, immediately followed by a `break`. This inclusion costs nothing – the compiler will not generate any code for this case – and it will be evident, from the source code, that the author intended to do nothing in such cases.

Since `enum` in C are simply integers, it is possible to discriminate over an `enum` type by enumerating its values (which constitute a finite set), as shown in example 6.39.

EXAMPLE 6.39.–

```
C
enum dir { North, East, South, West } ;

enum dir invert (enum dir d) {
  enum dir res ;
  switch (d) {
  case North: res = South ; break ;
  case East: res = West ; break ;
  case South: res = North ; break ;
  case West: res = East ; break ;
  }
  return (res) ;
}
```

In this example, there is no `default` case. Indeed, its utility is not evident since all possible values of the type have been considered. In practice, since the values of the `enum` are encoded by integers, the program does not cover all possible values in the computing domain, only those declared in the type. In C, the `enum` type is fully compatible with integer values: a function with a parameter of the `enum` type may be called using any integer. Hence, if no `default` case is present, erroneous calls to the `invert` function will go undetected. The addition of a `default` case is highly recommended as a means to increase robustness, and is essential when developing critical software. This seemingly regular case structure hides a more complex reality. The `break` instruction is used to cut off the instruction sequence for a case, just as it can be used to interrupt a loop. In example 6.40, if the `break` between two cases is omitted, the program will go on to execute the code for the second case straight after the first.

EXAMPLE 6.40.–

```
C
void f (int x) {
  switch (x) {
  case 0: printf ("x is 0\n") ;
  case 1: printf ("x is 1\n") ; break ;
  default: printf ("x is something else\n") ; break ;
  }
}
```

In example 6.40, the `break` at the end of `case 0` was removed. Hence, a call `f(0)` will display "x is 0", but also "x is 1".

Furthermore, cases may be seen as simple jump labels, as their order does not follow any particular structure. Cases can be arranged in any order (although `default` is, by pure convention, the last considered case, but this is not mandatory), and can be put anywhere within the scope of the curly brackets delimiting the `switch`. These points are illustrated in example 6.41.

EXAMPLE 6.41.–

```
C
void f (void) {
  int j = 24 ;
  switch (j) {
  case 23 : {
    printf ("case 23\n") ;
    while (j) {
    default :
      printf ("case default\n") ;
    }
    } break ;
  }
}
```

In this example, the `default` case appears in the code between the label for case 23 and the `break` terminating this case. Furthermore, it is located within the body of a loop. The program above is entirely legal in C, and compilation will not trigger any warning. On execution, as `j` is different from 23, the program will jump to the `default` case, i.e. to the middle of the `while` loop, and will continue to loop infinitely as the condition is always true.

6.5.2. *Pattern matching*

Certain languages, such as OCaml, Haskell, Scala, F# and C#, include a construct known as *pattern matching*. This is similar to the `switch/case` construct seen in section 6.5.1, but it is a lot more expressive as it permits the deconstruction of structured values (sum type values, records or objects). The `switch/case` construct tests the equality of the examined value with atomic constants. Pattern matching is a way of testing using the structure of a value. Consider the sum type `type t = I of int | B of bool` and v a value of `t`. We can state cases such as "v is of the form `I` of something or `B` of something", and use the value "something" in the treatment devoted to the case.

A *pattern* is a kind of skeleton built from value constructors, labels of records and variables, defining a shape. Such a pattern matches all values having the same shape and this matching binds the variables of the pattern to the corresponding subterms of the value. The names of these variables may be used by the instructions associated with the pattern and their bindings will be used when the instructions are executed. For the type `t` above, the pattern `I x` matches the value `I 46`: they are both built with the same value constructor `I`. This matching binds `x` to 46 when the code associated with the pattern `I x` is executed.

A more formal definition of pattern/value compatibility is given below. First, however, let us consider some examples, shown in OCaml and using the `match/ with` construct.

EXAMPLE 6.42.–

```
OCaml
type dir = North | East | South | West

let invert d =
  match d with North -> South | East -> West | South -> North | West -> East
```

This function, defined by cases, is identical to that of the example 6.39 written to illustrate the `switch/case` construct in C. As the type `dir` is made up of exclusively argument-free constructors, pattern matching operates in the same way as a simple `switch/case`. If the argument of `invert` is `West`, then the fourth branch of the matching will be selected during the execution and the value `East` will be returned.

EXAMPLE 6.43.–

```
OCaml
type intlist = End | Cell of (int * intlist)

let rec sum l = match l with Cell (i, remainder) -> i + sum remainder | End -> 0
```

The `intlist` type represents lists of integers, and the function `sum` returns the sum of the integers in a list. If `l = Cell (a, r)`, then the variable `i` is bound to `a` and the variable `remainder` is bound to `r`. The bindings of these two pattern variables are used in the body of the case, enabling the addition of the current integer value and the sum of the remainder list, computed recursively. If `l = End`, the recursion stops and the value 0 is returned.

Note that a "catch-all" pattern, denoted as `_`, may replace a pattern variable if the value matched by this variable is unused in the body of the case. This pattern is equivalent to a variable, but avoids introducing a new name. It provides a clear indication to readers that this part of the matched value is not used.

EXAMPLE 6.44.–

```
OCaml
let is_empty l = match l with Cell (_, _) -> false | _ -> true
```

A list is empty if the `Cell` constructor has no occurrence in its value, which is therefore `End`. We first consider the case of a value built with the constructor `Cell`, which will return `false` whatever is the integer and the remainder list arguments of this constructor. Thus, the corresponding pattern is `Cell(_,_)` with two occurrences of `_`. For all other cases (in our example, only `End`), the list is empty. Note that the final `_` "catches" the only remaining case `End`: as such, we could write `| End -> true` in a completely equivalent manner.

6.5.2.1. *ML style pattern matching: formalization*

Before establishing a formal semantics of pattern matching, we need to provide a more detailed explanation of the notion of *term*, which we have discussed briefly in the context of type inference in section 5.4 where the word *term* referred to type expressions. More generally, terms are constructed from a set Σ of symbols, which each have an *arity* (the number of "arguments" of the symbol) and a set of variables V, according to the following inductive definition.

DEFINITION 6.1.– *Term: the set* $\mathcal{T}(\Sigma, V)$ *of terms over* Σ *and* V *is defined by:*

– *every symbol of arity 0 is a term in* $\mathcal{T}(\Sigma, V)$*;*

– *every variable in* V *is a term in* $\mathcal{T}(\Sigma, V)$*;*

– *if* $t_1, \ldots, t_n$ *are terms in* $\mathcal{T}(\Sigma, V)$*, and if* C *is a symbol of arity* n*, then* $C(t_1, \ldots, t_n)$ *is a term in* $\mathcal{T}(\Sigma, V)$*.*

As we defined a substitution on types, replacing type variables by types, we define a substitution on terms as a partial finite domain function of the set of variables into the set of terms and extend it naturally into a function of terms into terms.

Substitutions can be used to establish a preorder of generality on terms. A term t_1 is *less general* than a term t_2 if there exists a substitution ϕ such that $t_1 = \phi(t_2)$ and it is noted $t_1 \leq t_2$.

Formally, matching a value v against a pattern m consists of considering the pattern m as a term and determining if $v \leq m$. So we have to compute the substitution ϕ associating the variables of the pattern with subterms (subvalues) of v. If such a substitution ϕ exists, then m is said to *match* v and ϕ is used to extend the execution environment of the pattern matching case.

6.5.2.1.1. Order of patterns

Pattern matching is done according to the patterns involved in the `match/with` construct and, before giving a formal semantics to this mechanism, we precise how the choice of the right pattern is done. Patterns are examined in the order in which they appear. If the first pattern does not match the examined value, the second case is tested, and so on for as long as necessary.

If no pattern is found that matches the value in question, an execution error is raised. Note that languages in the ML family analyze pattern exhaustiveness at compile-time and a warning is emitted by their compiler in case of non-exhaustive pattern matching.

If a value matches more than one pattern, the order in which patterns are listed is decisive: the first matching pattern is used, and the other matching patterns are ignored.

If several patterns always match the same values, then the principle of choosing the first one still holds; however, for example in OCaml, the compiler raises a warning to indicate the existence of this redundancy in the `match/with` construct.

These different situations – partially overlapping cases, totally overlapping cases and missing cases – are illustrated in example 6.45.

EXAMPLE 6.45.–

OCaml
```
type logic_op = Or of (bool * bool) | And of (bool * bool)

let eval e =
  match e with
  | Or (true, _) | Or (_, true) | Or (false, true) | And (true, true) -> true
```

```
  | And (_, _) -> false
Warning 8: this pattern-matching is not exhaustive.
Here is an example of a case that is not matched:
Or (false, false)
Warning 11: this match case is unused.
val eval : logic_op -> bool = <fun>
```

The type `logic_op` represents boolean conjunction and disjunction expressions. Note that `true` and `false` are exceptions to the rule in OCaml syntax stating that sum type constructors must always start with an uppercase letter. The function `eval` returns the boolean value of an expression. If we compute `eval (Or (false, true))`, the first pattern does not match, as the first argument of `Or` in the pattern is `true`. We then examine the second case: this pattern matches, as any value is accepted as the first argument, and `true` is the second argument in both the pattern and the matched value. The third pattern is redundant as all of the values that it captures have already been matched by the previous case. The compiler flags this issue in the second warning. The case where both arguments in the pair are `false`, However, the case of a value of the form `And (false, false)` is not considered: the compiler issues a warning, see the first message shown in the example.

6.5.2.1.2. Operational semantics

We shall extend the abstract syntax of $\mathbf{Exp_2}$, presented in section 3.1.1, as shown in Table 6.3.

$p ::=$	v	Pattern variable
$\mid$	$C(p_1, \ldots, p_n)$	Value constructor
$e ::=$	$\ldots$	Previous expressions
$\mid$	match e with $p \rightarrow e \mid \_ \rightarrow e$	Pattern matching

Table 6.3. *Extension of the syntax of* $\mathbf{Exp_2}$ *for pattern matching*

First, we need to define the syntax of patterns, which are essentially composed of constructors and variables (the catch-all pattern can be assimilated to a variable of which the name is irrelevant). We then add a construct for pattern matching with only two cases. The first case is a given pattern, and the second is the catch-all, which always succeeds. This structural limitation is used to simplify presentation, and it has no effect on the generality of the language. It is always possible to rewrite

```
match e with P1 -> e1 | P2 -> e2 | P3 -> e3
```

as

```
match e with
| P1 -> e1
| _ -> (
  match e with
  | P2 -> e2
  | _ -> e3
  )
```

The function `filter`, used to check if a pattern p matches a value v, is defined by recursively comparing the structures of p and v. It calls a function `filter'` using an empty initial substitution. The substitution that is used as input for `filter'` accumulates the bindings created by the comparison between the pattern and the value. Box 6.1 shows the definition of these two functions, providing only nominal cases. The matching process implicitly fails on all other cases.

$$\frac{x \notin \mathrm{dom}(\phi)}{\texttt{filter'}(\phi, x, v) = \phi \oplus (x, v)} \qquad \frac{\phi(x) = v}{\texttt{filter'}(\phi, x, v) = \phi}$$

$$\frac{\phi_1 = \texttt{filter'}(\phi, p_1, v_1) \quad \ldots \quad \phi_n = \texttt{filter'}(\phi_{n-1}, p_n, v_n)}{\texttt{filter'}(\phi, C(p_1, \ldots, p_n), C(v_1, \ldots, v_n)) = \phi_n}$$

$$\texttt{filter}(p, v) = \texttt{filter'}(\emptyset, p, v)$$

Box 6.1. *Matching function*

The rules of the operational semantics are shown in Box 6.2 and use the matching function of Box 6.1. The rule E_{match1} corresponds to the case where the first pattern p matches the value v of the expression e, returning a substitution ϕ. It is interesting to note that the structures of a substitution and an environment are identical, that is, lists of associations of a variable and a value. Thus, we may extend the environment with the substitution ϕ returned by `filter` to evaluate e_1, the expression associated with the first pattern. The rule E_{match2} corresponds to the case where the first pattern does not match the value v of e, indicating that a substitution ϕ cannot be found. The result of the computation is the evaluation of the expression e_2 associated with the second pattern, which captures any value.

$$(E_{match1}) \ \frac{\mathrm{Env} \vdash e \rightsquigarrow v \quad \phi = \texttt{filter}(p, v) \quad \phi \oplus \mathrm{Env} \vdash e_1 \rightsquigarrow v_1}{\mathrm{Env} \vdash \mathrm{match}\ e\ \mathrm{with}\ p\ \rightarrow\ e_1 \mid \_ \rightarrow\ e_2 \rightsquigarrow v_1}$$

$$(E_{match2}) \ \frac{\mathrm{Env} \vdash e \rightsquigarrow v \quad \nexists \phi = \texttt{filter}(p, v) \quad \mathrm{Env} \vdash e_2 \rightsquigarrow v_2}{\mathrm{Env} \vdash \mathrm{match}\ e\ \mathrm{with}\ p\ \rightarrow\ e_1 \mid \_ \rightarrow\ e_2 \rightsquigarrow v_2}$$

Box 6.2. *Operational semantics of pattern matching*

6.5.2.1.3. Typing rules

There are similarities between the typing of the `match/with` construct and that of the conditional, in that all branches must return values of the same type. The type of a value constructor is a functional type, of which the argument type is the product of the types of its arguments and the result type is t, (the type of the value built with this constructor). We assume given a function `TypeCstr` that provides the type scheme of a constructor. The value constructors of sum types are bound to their type scheme in the typing environment when the sum type is defined.

Rules T_{pvar} and T_{pcons} typecheck patterns and return an environment extended by the bindings of pattern variables to their respective types. The rule T_{match} ensures that the pattern and the examined value are of the same type. It typechecks the body of the first case in the environment extended by the bindings of pattern variables, and typechecks the body of the second case in the initial environment, verifying that the types of these bodies are the same.

$$(T_{pvar})\ \text{Env} \vdash x : \tau \Rightarrow (x, \tau) \oplus \text{Env}$$

$$(T_{pcons})\ \frac{(\tau_1 * \ldots * \tau_n) \rightarrow \tau \leq \texttt{TypeCstr}(C) \quad \text{Env} \vdash p_1 : \tau_1 \Rightarrow \text{Env}_1 \quad \ldots \quad \text{Env}_{n-1} \vdash p_n : \tau_n \Rightarrow \text{Env}_n}{\text{Env} \vdash C(p_1, \ldots, p_n) : \tau \Rightarrow \text{Env}_n}$$

$$(T_{match})\ \frac{\text{Env} \vdash e : \tau \quad \text{Env} \vdash p : \tau \Rightarrow \text{Env}' \quad \text{Env}' \vdash e_1 : \tau' \quad \text{Env} \vdash e_2 : \tau'}{\text{Env} \vdash \text{match } e \text{ with } p \ \rightarrow \ e_1 \mid \_ \rightarrow \ e_2 : \tau'}$$

Box 6.3. *Typing rules for pattern matching*

6.5.2.2. *Extensions and other forms of matching*

6.5.2.2.1. Guarded patterns

Many of the languages that use pattern matching associate a condition, known as a *guard*, with a pattern (the `when` construct in OCaml, C#, and F#, or `if` in Scala). The associated branch will only be taken at runtime if the pattern matches the examined value and if the condition is satisfied, as shown in example 6.46.

EXAMPLE 6.46.–

OCaml
```
card type = Jack | Queen | King | Number of int

let card_name c =
  match c with
  | Jack -> "jack"
  | Queen -> "queen"
  | King -> "king"
  | Number 1 -> "ace"
  | Number n when n > 1 && n < 11 -> string_of_int n
  | Number _ -> "illegal card!"
```

The `card_name` function handles the case `Number 1` separately, enabling it to return a more usual string, i.e. "ace" instead of "1". The other legal numerical values, from 2 to 10 inclusive, are handled separately. This is done using the match case `Number n`, which is limited to numerical values in this interval. Adding a guard to this pattern means that it cannot match all remaining values of the `card` type, even if they are structurally compatible. An additional case is therefore needed to handle values rejected by the guard.

6.5.2.2.2. Disjunction of patterns

Some situations require the same treatment to be applied to several cases. In this situation, instead of enumerating patterns one by one and repeating the same body each time, we may use a single pattern defined by a disjunction of (sub-)patterns, and a single processing may then be applied, as shown in example 6.47.

EXAMPLE 6.47.–

```
OCaml
let crowned_head c = match c with Queen -> true | King -> true | _ -> false
let crowned_head2 c = match c with Queen | King -> true | _ -> false
```

The function `crowned_head`, which specifies whether a crown is shown on a playing card, produces the same code twice: once for `Queen`, and once for `King`. In the second version, the first two patterns are replaced by a single pattern that accepts both `Queen` and `King`. This pattern may be read as "match `Queen` or `King`". Using "or"-patterns is particularly helpful in situations where the (identical) processing to apply to each case is a long piece of code. Sharing the code in these situations reduces the sizes of the source code and the executable.

Given that a disjunction pattern uses the same processing for each subpattern, if one of these subpatterns contains a variable, this one must also occur – and be of the same type – in the other subpatterns. This condition is necessary to ensure that this pattern variable will be well bound, independently of the specific subpattern used to catch the examined value, and will only have one type. This point is illustrated in example 6.48.

EXAMPLE 6.48.–

```
OCaml
type exparith =
  | Plus of (exparith * exparith) | Minus of (exparith * exparith)
  | Mult of (exparith * exparith) | Div of (exparith * exparith) | Num of int

let rec count_operators e =
  match e with
  | Plus (e1, e2) | Minus (e1, e2) | Mult (e1, e2) | Div (e1, e2) ->
      1 + (count_operators e1) + (count_operators e2)
  | Num _ -> 0
```

The purpose of the `count_operators` function is to count the number of operators present in an arithmetic expression. The four cases concerning $+, -, *, /$ all require two recursive calls, one for each subexpression. The treatment of the associated constructors can be factorized, as they allow the same variables `e1` and `e2` to be bound to the same type on every occasion.

6.5.2.2.3. Matching records

Patterns syntax also includes records with named fields, as presented in section 6.4.2. In a pattern, only those fields that are relevant to a treatment need to be named. This avoids the need to include all fields with a _ as a subpattern, as we see in example 6.49.

EXAMPLE 6.49.–

```
OCaml
type transport = Bus | Metro | Tram
type line = { number : int ; kind : transport ; company : string }

let rec filter_metro l =
  match l with
  | [] -> []
  | li :: rest -> (
      match li with
      | { number = n ; kind = Metro } -> n :: (filter_metro rest)
      | _ -> filter_metro rest
    )
```

The function `filter_metro` takes a list of public transport lines as input, and returns a list containing only the numbers of metro lines. Since the `company` field is irrelevant to this function, it is left out of the pattern. In practice, the final pattern _ has the record type `line`, implicitly signifying "any value for all fields of this type".

6.5.2.2.4. Matching on types

Some languages have the ability to test the type of the values encountered during execution. This ability is often linked to object mechanisms, which we shall present in greater detail in Volume 2, Chapter 4.

For example, `classes` of C# or Scala, can be used to define a hierarchy of structured types, enriched incrementally, along with functions used to test the type of the objects which they handle and to adapt their treatment accordingly.

EXAMPLE 6.50.–

```
Scala
import scala.math.sqrt

abstract class Shape

case class Point (x : Double, y : Double) extends Shape
case class Line (x1 : Double, y1 : Double, x2 : Double, y2 : Double)
  extends Shape
case class Circle (x : Double, y : Double, radius : Double) extends Shape

def perimeter (shape : Shape) : Double = {
  shape match {
    case Point (_, _) => 0
    case Line (x1, y1, x2, y2) =>
      sqrt ((x2 - x1) * (x2 - x1) + (y2 - y1) * (y2 - y1))
    case Circle (_, _, r) => 2.0 * 3.14159 * r
  }
}
```

This program, written in Scala, defines a general type `Shape` then two types `Point` and `Line` which "extend" the first type using their own fields. The `perimeter` function uses a value of the general type as input, and examines the precise type of this value. The body of each match case may then use the specific fields coming from the actual type of the examined value.

At first glance, `case class` could be perceived as a different syntactic approach to defining sum types; however, the construct is not simply limited to matching and provides several additional mechanisms used in object-oriented programming.

6.6. Equality

In section 6.5, we introduced constructs generalizing the conditional, first to deal with more than two cases, then to provide a notion of "equality" broadened by the pattern matching. While the concept of equality used in programming languages appears straightforward, it merits more detailed examination.

In a mathematical context, equality is defined as a reflexive, symmetric and transitive relation between two elements of the same set. The intuitive semantics of the equality of two elements consists of stating that if, in a mathematical expression, the first is replaced by the second (or vice versa), the value of the expression remains unchanged. The notion of equality was formalized in the 17th century by Gottfried Wilhelm Leibniz, who stated that two values x and y are equal if, and only if, for any predicate P, $P(x)$ is true if and only if $P(y)$ is true:

$$(x = y) \Leftrightarrow \forall P, P(x) \Leftrightarrow P(y)$$

The difficulty with this definition lies in the need for universal quantification on predicates, i.e. on functions returning a truth value. Hence, this definition cannot be directly applied in computer science.

It is often said that two "computer" values are considered to be equal if their representations in the memory are identical. In other terms, the sequences of bits encoding these values must be of the same length and contain two-by-two identical bits. Using the usual semantics for adding integers, it is easy to understand that $4 = 3 + 1$; in other terms, the *values* resulting from the evaluation of the expressions 4 and $3 + 1$ are equal, and both will be represented internally, in binary, as 100_b.

NOTE.– The syntax used for equality varies between languages. For example, in C, Java, Python and others, the notation == is used. In OCaml, on the other hand, equality is noted = (== has a different equality semantics) like in Basic, Pascal and Ada. Given that assignment is also noted = in many languages, users are strongly advised to study the syntax of any new language in order to avoid misinterpretations.

There are two possible interpretations when testing the equality of two identifiers x_1 and x_2: we may wish to check whether $\text{Env}(x_1) = \text{Env}(x_2)$, or whether $\text{Mem}(\text{Env}(x_1)) = \text{Mem}(\text{Env}(x_2))$.

6.6.1. *Physical equality*

In languages such as C, the equality operator == tests the identity of values in the compared expressions. These values must be scalar, otherwise the test will be rejected during the compilation. This point has significant implications for comparing data structures. Generally, the encoding used to construct evolved data structures involves pointers, which are scalar values. Equalities between data structures can relate to the addresses of these structures, in the case of *physical* equality, or to the values of the structures, known as *structural equality*, a notion presented in section 6.6.2. In C, when the equality == is used in relation to data structures, it denotes physical equality.

The most striking example of this concerns the comparison of strings. As we saw in section 6.3, strings in C are encoded as arrays of characters (terminating with the character `'\0'`) and the arrays are denoted by pointers (see section 7.3). In testing the equality of two variables representing strings, we are, in fact, testing the equality of the pointers representing these strings, and not equality of the data stored at these locations. This is illustrated in example 6.51.

EXAMPLE 6.51.–

```
C
#include <stdio.h>

int main () {
  char *s1 = "Foo" ;
  char s2[4] = { 'F', 'o', 'o', '\0' } ;
  if (s1 == s2) printf ("Equal\n") ;
  else printf ("Different\n") ;
  return (0) ;
}
```

This program declares the variable `s1`, containing the string `"Foo"` and the variable `s2` as a character array containing the same letters as the string `"Foo"` and ending with a `'\0'`. Thus, `s1` and `s2` "contain" identical strings. However, at runtime, this program will return `Different`, as the locations denoted by `s1` and `s2` differ.

The semantics of equality in C is even more complex when we consider the case of constant string declarations. The compiler operates by creating only one occurrence of each constant string in a program. Thus, different variables initialized with the same constant string will denote the same reference. A modified version of example 6.51 is shown below, in which `s1` and `s2` are initialized with the same string `"Foo"`.

EXAMPLE 6.52.–

```
C
#include <stdio.h>

int main () {
  char *s1 = "Foo" ;
  char *s2 = "Foo" ;
  if (s1 == s2) printf ("Equal\n") ;
  else printf ("Different\n") ;
  return (0) ;
}
```

At runtime, the program will now return `Equal` since the compiler initialized `s1` and `s2` using the location of a single occurrence of the string `"Foo"`, assigned once and for all in the program code.

As we saw in section 6.3, the difference between declaring a string as a pointer or an array adds another subtle level of difficulty. A declaration of an array will necessarily result in a distinct memory allocation, and will thus produce a negative result in an equality test, even if two variables of this type are initialized using the same string.

In the example below, a subtle change has been made to the program in 6.52. Instead of declaring `s2` as a `char*`, we have chosen to declare it as a `char[]`, i.e. as an array.

EXAMPLE 6.53.–

```
C
#include <stdio.h>

int main () {
  char *s1 = "Foo" ;
  char s2[] = "Foo" ;
  if (s1 == s2) printf ("Equal\n") ;
  else printf ("Different\n") ;
  return (0) ;
}
```

At runtime, this program will return `Different` since the array designated by `s2` is necessarily found at a separate location to the string `"Foo"` used to initialize both cases.

6.6.2. *Structural equality*

Structural equality relates to the contents of addresses. Two values that are physically equal will be structurally equal, but the reverse is not necessarily true.

Structural equality is only possible in languages with native data structures, and the size and form of the information composing the structure must be known at compile-time. This is the case in both OCaml and Python, which allow structural comparison of lists, arrays, etc. In these languages, the compiler chooses the representation of these structures, unlike in C, where encoding is left to the programmer. The internal implementation of equality draws on knowledge of the compiler's representation choices. Note that these languages also have the capacity to test for physical equality (== in OCaml, `is` in Python).

In languages such as C or C++, the compiler cannot automatically determine which semantics to use in comparing two arbitrary data structures. There is no means of knowing whether any given pointer represents either the address of a single entity, or that of an array (and thus of multiple values). This task can only be carried out by the developer, who knows the semantics of the structures which she or he has created. Thus, in the case of a linked list structure, the programmer must implement a comparison which follows the links in the list, knowing that each pointer only provides the address of the next element. In a situation involving arrays of linked lists, the programmer must develop an operation to compare consecutive objects, knowing that the pointer for an array designates the beginning of a sequence of elements. This point is illustrated in example 6.54.

EXAMPLE 6.54.–

```
C
#include <stdlib.h>
#include <stdbool.h>

struct list {
  int val ;               /* Cell value. */
  struct list *next ;  /* Address of the next cell. */
};

/* Equality check of two lists denoted by the address of their head
   cells. */
bool list_equal (struct list *l1, struct list *l2) {
  if (l1 == NULL) {
    if (l2 == NULL) return (true) ;
    else return (false) ;
  }
  else {
    if (l2 == NULL) return (false) ;
    else return ((l1->val == l2->val) && list_equal (l1->next, l2->next)) ;
  }
}

/* Equality check of two arrays of lists. The sizes of the arrays are used
   as arguments, as they are not included in the array structures. */
bool tab_list_equal (int long1, struct list *t1, int long2, struct list *t2) {
  int i ;
```

```
  if (long1 != long2) return (false) ;
  else {
    for (i = 0; i < long1; i++) {
      if (! list_equal (&t1[i], &t2[i])) return (false) ;
    }
    return (true) ;
  }
}
```

Although both functions `list_equal` and `tab_list_equal` use pointers over `struct list`, their behavior is radically different due to the fact that, semantically, these pointers do not represent the same data structure. Note that in C, the size of an array is not associated with its structure; the size of each array must therefore be supplied as an argument for function `tab_list_equal`. Since arrays `t1` and `t2` contain lists and not pointers over lists, function `list_equal` is called specifying the addresses of the elements in the array to compare. The decision to define `t1` and `t2` as arrays of lists instead of list addresses is arbitrary; our aim is essentially to show that, even if the two comparison functions are of similar types, their algorithms are fundamentally different. Note that in practice, where data structures of this kind are created dynamically, `t1` and `t2` would most likely contain pointers to lists, and would therefore be of the type `struct list**`.

For character strings, the standard library of C includes a function `strcmp` for structural comparison purposes. This function is easy to write, as the strings are arrays terminating with a sentinel marker (`'\0'`). In object-oriented languages, such as Java or C++, strings are encapsulated within classes from their standard library, which include a structural comparison function. Nevertheless, this function is not one of the primitives of these languages and was written by the library's developers. In languages that include an operator overloading mechanism, such as C++ (see section 5.7.2), it is possible to change the behavior of == when used with data structures in order to assign a dedicated semantics, for example structural equality. This redefinition is done during the implementation of the data structure. Once again, this approach relies on explicit development work carried out by the designer of the data structure based on a thorough knowledge of the semantics of this structure.

6.6.3. *Equality between functions*

We discussed the semantics of equality tests for functions at the start of Chapter 5, indicating that there cannot exist a satisfactory response to this problem. The choices used by a number of languages are described as follows.

6.6.3.1. *C, C++, Python*

These languages compare the addresses of the code of the compared functions. The equality is thus implicitly physical in nature. The only way to obtain a positive test result is to compare a function identifier with itself.

6.6.3.2. *OCaml*

In this case, any attempt to test the structural equality of two functions will result in an error message, even if the two operands are the same function identifier. Nevertheless, physical equality can still be tested, using the same semantics found in C or C++, as shown in example 6.55.

EXAMPLE 6.55.–

```
OCaml
# let f x = x ;;
val f : 'a -> 'a = <fun>
# f = f ;;
Exception: Invalid_argument "compare: functional value".
Raised by primitive operation at unknown location
Called from file "toplevel/toploop.ml", line 180, characters 17-56
# let g x = x ;;
val g : 'a -> 'a = <fun>
# f == f ;;
- : bool = true
# g == f ;;
- : bool = false
```

6.6.3.3. *Java*

Function comparison is illegal in Java, and this is indicated during the compilation. There is no implementation of the predefined == operator for functions, so the compiler simply rejects these expressions. Furthermore, functions (methods) are not objects belonging to classes, so the `equal` method, used to test equality between objects in a class, is not applicable.

6.6.3.4. *Coq*

This tool is both an environment for formal proof and a programming language. It is widely used to construct proofs for properties verified by computer, such as those concerning communication protocols [DUB 18]. Users may write expressions, including functions, and prove properties over them. Under certain conditions – which lie beyond the scope of this book – OCaml code can be extracted automatically from a development in Coq.

Coq can manipulate terms representing a program, and thus can rewrite certain terms and deduce an equality. This even holds true for certain functions, meaning that it is possible to prove an equality between functions which are *a priori* syntactically different, as shown in example 6.56.

EXAMPLE 6.56.–

```
Coq
Welcome to Coq 8.8.2 (November 2018)

Coq < Definition f (x : nat) := x.
f is defined

Coq < Definition g (y : nat) := if true then y else 0.
g is defined

Coq < Goal f = g.
1 subgoal

  ============================
   f = g

Unnamed_thm < compute.
1 subgoal

  ============================
   (fun x : nat => x) = (fun y : nat => y)

Unnamed_thm < auto.
No more subgoals.

Unnamed_thm < Qed.
Unnamed_thm is defined
```

We begin by defining two functions `f` and `g` and precise our goal which is to prove their equality (using the `Goal` command). We use a tactic that is native to Coq to compute and rewrite the terms found in the goal to prove (using the `compute` command). Coq is then able to eliminate the test (false) found in `g`, producing two function bodies that are identical, apart from the names of the bound variables. Finally, a conclusion of equality is reached automatically (`auto` command) despite this difference in the variable names.

Of course, this power comes both from the fact that the developer guides Coq through the proof process and from the fact that the functions being compared are relatively simple. The automatic process only works for relatively simple functions. In more complex cases, the responsibility for proof falls to the user, and may become arbitrarily complicated (if not impossible) depending on the way in which functions are defined. For example, the equality of the functions `f x = x + x` and `g y = 2 * y` cannot be verified automatically, and the proof can be done, but requires significant work.

7

Pointers and Memory Management

This chapter is dedicated to the notions of pointers and references, already mentioned in Chapter 4, with illustrations mostly written in C and C++. These languages include explicit pointers, which sometimes reveal technical details closely linked to computer architecture. Pointers and references are essential to the notion of *objects*, which will be discussed in greater detail in the context of object-oriented languages (Volume 2, Chapters 3 and 4) in relation to Java (see Volume 2, section 4.1), Python (see Volume 2, section 4.4) and C++ (see Volume 2, section 4.2).

7.1. Addresses and pointers

In terms of low-level implementation, a computer memory may be seen as a large "array" of bytes. This "array" is made up of zones corresponding to program variables and, more generally, of zones explicitly or silently allocated by the program. Some of these zones are created automatically by the program's execution mechanisms, some by compilation mechanisms and some by demands formulated explicitly in the source code. The information manipulated by a program is stored in the memory at a certain location, indicated by an *address*: this notion has already been discussed in Chapters 2 and 4. An address may thus be seen as the index in the memory "array" where an item of information is stored.

An address is an element of information like any other: it may be recorded, manipulated or used to read or write at the location it denotes. A value of the address type is called a *pointer*. A variable containing an address is therefore of address type. The term "pointer" is often used indifferently to designate an address type or an address value according to the context.

As we saw in our study of typing imperative features, a pointer type indicates the type of the pointed values. This rule holds true for languages such as C and C++, and guarantees that values stored at an address will be manipulated in a way that is consistent with their type. In C, for example, suppose that we are reading the value at a location denoted by a pointer `p`. If `p` denotes the address of an integer (`int`), there will be 4 bytes to read. In the case of a float (`double`), there will be 8 bytes to read. Figure 7.1 shows a schematic representation of the integer 23 in the memory, beginning at address 10. The way bytes are organized in the memory will be discussed in detail in section 7.2.

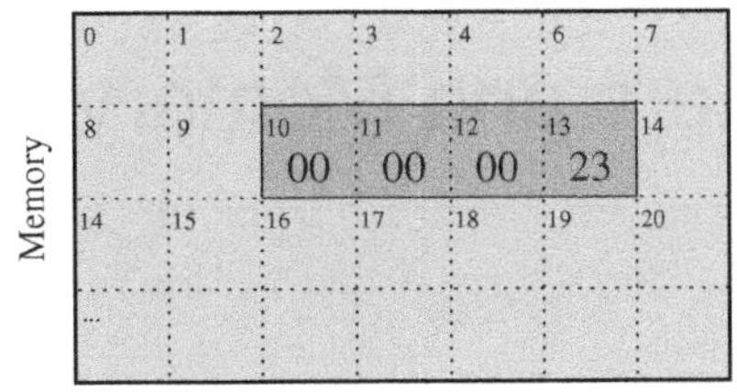

Figure 7.1. *Schematic representation of memory and address. For a color version of this figure, see www.iste.co.uk/hardin/programming1.zip*

In C, the pointer type is denoted by the constructor `*`. Thus, `int *` denotes a pointer over `int`, i.e. an address where integer type information will be found. The notation `int *x;` declares the variable `x` as a pointer to an integer: literally, the value pointed by `x` has type `int`. Variables may be created to record addresses, and these variables are also stored at an address in the memory. Thus, `int**` represents a pointer to a pointer to an integer. The information found at the address designated by such a pointer is itself an address.

Figure 7.2 shows the integer introduced in Figure 7.1 alongside a pointer, the value of which represents the address of this previous integer. For the purposes of this representation, the pointer was arbitrarily considered to be of the same size as an integer, but this is not always the case in current architectures.

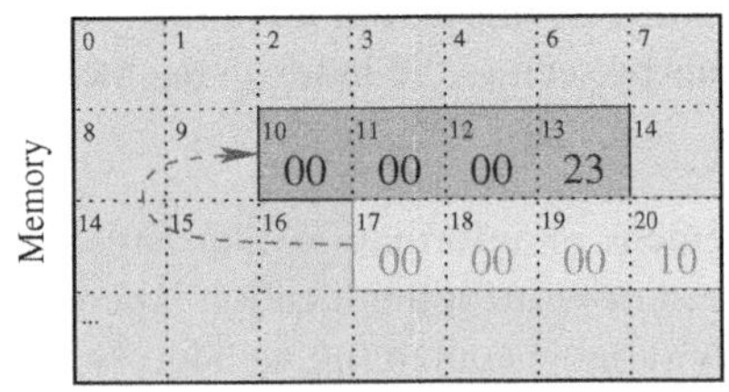

Figure 7.2. *Pointer to a pointer. For a color version of this figure, see www.iste.co.uk/hardin/programming1.zip*

7.2. Endianness

The way in which integers are stored in the memory varies from processor to processor. Storage of integers needs several bytes, which may be stored in the memory from left to right or from right to left. This order, known as *endianness*, is imposed by the processor's architecture. Processors that store the least important bytes of an integer at the lowest addresses are said to be "little endian". In an architecture of this kind, for the hexadecimal value `01020304`, the byte `01` would be placed at the address at the start of the integer, `02` would be placed at the following address, and so on. Intel x86 and ARM processors are little endian. Processors that store bytes in the reverse order, with the most important byte at the lowest address, are said to be "big endian". In this case, for the value `01020304`, the byte `04` would be stored at the address of the start of the integer, followed by `03`, and so on. The MC68000 family of Motorola processors are big endian. Certain processors, such as MIPS or PowerPC models, can be configured (using hardware, software, or both) to operate in either mode. Of course, endianness can lead to compatibility problems if data structures are directly extracted from memory and transmitted (over a network, or as a file on a disk) to a processor with a different endianness. In these cases, a program must be used to reorder the bytes in question before attempting to access the received integer values.

7.3. Pointers and arrays

As we have seen, an address represents a "location" in the memory. As such, it may be the location of a variable, or the start of a contiguous data zone. In certain languages, arrays – which are contiguous sequences of data of the same type – are simply represented by the address of the start of the memory zone in which they are located. Thus, as we saw in section 6.2, an element in an array is accessed by computing its address from the start of the array, the size of each element and the index. Computation using addresses is often referred to as *pointer arithmetic*. In C and C++, the addition operation $p + n$, using a pointer p over a type t and an integer n, is interpreted by the compiler as a "shift of n elements of type t from p" and this is the only permitted arithmetic operation on pointers.

For example, if p is declared by `int *p`, the memory access operation for expression `p[5]` will occur at address `p + 5 * 4`, since an integer is made up of 4 bytes. In practice, the bracketed notation `p[5]` is identical to the dereferencing of the address $p + 5$. In C, this is written `*(p + 5)` (as addition is commutative, the notation `i[tab]`, which is semantically equivalent to `tab[i]` in C, also denotes cell `i` in the array `tab` but should be avoided for readability reasons).

7.4. Passing parameters by address

In section 4.6, we examined different procedure call mechanisms, defining different modes of communication between the caller and the called procedure. In the case of passing by value, the caller transmits the value to the called procedure: this value is either a constant or a mutable variable, which is local to the procedure and initialized by the transmitted value. In the case of passing by reference, communication between the caller and the called procedure operates in both directions: the address of a mutable variable from the calling environment is transmitted to the procedure, and the procedure can then modify the value of this variable. Box 4.6 describes the possible semantics of these different call mechanisms. In this section, we shall present the use of pointers in passing arguments: this is known as *passing an argument by address*. This is a technique for implementing passing by reference but also, for certain compilers, a means of optimizing the transmission of large arguments. Several situations in which passing by address is helpful are described below.

By definition, a function returns a single value. However, we may require several values to be returned. One simple example is a function designed to take an array of integers and return the numbers of negative, null and positive values found in the array. In languages with native support for n-uples, a single triple value will be returned. Example 7.1 shows the definition of this function in Python.

EXAMPLE 7.1.–

```
Python
def count (t) :
    vneg = 0
    vnul = 0
    vpos = 0
    i = 0
    while i < len (t) :
        if t[i] < 0 : vneg = vneg + 1
        elif t[i] == 0 : vnul = vnul + 1
        else : vpos = vpos + 1
        i = i + 1
    return (vneg, vnul, vpos)
```

If the chosen language does not support n-uples – as in C – another mechanism must be used, such as creating a record type containing three integers and returning a value of this type, as shown in example 7.2.

EXAMPLE 7.2.–

```
C
struct three_ints { int vneg ; int vnul ; int vpos ; } ;

struct three_ints count (int *t, int size_t) {
  int i ;
  struct three_ints res = { 0, 0, 0 } ;
  for (i = 0; i < size_t; i++) {
    if (i < 0) res.vneg++ ;
    else if (i == 0) res.vnul++ ;
    else res.vpos++ ;
  }
  return res ;
}
```

The drawback to this solution is that we need to create a "utility" structure for every type of multiple values, which we wish to return. Thus, if another function needs to return two integers and a float, a new structure must be created for this purpose and so on, increasing the size of the code. Furthermore, the definitions of these structures "pollute" the source code and limit maintainability, to the detriment of those definitions that help to reveal the algorithmic architecture of the program.

Another solution is to create global variables dedicated to the function, in which the function writes the values to return. Evidently, a global variable will be needed for each value to return, just as multiple structures were required in the previous approach. This makes the program more complicated to maintain, as we need to ensure that each function writes to its own global variables, and that each calling function reads the results from the right global variables. The solution also limits reuse: the code for a function cannot be reused in another program without also importing the declarations of the associated global variables.

These issues can be avoided by *passing arguments by address*, an approach involving pointers. The function takes arguments representing the addresses where the results to transmit to the caller should be stored. The formal and actual parameters of this function are therefore of pointer type. The actual parameters transmitted by the caller are the addresses of the variables where the results of the function call will be stored. This point is illustrated in example 7.3.

EXAMPLE 7.3.–

```
C
void count (int *t, int size_t, int *vneg, int *vnul, int *vpos) {
  int i ;
  for (i = 0; i < size_t; i++) {
    if (t[i] < 0) (*vneg)++ ;
```

```
    else if (t[i] == 0) (*vnul)++ ;
    else (*vpos)++ ;
  }
}

void caller () {
  int t[6] = { 1, 1, 0, 0, 0, -2 } ;
  int vneg, vnul, vpos ;
  count (t, 6, &vneg, &vnul, &vpos) ;
  printf ("Negative: %d Null: %d Positive: %d\n", vneg, vnul, vpos) ;
}
```

The `caller` function transmits the addresses of the three local variables `vneg`, `vnul` and `vpos`, which will act as "mailboxes" to `count`. The `count` function will store the values returned by its computation in these locations. The computation increments the values found at the specified addresses (use of `*` in `(*vneg)++`, `(*vnul)++` and `(*vpos)++`). We have chosen not to use `return` in this function: all results will be transmitted using the passing by address mechanism.

In example 7.3, the `count` function uses variables passed by address in write-only mode. This mode is often known as "out", indicating that the parameters are the output of the function. These arguments may also be used in both reading and writing: this is known as "in-out" mode. In this case, the function begins by reading the values located at the addresses in order to carry out a computation; it then overwrites them in order to store the results of the computation. These modes, with a dedicated syntax, for passing arguments by address are also found in Ada, a language that is briefly presented in Volume 2, section 2.2. Consider the function in example 7.4, which multiplies a vector of dimension 2 by a scalar. The resulting vector may be re-written "over" the argument vector.

EXAMPLE 7.4.–

```
C
void mult_vect_scal (float *x, float *y, float n) {
  *x = *x * n ;
  *y = *y * n ;
}
```

The function `mult_vect_scal` must first retrieve the value of each component of the vector before multiplying it by the scalar, then writing the result of this product in the original vector.

Passing arguments by address is also justified when the argument to pass to a function is large. Instead of copying the entire contents of the structure to transmit to the called function – which can be costly in terms of both time and memory – we may simply transmit the address of the structure. However, this strategy is not without its risks if the contents of the structure must be left as given by the called function: if this function has access to the memory zone of the caller, there is nothing to prevent it from making modifications. Certain languages include type modifiers that can be used to declare a variable or formal parameter as *constant*. The code in the scope of the declaration prohibits itself any modification of the variable following initialization. This modifier is named `const` in C and C++, `in` in Ada, and `final` in Java.

In example 7.5, the function `f` attempts to write to the address designated by `i`, but this address is defined as a constant. The program is thus rejected during the compilation.

EXAMPLE 7.5.–

```
C
void f (const int *i) { *i = *i + 1 ; }
```

```
$ gcc foo.c -c
faux.c:5:6: error: read-only variable is not assignable
  *i = *i + 1 ;
  ~~ ^
1 error generated.
```

7.5. References

NOTE.– The notion of *reference*, as described here, arises from the term used in C++ and Java. It is not the same as the references used in the context of mutable variables in Chapters 2 and 4. Unfortunately, both uses of the term are firmly established in the context of programming languages, so it is important to be aware of the specific meaning used in any given language in order to prevent ambiguity.

As we saw in section 7.4, arguments can be passed to a function by address, allowing the function to modify the corresponding values in the caller. In C, this is done using pointers, and the called function must read or write to the addresses received as arguments. The notion of a *reference* produces the same effect without recourse to pointers, simplifying the way the source code is written.

7.5.1. *References in C++*

In the context of C++, a reference is a variable that is an alias of a pre-allocated memory zone. References are initialized using an expression that represents a memory zone, accessed by a name (a more precise definition of the *l-value* introduced in section 2.3.2). Such a name can be a variable name, a cell in an array, an object's attribute, etc., hence it cannot be an arbitrary expression, such as a constant or the result of an intermediate computation. Since the introduction of C++11, however, users can define references that are initialized by other expressions, known as *r-value references*; a description of this mechanism lies outside the scope of this book. Even in older standards, references may be initialized by an *r-value* if the reference is declared as a constant. Readers interested in further details on this topic may read specialized documents about C++.

Just as every pointer is a pointer "to a type", every reference is a reference "to a type". The syntax used in C++ to declare a reference to a type `t` is `t&`. Example 7.6 shows the declaration of a reference, and the effect of a modification on the variable to which it refers.

EXAMPLE 7.6.–

```
C++
#include <stdio.h>

int main () {
  int i = 46 ;
  int& j = i ;
  i = i + 1 ;
  j = j + 1 ;
  printf ("i = %d j = %d\n", i, j) ;
  return 0 ;
}
```

The variable `j` is declared as a reference to `int` and initialized with the address of `i`. When `i` is modified, `j` is also modified and vice versa. During execution, this program will thus display `i = 48 j = 48` since two incrementations have been made, one on the identifier i and the other on `j`.

Example 7.7 is a modified version of example 7.6 in which the reference is replaced by a (constant) pointer. Note the explicit manipulation of the pointer, both in extracting the address of `i` and in dereferencing `j`.

EXAMPLE 7.7.–

```
C
#include <stdio.h>

int main () {
  int i = 46 ;
  int *j = &i ;
  i = i + 1 ;
  *j = *j + 1 ;
  printf ("i = %d j = %d\n", i, *j) ;
  return (0) ;
}
```

The program in example 7.8, in which we attempt to initialize the reference `j` with a constant and the reference `k` with the value returned by a function call, does not compile: the constant 10 and the integer returned by `times2` do not represent allocated memory zones.

EXAMPLE 7.8.–

```
C++
#include <stdio.h>

int times2 (int v) {
  return (2 * v) ;
}

int main () {
  int i = 46 ;
  int& j = 10 ;
  int& k = times2 (i) ;
  return 0 ;
}
```

Attempts to compile this program using the g++ compiler will result in an error regarding the initializations of `j` and `k`:

```
$ g++ refs2.cpp
refs2.cpp:11:8: error: non-const lvalue reference to type 'int' cannot bind to
      a temporary of type 'int'
  int& j = 10 ;
       ^   ~~
refs2.cpp:12:8: error: non-const lvalue reference to type 'int' cannot bind to
      a temporary of type 'int'
  int& k = times2 (i) ;
       ^   ~~~~~~~~~
2 errors generated.
```

Using an array to represent a memory zone, it is possible to create a reference to a cell, as shown in example 7.9.

EXAMPLE 7.9.–

```
C++
#include <stdio.h>

int main () {
  int t[2] = { 12, 21 } ;
  int& i = t[1] ;
  i = i + 1 ;
  printf ("i = %d t[1] = %d\n", i, t[1]) ;
  return 0 ;
}
```

The variable `i` is a reference to an integer representing the second cell in the array `t`. As `t` denotes a zone in the memory, the second cell will also denote a zone in the memory. Running this program will return the expected result `i = 22 t[1] = 22`.

References can also be used as function parameters in order to "silently" pass a parameter by address, as shown in example 7.10.

EXAMPLE 7.10.–

```
C++
#include <stdio.h>

void f (int& v) { v = v + 1 ; }

int main () {
  int i = 46 ;
  f (i) ;
  printf ("i = %d\n", i) ;
  return 0 ;
}
```

The formal parameter `v` of `f` is a reference to an integer. Thus, during the call `f(i)`, `v` becomes an alias of `i`; in other terms, `i` and `v` denote the same location in the memory, although the two identifiers are not in the same scope. Thus, the incrementation of `v` in `f` is visible in `i` in the caller function, and the program returns `i = 47`.

Given that a reference is an alias, it is entirely possible to create a reference to a reference, or to pass a reference to a function accepting a reference as an argument; the reference always refers to the same initial variable. This is illustrated in example 7.11.

EXAMPLE 7.11.–

```
C++
#include <stdio.h>

void f (int& v) { v = v + 1 ; }

int main () {
  int i = 46 ;
  int &j = i ;
  int &k = j ;
  f (i) ;
  f (j) ;
  f (k) ;
  printf ("i = %d j = %d k = %d\n", i, j, k) ;
  return (0) ;
}
```

Creating `j` as a reference to `i`, and `k` as a reference to `j`, identifiers `j` and `k` are both aliases of `i`. In the context of calls to `f` with each of these identifiers as arguments, the formal parameter `v` also becomes an alias of `i`. Three incrementations are therefore applied, and the program returns `i = 49 j = 49 k = 49`.

The same applies in the context of a function call: the argument passed during the call for a parameter of reference type must designate a memory area. Once again, the purpose of this mechanism is to create aliases for identifiers. Thus, example 7.12 will be rejected by the compiler.

EXAMPLE 7.12.–

```
C++
#include <stdio.h>

void f (int& v) { v = v + 1 ; }

int main () {
  f (67) ;
  return 0 ;
}
```

The call to `f` uses an argument that does not represent a zone in the memory, resulting in an error message during the compilation.

```
$ g++ refs.cpp
refs.cpp:8:3: error: no matching function for call to 'f'
  f (67) ;
  ^
refs.cpp:3:6: note: candidate function not viable: expects an l-value for 1st
      argument
void f (int& v)
     ^
1 error generated.
```

As in the case of passing by address, references may be used to transmit large arguments of a call, avoiding the need to copy them. A constant reference may be declared (in the same way as a constant pointer) to prevent the called function from modifying the memory zone transmitted by the caller function.

7.5.2. *References in Java*

While Java does not include an explicit reference mechanism, some authors consider that the language passes an argument by value if the argument is of a primitive type (integer, boolean or float), and by reference if this argument has an object type (including character strings, seen here as objects). This is not strictly true: in reality, Java always passes by value. The difference between a primitive value and an object value is that objects are always dynamically allocated, and are thus represented by pointers. When the argument of a call by value is an object value, the value which is actually passed as the argument is a copy of the pointer. As Java does not use explicit pointers, each time an object is manipulated, it is done silently through its address.

7.6. Memory management

In this section, we shall discuss memory allocation and disallocation using manual or automatic approaches. In high-level languages, data are stored in the memory by the execution process, without direct intervention from the developer. In certain languages, developers also have the option to use lower level features to obtain full or partial control over memory storage. A relatively brief presentation of memory management by the execution process, sometimes with the participation of the programmer, is now discussed.

7.6.1. *Memory allocation*

When a compiler analyzes the definition of a variable, it assigns an area of memory in which to store the value bound to this variable. This operation is known as *memory allocation*. For variables, and arrays whose size can be determined at compile-time, the memory requirements are known and the allocation can be planned during the compilation. This is known as *static allocation*.

In example 7.13, the size of variables `i` and `s` cannot vary from one execution to the next: their allocation is static.

EXAMPLE 7.13.–

```
C
int i = 42 ;
char s[20] ;
```

7.6.1.1. *Dynamic allocation*

In other cases, the quantity of memory required for all possible executions of a program cannot be determined in advance. For example, the designers of the text editor used to write this book had no means of knowing how many characters each text file would contain, meaning that they had no knowledge of the amount of memory required during the writing process. In other words, memory requirements are unknown at compile-time.

The programmer can certainly reserve a "fairly large" amount of memory (for example using a very large array), and say to themselves that "that will be enough". This solution is unsatisfactory for several reasons. First of all, when an execution uses less memory than expected at the time of allocation, the unused space cannot be used for the execution of another program. Recall that, as a resource, memory is shared by all programs, and its size is necessarily limited. Then, if the default memory allocation is insufficient, all that remains is to modify the source code to request a larger size. This is unacceptable because, on the one hand, a software user does not necessarily know (or need to know) how to modify this source code; on the other hand, this source code is not always available, especially when it comes from proprietary and paid software. The execution system (in collaboration with the operating system) must therefore include a mechanism, which allocates memory during program execution in order to satisfy the actual needs of each execution. This mechanism is called *dynamic allocation*. The result of a dynamic allocation is the start address of the area just allocated, which must be stored in a pointer.

A memory allocation may be rejected, for example if there is insufficient space. Usually in this case, the primitive used for memory allocation returns an "illegal" address, often referred to as the `NULL` *pointer* (from the name of the constant used to define this address in many languages). This raises two important points. You have to check whether memory allocation has been successful before attempting to access the returned address. In addition, any pointer in a program is likely to contain (for whatever reason) this illegal `NULL` value. This value must be detected, because accessing this illegal address interrupts the execution of the program.

7.6.1.2. *Allocation for large structures*

Memory allocation for very large data, e.g. of multiple megabytes, can raise certain issues which differ according to the scope of the binding of the data.

Consider the case of a local variable bound to such data. Memory allocation is often done on the execution stack, for example, this is the case in C. As the size of the stack is limited, it can become totally full and overflow, resulting in an execution error. This is illustrated in example 7.14, despite the 16 gigabytes of RAM available in the machine.

EXAMPLE 7.14.–

```
C
#include <stdio.h>
#include <stdlib.h>

#define SIZE (10000000L)

int main () {
  int t[SIZE] ;
  t[0] = 5 ;
  printf ("%d\n", t[0]) ;
  return 0 ;
}
```

It is not possible to allocate 10 million integers to the stack, and so executing this program generates a segmentation error.

```
$ gcc foo.c
$ ./a.out
Segmentation fault: 11
```

Now, consider the case of a global variable bound to a very large data element. If this variable is not initialized, most compilers will generate a memory allocation to be performed when loading the executable. If the data are initialized, it will appear *in extenso* in the code generated by the compiler; this significantly increases the size of the executable (and the compilation time), as we see from example 7.15.

EXAMPLE 7.15.–

```
C
#include <stdio.h>
#include <stdlib.h>

#define SIZE (10000000L)
int t[SIZE] = { 1 } ;
int main () {
  t[0] = 5 ;
  printf ("%d\n", t[0]) ;
  return (0) ;
}
```

The executable generated by example 7.15 contains the sequence of statically allocated bytes, initialized for `t`, which is added to the code produced for the functions. The displayed size of the binary produced in this case is clearly different to that of the same program using dynamic allocation that is smaller, around only 10 kB.

```
$ gcc foo.c
$ ls -l a.out
-rwxr-xr-x  1 user  user  40005896  8 mar 09:34 a.out
```

Finally, consider the case of a function, transmitting a local memory zone containing results to its caller, when this result is large (see section 7.4). In example 7.16, `f` creates a local variable that is an array, applies a treatment to this array and returns the array to the program, which is called `f`.

EXAMPLE 7.16.–

```
C
int* f () {
  int t[10] ;
  ...    /* Processing. */
  return t ;
}
int* g () {
  ... f () /* Call to f */
}
```

Example 7.16 is perfectly acceptable for languages with automatic memory management systems, or where arrays are native data structures. In languages such as C and C++, however, such a program is incorrect. Since the local variables of a function are allocated on the stack, the local variable `t` would be "stacked" at the beginning of the execution of `f`. In practice, this means that the 10 integers making up the array are placed on the stack, and `t` contains the address (in the stack) of the first element of the array. At the end of the execution of `f`, the local variables of `f` are removed from the stack. However, the `return t` instruction does not return the contents of the array `t` but simply the address of its first element in the stack. If the function g that calls `f` uses this address, there is no guarantee that the current data at this address still has any link with `t` as the stack may have been modified in the meantime.

These problems can be avoided through the use of dynamic allocation, even if the size of the memory allocation is known at compile-time. In example 7.16, the array `t` may be allocated dynamically. At the end of the execution of `f`, the local variables will disappear but the allocated zone remains accessible, as long as it has not been explicitly freed.

7.6.1.3. *Dynamic allocation operators*

Dynamic memory allocation is performed by calling a dedicated function in the language. In C, this function is called `malloc`, and takes as the argument the number of bytes to allocate. It returns an untyped pointer (`void*`), which is the start address of the area allocated by the memory allocator. The `malloc` function also exists in C++, but the more elaborate `new` operator is often used instead. Unlike `malloc`, `new` calls the constructor of the created object (studied in Volume 2, section 4.2). The number of bytes to allocate is not required as the argument by `new`, since the compiler handles this computation using the type of the object being allocated. Note that, as an operator, `new` can be overloaded (see section 5.7.2).

7.6.2. *Freeing memory*

When a dynamically allocated memory zone is no longer needed, it must be given back to the system for reuse. In certain languages, the programmer is responsible for this restitution mechanism and must specify points in the execution where some memory zones will no longer be used and can be freed. This form of memory management can be extremely complex in the case of large programs, and different kinds of errors may occur.

First of all, the programmer may misjudge when an area is no longer needed and release it too soon. Therefore, the program continues to run instructions using values stored in that zone. But, since this zone has been deallocated, it no longer "belongs" to the running program. If it attempts to access it, the program will be interrupted for violating memory space. Indeed, the operating system prohibits programs from accessing any areas of the memory which have not been assigned to them.

Premature freeing of memory can result in a significantly worse version of this error, when the memory area has been reallocated to the program for another use. In this case, the program continues to access this area that belongs to it, so the operating system does not interrupt it. However, the processing using the area released too early and the one allocated this area after this release, read and write their data concurrently in this area, leading to unpredictable data modifications. This silent data corruption leads to errors that can be difficult to find, and correcting the problem is particularly difficult.

Memory may also be freed too late, or not at all. In this case, the program "hogs" memory resources for no good reason. The worst-case scenario is that there will not be enough memory available to satisfy other requirements, resulting in an execution error. To a lesser extent, modern operating systems have the ability to *virtualize* memory: external storage peripherals (hard disks) act as buffers to temporarily free up memory from programs or processes that are not currently using it. The result is an overall slowdown of the machine and a performance collapse, which can lead to more or less total paralysis of the entire system.

For memory to be freed, the address of each allocated zone must be saved up until the point of release. If this address is lost, it will no longer be possible to free the zone, resulting in the effects described above. This is known as a *memory leak*.

In example 7.17, the zone allocated by the primitive `malloc` is not freed at the end of the function `f`. The address of this zone, memorized in the local variable `p`, disappears at the end of the execution of `f`. The address thus becomes unknown and, consequently, it is impossible to free the space.

EXAMPLE 7.17.–

```
C
int f (int size) {
  int *p = malloc (size) ;
  for (int i = 0; i < size; i++) {
    do_something (p[i]) ;
  }

  return (...) ;
}
```

Other common memory freeing errors include releasing a zone that has not been dynamically allocated, and making multiple attempts to free the same zone.

Just like allocation, memory release is achieved by calling a construct of the language. In C, this construct is the `free` function, whose argument is the start address of a zone allocated by `malloc`. The size to be freed is not to be specified by the developer, as this information is memorized by the memory manager during allocation. This is all the more important as `malloc` does not necessarily allocate exactly the requested amount of memory: the actual allocation may be somewhat larger, for reasons we will not go into here. The programmer cannot know exactly how much memory to free up, hence the importance of offloading this issue. In C++, all memory allocated using `new` must be freed up using the `delete` operator. During a call to `delete`, the destructor of the allocated object (described in Volume 2, section 4.2) is called automatically; this would not be the case using `free`. As an operator, `delete` may also be overloaded.

7.6.3. *Automatic memory management*

To overcome the difficulty of memory release and the errors it induces, languages such as Java, OCaml, Python and Lisp include a mechanism in their runtime; this automatically retrieves and frees areas of memory that are no longer accessible and are therefore useless. This mechanism, called the *garbage collector* (GC), increases language safety as users are relieved of the responsibility of memory management.

To manage memory automatically, the execution manager needs to know the size of each allocated memory zone. The set of memory *blocks* (zones) used by the program (known as *live blocks*) is organized into a memory graph. Its starting points, called *roots*, are the registers, the execution stack and the global variables. The edges of this graph are the pointers located in the blocks. Any live block can be accessed by transitively following pointers from the roots: this block is still useful at execution-time. The so-called *dead blocks* are inaccessible from the roots so they can be retrieved. Thus, the essential task of a GC is to browse the memory graph during execution, looking for dead blocks. It must therefore be able to differentiate pointers from other types of data.

The efficiency of a GC is measured by the extent to which it increases the execution-time, by the amount of memory space it requires and by its ability to free up all the memory that can be freed. It depends on the frequency of allocations and the frequency of assignments within programs. Different GCs use different techniques, none of which is notably better than the others. A basic outline is provided in the following sections.

7.6.3.1. *Reference counting GC*

A *reference counting GC* associates a counter with each allocated memory block. This counter is incremented each time a pointer to the block is created, and decremented each time a pointer is destroyed. If the counter reaches 0, this means that the block is no longer used and can be freed; this may result in the deletion of pointers to other blocks. If this is the case, the counters of these latter blocks will be decremented; if any of these counters reach 0, a new pass of the GC is triggered. Pointer management in blocks comes at a cost, which may be significant. This GC mechanism is not able to free cyclic data structures, that is which reference themselves, as their counters never reach zero: this is shown in Figure 7.3. This phenomenon results in memory leaks.

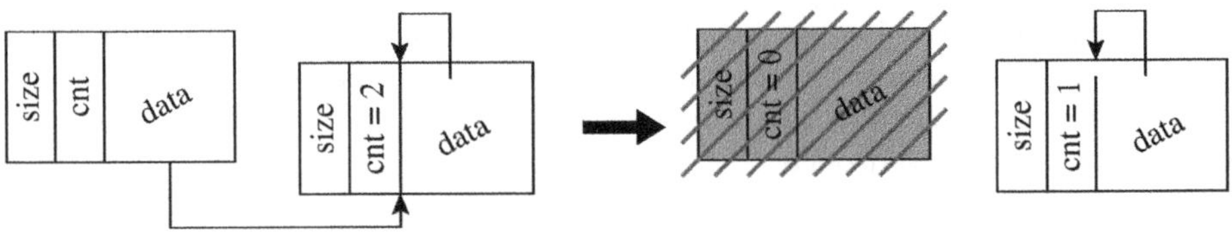

Figure 7.3. *Cycle and reference counter. For a color version of this figure, see www.iste.co.uk/hardin/programming1.zip*

7.6.3.2. *Mark-and-sweep GC*

A *mark-and-sweep GC* goes through the memory graph, marking live blocks. Unmarked blocks are therefore dead and the GC adds them to the free memory, represented by a linked list of free blocks.

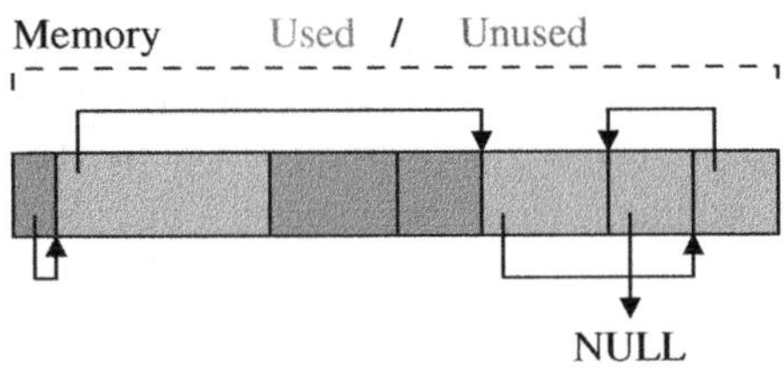

Figure 7.4. *List of free blocks. For a color version of this figure, see www.iste.co.uk/hardin/programming1.zip*

Mark-and-sweep GCs have the ability to free cyclic structures. However, they are not able to merge consecutive free blocks, as each block is reinserted into the list of free blocks independently. This results in fragmentation of the memory, as shown in Figure 7.5, which can prevent the allocation of a structure that is bigger than any individual free block, even if the overall quantity of free memory is sufficient.

7.6.3.3. *Copying GC*

A *copying GC* uses two memory zones which are equal in size. Only one of these zones is used to allocate memory at any time. When the quantity of available memory becomes insufficient, the GC goes through the memory graph and copies all live blocks, consecutively, into the unused zone. At the end of this process, the previously unused zone contains all live blocks and the GC reverses the roles of the two zones. Block copying prevents memory fragmentation, but the pointers in the copied blocks must be updated. There are two main drawbacks to copying GC: first, program execution stops during the GC phase, and second, it doubles the amount of memory needed.

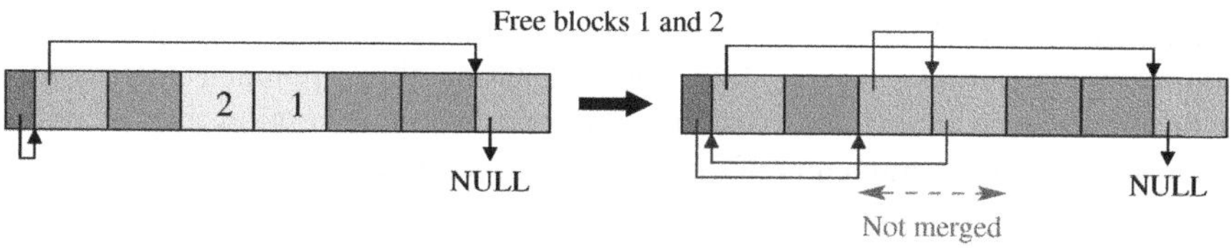

Figure 7.5. *Freeing and fragmentation. For a color version of this figure, see www.iste.co.uk/hardin/programming1.zip*

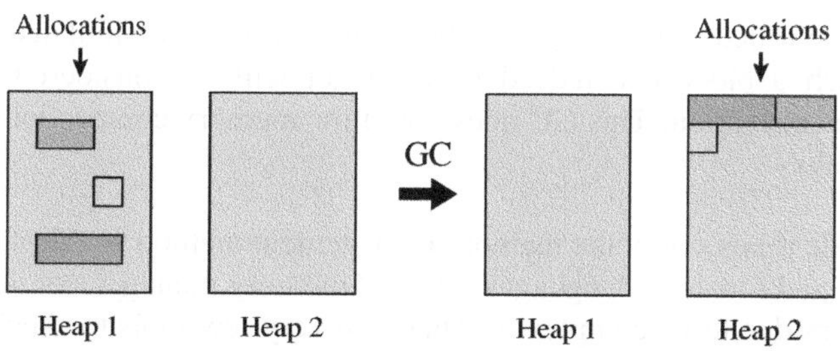

Figure 7.6. *Copying and compacting in a copying GC. For a color version of this figure, see www.iste.co.uk/hardin/programming1.zip*

7.6.3.4. *Generational GC*

Generational GC is based on the empirical observation that recently allocated (so-called *young*) blocks tend to stay alive for a short period of time. These young blocks mostly store values of temporary variables, of local variables of blocks or functions that disappear once the block or function has been executed. A global variable, which must remain accessible throughout the whole execution, will have its value stored in a block that will "age". Memory is thus split into several areas containing blocks of similar ages, each referred to as a *generation*. Allocation is done in the zone containing the youngest blocks. If this zone is not large enough to satisfy an allocation request, the GC scans the graph for the living blocks of this younger generation and copies them into the slightly older generation. Once this older generation becomes saturated by the copying, the GC looks for dead blocks to eliminate them. In a two-generation GC, the processings of the young and old generations are often referred to as *minor* and *major* GC, respectively.

As in the case of copy-based GC, memory is automatically compacted. Furthermore, if the language causes a large number of temporary allocations, the minor GC will be able to free large amounts of memory simply by examining the young generation. The execution of generational GC implies recording and managing pointers from old blocks to young blocks, resulting in additional costs during assignment operations. Again, like copying GCs, this approach interrupts the execution of the program; however, the duration of the interruption may be reduced by adjusting the size parameters of the different generations.

7.6.3.5. *Conservative GC*

To add an automatic memory recovery mechanism to a language relying on "manual" allocation/deallocation, a *conservative GC* can be used. It scans the memory assuming that any sequence of bytes that can represent an address value in the memory is a pointer. It frees all blocks that are definitely dead. But some blocks are wrongly identified as alive because they contain a value assimilated by the GC to a pointer. Such a block can indeed be dead but will be considered alive for the duration of the execution. This GC does not allow memory compaction and can lead to memory leaks.

To conclude this section, the memory management method is a demarcation point between high and low level languages. Manual memory management is a recognized source of errors due to its complexity. There are very few tools to check this manual management. In this section, we have provided a brief overview of GCs, which partly remedy these difficulties. Other more complex solutions, such as concurrent or parallel GCs operating over multi-processor architectures [DOL 93, DOL 94] will not be addressed here, as they lie outside the scope of this book.

8

Exceptions

8.1. Errors: notification and propagation

Errors may occur when executing a program. They may result from choices made by the programmer who may specify certain conditions, which must stop the execution. For example, a program to compute the average of a list of ages will not be able to handle a negative age value in the same way as a positive value. Other kinds of errors may occur that are not directly related to the programmer's intent, for example an attempt to divide by 0 or an attempt to illegally access a file due to missing permissions.

To produce a robust program, any programmer must be aware of these possible errors and should prevent their consequences when they occur. The most basic solution is to include multiple checks throughout a program and to implement a mechanism to highlight incorrect results, from error detection to management. The result obtained from each function call should be checked and if this value indicates that an error occurred during the execution of this call, then an error must be returned. A number of possible approaches for error notification and handling are outlined below using a simple pseudo-language.

```
f ()
  ...
  if ... then return error
  ...

g ()
  r = f ()
  if r is error then return error
  ...
h ()
  r = g ()
```

```
  if r is error then return error
  ...
```

This mechanism, which is entirely reliant on the programmer, presents several drawbacks. Error propagation is explicit in the algorithm. Furthermore, forgetting to check the validity of the result of a function call could go unnoticed. In addition, it must be possible to notify an error in the returned value. This is easy when the codomain of a function is partial: just encode the error in a value outside the codomain. For example, a function computing the absolute value of the difference between its arguments, which are assumed to be strictly positive, can only return positive or null integer values. Hence, errors can be represented using -1, -2, etc., as follows:

```
abs_diff_positive_ints (x, y)
  if x = 0 or y = 0 then return (-1)
  if x < 0 or y < 0 then return (-2)
  if x > y then return (x - y)
  else return (y - x)
```

This mechanism for encoding error values becomes impossible if the codomain of the function is total: all the values that can be returned are legal values. In other words, there is no available value to represent an error. In the following (simplistic) example computing the acceleration of a mass subjected to a force, if the mass is null, we do not know what value to return.

```
accel (force, mass)
  if mass = 0 then return ???
  else return (force / mass)
```

Some languages natively include n-uples. In this case, a pair may be returned, the first component being a flag stating whether or not this result is meaningful, the other one being the value of the result. If the flag indicates an error, any value may be chosen for the result, as it is known to be meaningless.

```
accel (force, mass)
  if mass = 0 then return (false, 42)
  else return (true, (force / mass))
```

In languages with sum types, the type of returned values can also be modified to include a notion of validity or invalidity.

```
type result = Error | Valid (float)

accel (force, mass)
  if mass = 0 then return (Error)
  else return (Valid (force / mass))
```

For languages without native n-uples or sum types, there are three possible solutions.

8.1.1. *Global variable*

The first solution, which is extremely difficult to maintain, consists of using a global variable to act as a validity flag.

```
var validity

accel (force, mass)
  if mass = 0 then
    validity = false
    return 42
  else
    validity = true
    return (force / mass)

f ()
  ...
  v = accel (...)
  if validity then ...
  else ...
```

As we saw in section 7.4, this solution reduces maintainability and readability. Defining as many global variables as there are functions needing to return errors, or sharing a same global variable between several functions does not ease code review and maintenance. Nevertheless, this mechanism is used in the C standard library to detail errors in certain system calls and functions. The global variable `errno` is modified by these functions as a side effect, adding error information.

8.1.2. *Record definition*

As we saw in section 7.4, even if a language does not natively support n-uples, it remains possible to create a two-field record. However, a record type must be created for every possible type of returned value, so this solution to the error notification problem is also unsatisfactory.

8.1.3. *Passing by address*

Some languages, such as C, allow arguments to be passed by address (see section 7.4). Instead of passing a "conventional" value, we may pass the address of a memory zone, for example the address of a variable of the calling function. Thus, the called function will write at this address. In this way, information can be transferred from the called function to the caller, both for error-free results and for error notification, as shown in example 8.1.

EXAMPLE 8.1.–

```
C
float accel (float force, float mass, bool *valid) {
  if (mass == 0) {
     *valid = false ;
     return 42 ;
  }
  *valid = true ;
  return (force / mass) ;
}
```

8.1.4. *Introducing exceptions*

All of the techniques for reporting and propagating errors presented above have the common disadvantage of being explicitly the responsibility of the programmer. Many languages (Ada, Modula, ML, Java, C++, Python, etc.) use a different mechanism, known as *exceptions*. Exceptions are used to interrupt a computation in order to take account of an error or an abnormal configuration. When a running program encounters a situation of this kind, it raises an exception to warn that a valid result will not be supplied. If the program fails to handle the exception, its execution is immediately stopped.

Taking into account an exception consists of explicitly foreseeing in the program the case where such an exception would occur and writing the algorithm to handle this case. This is what is called *catching an exception* by an *exception handler*. Raising an exception at a point in the program transfers control to the closest handler for that exception in the program's call stack, as will be explained in example 8.2.

EXAMPLE 8.2.–

```
exception NullMass

accel (force, mass)
  if mass = 0 then raise NullMass
  return (force / mass)
f ()
  ...
  try {
    v = accel (...)
    ...
  }
  with x {
    if x = NullMass then error ("Error, mass is null.")
    else raise x
  }
```

In example 8.2, `raise` is the construct for raising an exception. If it is raised, the execution of `accel` is interrupted. The exception will propagate through the chain of function calls that led up to `accel`, until a dedicated handler is found. There is a handler (the `try ... with` construct) in the immediate calling function, `f`, which catches the exception and executes the prescribed treatment. If there had been no handler for `NullMass` in `f`, the exception would have propagated to the function that called `f`, and so on as long as no handler is found, possibly as far as the root function of the program, interrupting it as well, thus killing the execution of the global program. The handler in function `f` catches a possible exception, which is then bound to the variable `x`. If the caught exception is not the expected one, the program explicitly raises it again to ensure its detection. As we shall see, many languages automatically handle the raising of an exception if it is not one of the cases expected by the handler. We have chosen here a simplified construction of the handlers, without any loss of generality, in order to ease the presentation.

Exceptions provide a simple and flexible way of handling errors in software. If an exception is not explicitly handled by the call which raised it, it is automatically propagated upstream in the functions calls graph until a function which "knows" how to handle the exception is found. If there is no handler for this exception, then the execution of the program is interrupted, highlighting the existence of a configuration not foreseen in the algorithm. We can see that this mechanism contrasts radically with the one previously used to report errors in the traditional style, where the "impossible values" returned by functions must be encoded and propagated manually.

Exceptions can be "entities" of different kinds, depending on the language. Nevertheless, the same two functions – raising and catching exceptions – are always present.

8.2. A simple formalization: ML-style exceptions

In this section, we shall give a formal presentation of the syntax, the execution semantics and the exceptions typechecking for our functional language $\mathbf{Exp_2}$. This presentation is based on the model used by the ML family of languages. We deliberately consider only constant exceptions in order to simplify the presentation. We will briefly discuss the approaches used in certain other languages in section 8.3, highlighting the main differences between these approaches and the model presented here. These differences do not affect the fundamental principle of exceptions.

8.2.1. *Abstract syntax*

Let us extend the abstract syntax of $\mathbf{Exp_2}$ given in section 3.1.1. Table 8.1 presents three additional rules, added to those in Table 3.1, to define constructs relating to exceptions:

1) declaration of a constant exception by its name Exc;

2) raising an exception expression e;

3) catching an exception, binding the identified exception to the variable x.

$e ::=$ ... previous expressions		
|	*exception* Exc *in* e	Declaration of an exception
|	*raise* e	Raising an exception
|	*try* e_1 *with* $x \to e_2$	Catching an exception

Table 8.1. *Extension of the syntax of* $\mathbf{Exp}_2$ *to include exceptions*

When executing a `try/with` construct, if the evaluation of e_1 raises an exception Exc, then the result of the execution is the evaluation of e_2 in which x is bound to Exc.

8.2.2. *Values*

In $\mathbf{Exp}_2$, an exception is a constant value just like an integer in $\mathbb{Z}$, a boolean in $\mathbb{B}$ or a closure in $\mathbb{F}$. The fact that exceptions are values means that they can be results of evaluating arbitrary expressions, before – possibly – being raised. We must therefore differentiate between a result being an exception value and one being an explicitly raised exception. In the first case, no error is reported; the result is simply a value allowing a later error notification. In the second case, the result is an actual and characterized error.

In section 3.2.2, "error results" (denoted by $\mathbb{E}$) were "highlighted" errors and not values. In addition to the existing values, we must introduce values for non-raised exceptions. Let $\mathbb{E}$ be the set of exception values and $\widehat{\mathbb{E}}$ the set of raised exceptions. Thus, the set of values in our language is now defined by $\mathbf{V} = \mathbb{Z} \cup \mathbb{B} \cup \mathbb{F} \cup \mathbb{E}$. The set containing all possible results of the evaluation of an expression, that is either a value or a raised exception, is now defined as $\mathbb{V} = \mathbf{V} \cup \widehat{\mathbb{E}}$.

8.2.3. *Type algebra*

Since exceptions are values, they must have a type. We have chosen to name this type exn. It extends the type algebra presented in Table 5.1.

8.2.4. *Operational semantics*

Now, let us extend the rules of the big-step operational semantics using call by value presented in Box 3.3. Intuitively, evaluation results can now be represented by a sum type as follows:

```
type 'a result = Raised of exn | Value of 'a
```

To verify whether a result belongs to the "error" or "value" sets, it suffices to look at the result constructor.

When an exception is declared, its existence is recorded in the environment by binding its name to a value. For simplicity's sake, we shall consider that this value is simply the name of the exception.

In the languages of the ML family, the exceptions are only the constructors of a particular sum type that can be extended throughout the program. For this reason, exceptions can be parameterized using arguments, just like any other sum type constructor. For example, an exception declared as

```
exception SyntaxError of (string * int * int)
```

may indicate the presence of a syntax error by specifying the name of the file, the line number and the number of the character in the line where the propagated error was detected.

$$(E_{dexc})\ \frac{(Exc, \text{EXC}) \oplus \text{Env} \vdash e \rightsquigarrow v}{\text{Env} \vdash \text{exception } Exc \text{ in } e \rightsquigarrow v} \qquad (E_{raise})\ \frac{\text{Env} \vdash e \rightsquigarrow v}{\text{Env} \vdash \text{raise } e \rightsquigarrow \widehat{v}}(v \in \mathbb{E})$$

$$(E_{try})\ \frac{\text{Env} \vdash e_1 \rightsquigarrow v_1}{\text{Env} \vdash \text{try } e_1 \text{ with } x \rightarrow e_2 \rightsquigarrow v_1}$$

$$(E_{catch})\ \frac{\text{Env} \vdash e_1 \rightsquigarrow \widehat{v_1} \quad (x, v_1) \oplus \text{Env} \vdash e_2 \rightsquigarrow v_2}{\text{Env} \vdash \text{try } e_1 \text{ with } x \rightarrow e_2 \rightsquigarrow v_2}$$

Box 8.1. *Operational semantics of exceptions*

The rule E_{dexc} is used to declare an exception by binding its name to a new value of the same name in the environment. The rule E_{raise} describes the evaluation of an error raising expression. The result of this evaluation is not a value, but a raised expression belonging to $\widehat{\mathbb{E}}$. Rules E_{try} and E_{catch} are used for error catching. Rule E_{try} makes the hypothesis that the evaluation of the expression e_1 in the environment Env does not raise an exception, so the exception handler is not used and the returned value is the one of e_1. Rule E_{catch} makes the hypotheses that the evaluation of the expression e_1 in the environment Env raises an exception ($\widehat{v_1}$) and that this exception v_1 is caught by the handler. Thus, the handler expression e_2 is executed in Env extended by a binding of x to the intercepted exception value.

The environment in which the exception $\widehat{v_1}$ is raised for the first time, following the evaluation of an expression e required in order to evaluate e_1, is not recorded in this rule. As mentioned in section 8.1.4, if an exception is not explicitly handled by the call which raised it, then the exception mechanism is in charge to propagate it as a result to calls, figuring into the graph of function calls, until finding a function which handles its treatment. Note that this transmission of $\widehat{v_1}$ is described by the sub-tree of which the conclusion is the left premise of this rule E_{catch}; this sub-tree is built with the evaluation rules of function applications or operations seen in Chapter 3.

8.2.5. *Typing*

The typing rules presented in Box 5.1 can now be extended by three new rules, as shown in Box 8.2.

$$(T_{\text{catch}})\ \frac{(Exc, \forall\,\emptyset.\,\mathtt{exn}) \oplus \text{Env} \vdash e : \tau}{\text{Env} \vdash \text{exception } Exc \text{ in } e : \tau} \qquad (T_{\text{raise}})\ \frac{\alpha \text{ fresh}}{\text{Env} \vdash \text{raise } e : \alpha}$$

$$(T_{\text{try}})\ \frac{\text{Env} \vdash e_1 : \tau \quad (x, \forall\,\emptyset.\,\mathtt{exn}) \oplus \text{Env} \vdash e_2 : \tau}{\text{Env} \vdash \text{try } e_1 \text{ with } x \ \rightarrow\ e_2 : \tau}$$

Box 8.2. *Typing rules for exceptions in* $\mathbf{Exp_2}$

The rule T_{catch} simply extends the environment with the name of the exception (which is also its value) bound to the type exn of all the exceptions. The rule T_{raise} is used to type the raising of an exception. Since this construct raises an error instead of returning a value, it does not have, strictly speaking, a true type. The rule assigns it a polymorphic type denoted by α, which is a fresh type variable. This choice keeps the properties of the typing algorithm and allows an exception to be raised in any type context without type constraints. Thus, `if ... then 1 else raise Exc` is well-typed, as is `if ... then "Ok" else raise Exc`. Finally, the rule T_{try} requires the type of the result of the expression to be the same, whether or not the exception handler is used. In other words, whether or not the evaluation of e_1 ends in an error, the result of the expression must be of the same type.

Researchers have extended the type system of an ML language (OCaml) to include static detection of uncaught exceptions [PES 99]. The type algebra in this case is more complex, as it is annotated using exception expressions, and requires the implementation of cyclic types. To obtain more precise analytical results, the type inference requires the identifiers of recursive functions to be generalized in order to type their bodies (see section 5.3.3). A heuristic on the size of type terms allows the decidability of the inference process to be maintained.

8.3. Exceptions in other languages

In this section, we shall provide a brief overview of the way in which exceptions are handled in different languages. In C++, Python and Java, for example, exceptions are closely linked to the object mechanisms of the language. An exception is defined by a class inheriting from a base “exception” class. Objects in this class may be raised and caught by constructs similar to those presented in section 8.1.4. The option to create exceptions inheriting from other exceptions introduces subtle differences in handlers' ability to catch raised exceptions. A handler able to catch a given exception will also be able to catch all exceptions inheriting from this exception. Thus, the order of exception matching cases in a handler is significant in defining the case which effectively catches an exception.

8.3.1. *Exceptions in OCaml*

The `match/with` construct presented earlier is a simplified version of the construct used in OCaml. Instead of just one catching case, any number of cases may be used, as in a pattern matching. Thus, for a handler to catch different exceptions, instead of writing:

EXAMPLE 8.3.–

```
OCaml
try e with
| x -> (
  match x with
  | Ex1 (v) -> ...
  | Ex2 -> ...
)
```

we may write, more directly:

EXAMPLE 8.4.–

```
OCaml
try e with
| Ex1 (v) -> ...
| Ex2 -> ...
```

8.3.2. *Exceptions in Python*

In example 8.5, the class `MyExc` inherits directly from the `Exception` base class in Python and introduces an exception.

EXAMPLE 8.5.–

```
Python
class MyExc (Exception) :
  def what (self) : return ("MyExc")

class MyDerExc (MyExc) :
  def what (self) : return ("MyDerExc")
```

These two class definitions introduce two exceptions defined by the developer. `MyDerExc` inherits from `MyExc`. These two classes use a `what` method, returning a message associated with the expression that they represent.

Let us continue to examine example 8.5, defining a handler to catch the exceptions of `Exception` (after raising them).

```
Python
try : raise (MyExc)
except Exception as e : print (e.what ())

try : raise (MyDerExc)
except Exception as e : print (e.what ())
```

Since `MyExc` and `MyDerExc` are `Exceptions`, the handler catches them and displays the message bound to each exception:

```
MyExc
MyDerExc
```

Now, let us define a two-case handler with the same example; the first case catches `MyDerExc` and the second catches `MyExc`. We raise the exception `MyDerExc`.

```
Python
try :
  raise (MyDerExc)
except MyDerExc as e :
  print ("MyDerExc", e.what ())
except MyExc as e :
  print ("MyExc", e.what ())
```

The first case in the handler is able to catch `MyDerExc`; thus, this case is used, and the program displays:

```
MyDerExc MyDerExc
```

Now, let us invert the order of cases in the handler. Case `MyExc` is encountered first.

```
Python
try :
  raise (MyDerExc)
except MyExc as e :
  print ("MyExc", e.what ())
except MyDerExc as e :
  print ("MyDerExc", e.what ())
```

The program now displays:

```
MyExc MyDerExc
```

showing that the first case in the handler was used: this is the case used to catch `MyExc` but the printed value of `e.what()` is the name of the catched exception `MyDerExc`. The inheritance relation means that the exception `MyDerExc` can be caught by a handler of `MyExc`. Thus, the second case in the handler is redundant and will never be accessible.

8.3.3. *Exceptions in Java*

Writing the program from example 8.5 in Java results in a compilation error, and the program is rejected, as shown in example 8.6.

EXAMPLE 8.6.–

```
Java
class MyExc extends Exception {
  public String what () { return ("MyExc") ; }
} ;

class MyDerExc extends MyExc {
  public String what () { return ("MyDerExc") ; }
} ;

public class exn_class {
  public static void main (String[] args) {
    try { throw new MyExc () ; }
    catch (Exception e) { e.printStackTrace () ; }
    try { throw new MyDerExc () ; }
    catch (Exception e) { e.printStackTrace () ; }
    try { throw new MyDerExc () ; }
    catch (MyDerExc e) { System.out.println (e.what ()) ; }
    catch (MyExc e) { System.out.println (e.what ()) ; }
    try { throw new MyDerExc () ; }
    catch (MyExc e) { System.out.println (e.what ()) ; }
    catch (MyDerExc e) { System.out.println (e.what ()) ; }
  }
}
```

```
$ javac exn_class.java
exn_class.java:20: exception MyDerExc has already been caught
    catch (MyDerExc e) { System.out.println (e.what ()) ; }
    ^
1 error
```

The catch occurring in the last line of the source code has been recognized as redundant by the compiler, any exception inheriting from `Myexc` being catched by a handler of `Myexc`.

8.3.4. *Exceptions in C++*

In C++, using example 8.5, a warning is generated during compilation, but the resulting program can be executed and behaves in the same way as it did in Python, as we see from example 8.7.

EXAMPLE 8.7.–

```
C
#include <iostream>
#include <exception>
using namespace std ;

class MyExc : public exception {
public :
  virtual const char* what () const throw () { return ("MyExc") ; }
} ;

class MyDerExc : public MyExc {
public :
  virtual const char* what () const throw () { return ("MyDerExc") ; }
} ;

class MyExc exc ;
class MyDerExc dexc ;

int main () {
  try { throw exc ; }
  catch (exception& e) { cout << e.what () << '\n' ; }
  try { throw dexc ; }
  catch (exception& e) { cout << e.what () << '\n' ; }
  try { throw dexc ; }
  catch (MyDerExc& e) { cout << "MyDerExc " << e.what () << '\n' ; }
  catch (MyExc& e) { cout << "MyExc " << e.what () << '\n' ; }
  try { throw dexc ; }
  catch (MyExc& e) { cout << "MyExc " << e.what () << '\n' ; }
  catch (MyDerExc& e) { cout << "MyDerExc "<< e.what () << '\n' ; }
  return (0) ;
}
```

```
$ g++ exn_class.cpp
exn_class.cpp:28:10: warning: exception of type 'MyDerExc &' will be caught by
      earlier handler [-Wexceptions]
  catch (MyDerExc& e) {
         ^

exn_class.cpp:27:10: note: for type 'MyExc &'
  catch (MyExc& e) {
         ^

1 warning generated.
$ ./a.out
MyExc
MyDerExc
MyDerExc MyDerExc
```

```
MyExc MyDerExc
```

C++ has another special feature: exceptions are not necessarily class objects. Any value of any type can be raised as an exception, as shown in example 8.8.

EXAMPLE 8.8.–

```
C++
#include <iostream>
#include <exception>
using namespace std ;

struct pair_t { int left ; int right ; } ;

int main () {
  try { throw 42 ; }
  catch (int e) { cout << e << '\n' ; }
  try { throw "Catch me" ; }
  catch (char const *e) { cout << e << '\n' ; }
  try { throw 3.14159 ; }
  catch (double e) { cout << e << '\n' ; }
  try {
    struct pair_t pair = { 0, 1 } ;
    throw pair ;
  }
  catch (struct pair_t e) { cout << e.left << ", " << e.right << '\n' ; }
  return (0) ;
}
```

```
$ ./a.out
42
Catch me
3.14159
0, 1
```

The first handler catches an exception, which is an integer, and the second handler catches an exception, which is a string. It is important to note that the declared type of the exception is `char const *` and not simply `char *`. As the string `"Catch me"` is constant, it is implicitly qualified as `const` by the compiler. If this detail is omitted, the handler will not catch the exception. The third handler concerns an exception of type float. Once again, the constant `3.14159` is considered to be of type `double`, not *float*, by the compiler. The declared type of the exception must therefore be `double`, otherwise the exception will not be caught. The final handler shows that it is possible to raise values constructed from user data structures, and not only values of basic scalars, as exceptions.

Conclusion

The purpose of the two volumes of this book is to present the concepts, semantics and tools of the main sequential features of programming languages. In this first volume, after a brief tour of hardware and coding, we have considered functional and imperative features. These concepts were presented using a uniform approach, defining and comparing several different execution semantics for the two paradigms. We have also studied typing semantics and its applications, verification of program consistency, polymorphism and type inference, which are so important to lighten developers' work. The corresponding Chapters 3, 4 and 5 may seem difficult on first reading. They can be browsed first and then reworked as their content is referenced and used throughout the book. In addition, the OCaml and Python coding of the semantic notions is designed to facilitate their acquisition by reading and executing the codes provided.

We have also given a detailed presentation of data types and of automatic or explicit memory management, with examples mostly written in C and C++. We have highlighted certain characteristics of hardware components, which affect the semantics of these features and the program execution. The last chapter was devoted to exceptions, their semantics and their treatment in the languages OCaml, Java, C++ and Python.

Volume 2 is based on the foundations laid out in this volume. Starting with the analysis of certain needs of developers, we will develop step-by-step an original semantic model encompassing both modular and object features. This model is intended to allow a uniform point of view in the presentations of these constructs. However, for justified reasons, it does not have the rigor of the semantic models of this current volume. The modules of Ada and OCaml and the header files used in C will be studied as the first examples of this model. It will then be further developed to present the general mechanisms of object-oriented languages and will serve as a basis for a study of classes and packages in the languages Java, C++, OCaml and Python.

Appendix

Solutions to the Exercises

A.1. Introduction to language semantics

Solution to exercise 2.1

1) $\mathbf{X}^{\text{ref}}_{\text{Env}_0} = \{y, z\}$ and $\mathbf{X}^{\text{cst}}_{\text{Env}_0} = \{x\}$.

2) The evaluation of the sequence of definitions $\mathbf{d}$ = [var $z = 2$;let $x = 8$; var $y = 5$;let $x = 3$;] leads to the state etat_0 as follows:

$$([\,],[\,]) \xrightarrow{\text{var } z=2;}_{\mathbf{Def_1}} ([(z,r_z)],[(r_z,2)]) \xrightarrow{\text{let } x=8;}_{\mathbf{Def_1}} ([(x,8),(z,r_z)],[(r_z,2)])$$
$$\xrightarrow{\text{var } y=5;}_{\mathbf{Def_1}} ([(y,r_y),(x,8),(z,r_z)],[(r_y,5),(r_z,2)]) \xrightarrow{\text{let } x=3;}_{\mathbf{Def_1}} \text{etat}_0$$

There are several sequences of definitions, which lead to state etat_0. For example, in the sequence $\mathbf{d}$, the definition let $x = 8$ may be replaced by let $x = !z + 6$ since, in state $([(z,r_z)],[(r_z,2)])$, the expression $!z + 6$ is evaluated as 8.

3) Let us develop this expression as follows:

$$\begin{aligned}
&[\![(!y + !z) + x]\!]^{\text{Mem}_0}_{\text{Env}_0} \\
&= [\![!y + !z]\!]^{\text{Mem}_0}_{\text{Env}_0} + [\![x]\!]^{\text{Mem}_0}_{\text{Env}_0} = ([\![!y]\!]^{\text{Mem}_0}_{\text{Env}_0} + [\![!z]\!]^{\text{Mem}_0}_{\text{Env}_0}) + \text{Env}_0(x) \\
&= (\text{Mem}_0(\text{Env}_0(y)) + \text{Mem}_0(\text{Env}_0(z))) + 3 = (\text{Mem}_0(r_y) + \text{Mem}_0(r_z)) + 3 \\
&= (5 + 2) + 3 = 10
\end{aligned}$$

4) In state etat_0, the evaluation of the expression $!x$ triggers an error, as x is not a mutable variable in Env_0; the evaluation of $y + x$ triggers an error, as y is evaluated as a reference to which addition cannot be applied, and the evaluation of the identifier w triggers an error, as it is unbound in Env_0.

5) By applying the rules given in the statement, we obtain the following states and environments:

$$\begin{array}{l}\text{etat}_1 = ([(x,3),(y,r_y),(x,8),(z,r_z)],[(r_z,2),(r_y,3)])\\ \text{etat}_2 = ([(y,3),(x,3),(y,r_y),(x,8),(z,r_z)],[(r_z,2),(r_y,3)])\\ \mathbf{X}^{\text{ref}}_{\text{Env}_1} = \{y,z\} \quad \mathbf{X}^{\text{cst}}_{\text{Env}_1} = \{x\} \quad \mathbf{X}^{\text{ref}}_{\text{Env}_2} = \{z\} \quad \mathbf{X}^{\text{cst}}_{\text{Env}_2} = \{x,y\}\end{array}$$

6) Applying the rules in the statement, we obtain the following results:

$$\begin{array}{l}\text{etat}_3 = ([(x,r_y),(x,3),(y,r_y),(x,8),(z,r_z)],[(r_z,2),(r_y,5)])\\ \text{etat}_4 = ([(x,r_y),(x,3),(y,r_y),(x,8),(z,r_z)],[(r_z,2),(r_y,4)])\\ [\![!y]\!]^{\text{Mem}_0}_{\text{Env}_0} = 5 \neq 4 = [\![!y]\!]^{\text{Mem}_4}_{\text{Env}_4}\end{array}$$

Solution to exercise 2.2

1) Depending on the order in which expressions e_1 and e_2 are evaluated and the state in which these evaluations are carried out, there are several possible options for evaluating the expression $e_1 + e_2$:

– $[\![k]\!]^{\text{Mem}}_{\text{Env}} = (k, (\text{Env}, \text{Mem}))$ (for $k \in \mathbb{Z}$);

– $[\![x]\!]^{\text{Mem}}_{\text{Env}} = (\text{Env}(x), (\text{Env}, \text{Mem}))$ if $x \in \mathbf{X}$ and $x \in \text{dom}(\text{Env})$;

– $[\![x]\!]^{\text{Mem}}_{\text{Env}} = (\text{Err}, (\text{Env}, \text{Mem}))$ if $x \in \mathbf{X}$ and $x \notin \text{dom}(\text{Env})$;

– $[\![e_1 + e_2]\!]^{\text{Mem}}_{\text{Env}} = (v_1 + v_2, (\text{Env}_2, \text{Mem}_2))$ if $[\![e_1]\!]^{\text{Mem}}_{\text{Env}} = (v_1, (\text{Env}_1, \text{Mem}_1))$ and $[\![e_2]\!]^{\text{Mem}_1}_{\text{Env}_1} = (v_2, (\text{Env}_2, \text{Mem}_2))$ and $v_1, v_2 \in \mathbb{Z}$;

– $[\![e_1 + e_2]\!]^{\text{Mem}}_{\text{Env}} = (\text{Err}, (\text{Env}_2, \text{Mem}_2))$ if $[\![e_1]\!]^{\text{Mem}}_{\text{Env}} = (v_1, (\text{Env}_1, \text{Mem}_1))$ and $[\![e_2]\!]^{\text{Mem}_1}_{\text{Env}_1} = (v_2, (\text{Env}_2, \text{Mem}_2))$ and $v_1 \in \mathbb{Z}$ and $v_2 \notin \mathbb{Z}$;

– $[\![e_1 + e_2]\!]^{\text{Mem}}_{\text{Env}} = (\text{Err}, (\text{Env}_1, \text{Mem}_1))$ if $[\![e_1]\!]^{\text{Mem}}_{\text{Env}} = (v_1, (\text{Env}_1, \text{Mem}_1))$ and $v_1 \notin \mathbb{Z}$;

– $[\![!x]\!]^{\text{Mem}}_{\text{Env}} = (\text{Mem}(\text{Env}(x)), (\text{Env}, \text{Mem}))$ if $x \in \mathbf{X}^{\text{ref}}_{\text{E}}\text{nv}$;

– $[\![!x]\!]^{\text{Mem}}_{\text{Env}} = (\text{Err}, (\text{Env}, \text{Mem}))$ if $x \notin \mathbf{X}^{\text{ref}}_{\text{E}}\text{nv}$;

– $[\![x{+}{+}]\!]^{\text{Mem}}_{\text{Env}} = (\text{Mem}(\text{Env}(x)), (\text{Env}, \text{Mem}[\text{Env}(x) := \text{Mem}(\text{Env}(x)) + 1]))$ if $\text{Env}(x) \in \mathbf{R}$ and $\text{Mem}(\text{Env}(x)) \in \mathbb{Z}$;

– $[\![x{+}{+}]\!]^{\text{Mem}}_{\text{Env}} = (\text{Err}, (\text{Env}, \text{Mem}))$ if $\text{Env}(x) \notin \mathbf{R}$ or $\text{Mem}(\text{Env}(x)) \notin \mathbb{Z}$;

– $[\![{+}{+}x]\!]^{\text{Mem}}_{\text{Env}} = (\text{Mem}(\text{Env}(x)){+}1, (\text{Env}, \text{Mem}[\text{Env}(x) := \text{Mem}(\text{Env}(x)){+}1]))$ if $\text{Env}(x) \in \mathbf{R}$ and $\text{Mem}(\text{Env}(x)) \in \mathbb{Z}$;

– $[\![++x]\!]_{\text{Env}}^{\text{Mem}} = (\text{Err}, (\text{Env}, \text{Mem}))$ if $\text{Env}(x) \notin \mathbf{R}$ or $\text{Mem}(\text{Env}(x)) \notin \mathbb{Z}$.

2) We have $[\![!x]\!]_{\text{Env}}^{\text{Mem}} = (6, (\text{Env}, \text{Mem}))$, then $[\![x++]\!]_{\text{Env}}^{\text{Mem}} = (6, (\text{Env}, [(r_x, 7)]))$ hence $[\![!x + (x++)]\!]_{\text{Env}}^{\text{Mem}} = (12, (\text{Env}, [(r_x, 7)]))$. Furthermore, since $[\![!x]\!]_{\text{Env}}^{[(r_x,7)]} = (7, (\text{Env}, [(r_x, 7)]))$, then $[\![(x++) + !x]\!]_{\text{Env}}^{\text{Mem}} = (13, (\text{Env}, [(r_x, 7)]))$ and so $[\![!x + (x++)]\!]_{\text{Env}}^{\text{Mem}} \neq [\![(x++) + !x]\!]_{\text{Env}}^{\text{Mem}}$.

3) If x is not a mutable variable or if the value stored in the memory for x is not an integer, then:

$$[\![++x]\!]_{\text{Env}}^{\text{Mem}} = [\![(x++) + 1]\!]_{\text{Env}}^{\text{Mem}} = (\text{Err}, (\text{Env}, \text{Mem}))$$

Otherwise, by definition, we have:

$$[\![++x]\!]_{\text{Env}}^{\text{Mem}} = (\text{Mem}(\text{Env}(x)) + 1, (\text{Env}, \underbrace{\text{Mem}[\text{Env}(x) := \text{Mem}(\text{Env}(x)) + 1]}_{\text{Mem}_1}))$$

Furthermore, and also by definition, $[\![x++]\!]_{\text{Env}}^{\text{Mem}} = (\text{Mem}(\text{Env}(x)), (\text{Env}, \text{Mem}_1))$ and since $[\![1]\!]_{\text{Env}}^{\text{Mem}_1} = (1, (\text{Env}, \text{Mem}_1))$, we obtain:

$$[\![(x++) + 1]\!]_{\text{Env}}^{\text{Mem}} = (\text{Mem}(\text{Env}(x)) + 1, (\text{Env}, \text{Mem}_1))$$

and thus $[\![++x]\!]_{\text{Env}}^{\text{Mem}} = [\![(x++) + 1]\!]_{\text{Env}}^{\text{Mem}}$.

4) The extension of the relation is as follows:

$$(\text{Env}, \text{Mem}) \xrightarrow{x:=e}_{\mathbf{Lang}_1} (\text{Env}', \text{Mem}'[\text{Env}(x) := v])$$

if:

$$[\![e]\!]_{\text{Env}}^{\text{Mem}} = (v, (\text{Env}', \text{Mem}')),\ v \neq \text{Err and } x \in \mathbf{X}_{\text{Env}}^{\text{ref}}$$

$$(\text{Env}, \text{Mem}) \xrightarrow{x+:=e}_{\mathbf{Lang}_1} (\text{Env}', \text{Mem}'[\text{Env}(x) := v + \text{Mem}(\text{Env}(x))])$$

if:

$$[\![e]\!]_{\text{Env}}^{\text{Mem}} = (v, (\text{Env}', \text{Mem}')),\ v \in \mathbb{Z},\ \text{Mem}(\text{Env}(x)) \in \mathbb{Z} \text{ and } x \in \mathbf{X}_{\text{Env}}^{\text{ref}}$$

5) By definition

$$\begin{aligned} [\![++x]\!]_{\text{Env}_0}^{\text{Mem}_0} &= \begin{pmatrix} \text{Mem}_0(\text{Env}_0(x)) + 1, \\ (\text{Env}_0, \text{Mem}_0[\text{Env}_0(x) := \text{Mem}_0(\text{Env}_0(x)) + 1]) \end{pmatrix} \\ &= (3, \text{Env}_0, \underbrace{[(r_x, 3)]}_{\text{Mem}'_0}) \end{aligned}$$

and so:

$$(\mathrm{Env}_0, \mathrm{Mem}_0) \xrightarrow{x+:=++x}_{\mathbf{Lang}_1} (\mathrm{Env}_0, \mathrm{Mem}'_0[\mathrm{Env}_0(x) := 3 + \mathrm{Mem}_0(\mathrm{Env}_0(x))])$$

We therefore obtain $\mathrm{etat}_1 = ([(x, r_x)], \mathrm{Mem}'_0[r_x := 3{+}2]) = ([(x, r_x)], [(r_x, 5)])$.

6) Let $(\mathrm{Env}_0, \mathrm{Mem}_0)$ be an arbitrary state. If $x \notin \mathbf{X}^{\mathrm{ref}}_{\mathrm{Env}_0}$, or if the value stored in Mem_0 for x is not an integer, the three assignments lead to the same state $(\mathrm{Env}_0, \mathrm{Mem}_0)$. Otherwise, by definition:

$$\begin{aligned}
[\![x++]\!]^{\mathrm{Mem}_0}_{\mathrm{Env}_0} &= \begin{pmatrix} \mathrm{Mem}_0(\mathrm{Env}_0(x)), \\ (\mathrm{Env}_0, \underbrace{\mathrm{Mem}_0[\mathrm{Env}_0(x) := \mathrm{Mem}_0(\mathrm{Env}_0(x)) + 1]}_{\mathrm{Mem}_1}) \end{pmatrix} \\
[\![++x]\!]^{\mathrm{Mem}_0}_{\mathrm{Env}_0} &= \begin{pmatrix} \mathrm{Mem}_0(\mathrm{Env}_0(x)) + 1, \\ (\mathrm{Env}_0, \underbrace{\mathrm{Mem}_0[\mathrm{Env}_0(x) := \mathrm{Mem}_0(\mathrm{Env}_0(x)) + 1]}_{\mathrm{Mem}_2}) \end{pmatrix} \\
[\![1]\!]^{\mathrm{Mem}_0}_{\mathrm{Env}_0} &= (1, (\mathrm{Env}_0, \mathrm{Mem}_0))
\end{aligned}$$

and thus:

$$\begin{aligned}
(\mathrm{Env}_0, \mathrm{Mem}_0) &\xrightarrow{x:=x++}_{\mathbf{Lang}_1} (\mathrm{Env}_0, \underbrace{\mathrm{Mem}_1[\mathrm{Env}_0(x) := \mathrm{Mem}_0(\mathrm{Env}_0(x))]}_{=\mathrm{Mem}_0}) \\
(\mathrm{Env}_0, \mathrm{Mem}_0) &\xrightarrow{x:=++x}_{\mathbf{Lang}_1} (\mathrm{Env}_0, \underbrace{\mathrm{Mem}_2[\mathrm{Env}_0(x) := \mathrm{Mem}_0(\mathrm{Env}_0(x)) + 1]}_{=\mathrm{Mem}_2=\mathrm{Mem}_0[\mathrm{Env}_0(x):=1+\mathrm{Mem}_0(\mathrm{Env}_0(x))]}) \\
(\mathrm{Env}_0, \mathrm{Mem}_0) &\xrightarrow{x+:=1}_{\mathbf{Lang}_1} (\mathrm{Env}_0, \underbrace{\mathrm{Mem}_0[\mathrm{Env}_0(x) := 1 + \mathrm{Mem}_0(\mathrm{Env}_0(x))]}_{=\mathrm{Mem}_2})
\end{aligned}$$

The two assignments $x := ++x$ and $x+ := 1$ thus result in the same state when they are executed from the same state, but this state is different to that which would be reached by executing the assignment $x := x++$, which does not modify the state in which it is executed in any way.

A.2. Semantics of functional features

Solution to exercise 3.1

$$\begin{aligned}
[\![\mathsf{fun}\ x \mapsto \mathsf{let}\ y = 3\ \mathsf{in}\ x + y]\!]_{\mathrm{Env}} &= \langle x, \mathsf{let}\ y = 3\ \mathsf{in}\ x + y, \mathrm{Env}\rangle \\
[\![\mathsf{let}\ y = 3\ \mathsf{in}\ \mathsf{fun}\ x \mapsto x + y]\!]_{\mathrm{Env}} &= \langle x, x + y, (y, 3) \oplus \mathrm{Env}\rangle
\end{aligned}$$

Solution to exercise 3.2

1) We have $[\![\text{let } f = \text{fun } y \mapsto e \text{ in } (f\ x)]\!]_{\text{Env}} = [\![(f\ x)]\!]_{\text{Env}_1} = [\![e]\!]_{\text{Env}_2}$ and $[\![(f\ x)]\!]_{\text{Env}} = [\![e]\!]_{\text{Env}_3}$ where:

$$\begin{aligned}
\text{Env}_1 &= (f, \langle y, e, \text{Env}\rangle) \oplus \text{Env} \\
\text{Env}_2 &= (y, [\![x]\!]_{\text{Env}_1}) \oplus \text{Env} = (y, 4) \oplus \text{Env} \\
\text{Env}_3 &= (y, [\![x]\!]_{\text{Env}}) \oplus (x, 3) \oplus \text{Env}_0 = (y, 4) \oplus (x, 3) \oplus \text{Env}_0
\end{aligned}$$

We have $\text{Env}_2(y) = \text{Env}_3(y) = 4$ and $\text{Env}_2(x) = 4 \neq \text{Env}_3(x) = 3$. In order that $[\![e]\!]_{\text{Env}_2} = [\![e]\!]_{\text{Env}_3}$, the expression e must not contain any free occurrences of x (since f is not a recursive function, expression e will not contain any free occurrences of f).

2) When $e = y + x$, we have:

$$\begin{aligned}
&[\![\text{let } f = \text{fun } x \mapsto x + (f\ x) \text{ in } (f\ x)]\!]_{\text{Env}} \\
&= [\![(f\ x)]\!]_{\text{Env}_1} = [\![x + (f\ x)]\!]_{\text{Env}_2} = [\![x]\!]_{\text{Env}_2} + [\![(f\ x)]\!]_{\text{Env}_2} \\
&= [\![x]\!]_{\text{Env}_2} + [\![y + x]\!]_{\text{Env}_3} = [\![x]\!]_{\text{Env}_2} + ([\![y]\!]_{\text{Env}_3} + [\![x]\!]_{\text{Env}_3}) = 4 + (4 + 3) = 11
\end{aligned}$$

where:

$$\begin{aligned}
\text{Env}_1 &= (f, \langle x, x + (f\ x), \text{Env}\rangle) \oplus \text{Env} \\
\text{Env}_2 &= (x, [\![x]\!]_{\text{Env}_1}) \oplus \text{Env} = (x, 4) \oplus \text{Env} \\
\text{Env}_3 &= (y, [\![x]\!]_{\text{Env}_2}) \oplus (x, 3) \oplus \text{Env}_0 = (y, 4) \oplus (x, 3) \oplus \text{Env}_0
\end{aligned}$$

Solution to exercise 3.3

1) We note:

$$\text{let } x = 3 \text{ in } \underbrace{(\text{let } f = e_1 \text{ in } \underbrace{(\text{let } g = e_2 \text{ in } \underbrace{(\text{let } h = e_3 \text{ in } e_4)}_{e'_3})}_{e'_2})}_{e'_1}$$

Evaluation tree for e in environment Env_0:

$$\dfrac{(F_1)\dfrac{}{\text{Env}_0 \vdash 3 \rightsquigarrow 3} \quad (F_{13})\dfrac{(F_{15})\dfrac{}{\text{Env}_1 \vdash e_1 \rightsquigarrow F_f} \quad (F_{13})\dfrac{\nabla_1 \quad \nabla_2}{\underbrace{(f, F_f) \oplus \text{Env}_1}_{\text{Env}_2} \vdash e'_2 \rightsquigarrow v}}{\underbrace{(x, 3) \oplus \text{Env}_0}_{\text{Env}_1} \vdash e'_1 \rightsquigarrow v}}{\text{Env}_0 \vdash e \rightsquigarrow v}\ (F_{13})$$

where $F_f = \langle x, \mathsf{let}\ y = x\ \mathsf{in}\ y, \mathrm{Env}_1\rangle$ and ∇_1 and ∇_2 are the evaluation trees:

$$\underset{1}{\nabla}:\quad (F_{13})\dfrac{(F_{11})\dfrac{}{\mathrm{Env}_2 \vdash x \rightsquigarrow 3}\quad (F_{15})\dfrac{}{(y,3)\oplus \mathrm{Env}_2 \vdash \mathsf{fun}\ x \mapsto y \rightsquigarrow F_g}}{\mathrm{Env}_2 \vdash e_2 \rightsquigarrow F_g}$$

$$\underset{2}{\nabla}:\quad (F_{13})\dfrac{(F_{15})\dfrac{}{\mathrm{Env}_3 \vdash e_3 \rightsquigarrow F_h}\quad ()\dfrac{\nabla}{\underbrace{(h,F_h)\oplus \mathrm{Env}_3}_{\mathrm{Env}_4} \vdash e_4 \rightsquigarrow v}}{\underbrace{(g,F_g)\oplus \mathrm{Env}_2}_{\mathrm{Env}_3} \vdash e_3' \rightsquigarrow v}$$

where $F_g = \langle x, y, (y,3)\oplus \mathrm{Env}_2\rangle$ and $F_h = \langle k, \mathsf{fun}\ x \mapsto (k\,x), \mathrm{Env}_3\rangle$.

2) Evaluation tree for $e_4 = ((h\,f)\,2)$ in environment Env_4:

$$(F_{17})\dfrac{(F_{17})\dfrac{\underset{1}{\nabla}\ \underset{2}{\nabla}\ \underset{3}{\nabla}}{\mathrm{Env}_4 \vdash (h\,f) \rightsquigarrow F}\quad (F_1)\dfrac{}{\mathrm{Env}_4 \vdash 2 \rightsquigarrow 2}\quad (F_{17})\dfrac{\underset{4}{\nabla}\ \underset{5}{\nabla}\ \underset{6}{\nabla}}{\underbrace{(x,2)\oplus \mathrm{Env}_5}_{\mathrm{Env}_6} \vdash (k\,x) \rightsquigarrow 2}}{\mathrm{Env}_4 \vdash e_4 \rightsquigarrow 2}$$

where:

$$\nabla_1 : (F_{11})\dfrac{}{\mathrm{Env}_4 \vdash h \rightsquigarrow F_h} \qquad \nabla_2 : (F_{11})\dfrac{}{\mathrm{Env}_4 \vdash f \rightsquigarrow F_f}$$

$$\nabla_3 : (F_{15})\dfrac{}{\underbrace{(k,F_f)\oplus \mathrm{Env}_3}_{\mathrm{Env}_5} \vdash \mathsf{fun}\ x \mapsto (k\,x) \rightsquigarrow F}$$

with $F = \langle x, (k\,x), \mathrm{Env}_5\rangle$ and:

$$\nabla_4 : (F_{11})\dfrac{}{\mathrm{Env}_6 \vdash k \rightsquigarrow F_f} \qquad \nabla_5 : (F_{11})\dfrac{}{\mathrm{Env}_6 \vdash x \rightsquigarrow 2}$$

$$\nabla_6 : (F_{13})\dfrac{(F_{11})\dfrac{}{\mathrm{Env}_7 \vdash x \rightsquigarrow 2}\quad (F_{11})\dfrac{}{(y,2)\oplus \mathrm{Env}_7 \vdash y \rightsquigarrow 2}}{\underbrace{(x,2)\oplus \mathrm{Env}_1}_{\mathrm{Env}_7} \vdash \mathsf{let}\ y = x\ \mathsf{in}\ y \rightsquigarrow 2}$$

3) Evaluation tree for $e_4 = ((h\,g)\,2)$ in environment Env_4:

$$(F_{17})\dfrac{(F_{17})\dfrac{\underset{1}{\nabla}\ \underset{2}{\nabla}\ \underset{3}{\nabla}}{\mathrm{Env}_4 \vdash (h\,g) \rightsquigarrow F}\quad (F_1)\dfrac{}{\mathrm{Env}_4 \vdash 2 \rightsquigarrow 2}\quad (F_{17})\dfrac{\underset{4}{\nabla}\ \underset{5}{\nabla}\ \underset{6}{\nabla}}{\underbrace{(x,2)\oplus \mathrm{Env}_5}_{\mathrm{Env}_6} \vdash (k\,x) \rightsquigarrow 3}}{\mathrm{Env}_4 \vdash e_4 \rightsquigarrow 3}$$

where:

$$\nabla_1 : (F_{11})\dfrac{}{\mathrm{Env}_4 \vdash h \rightsquigarrow F_h} \qquad \nabla_2 : (F_{11})\dfrac{}{\mathrm{Env}_4 \vdash g \rightsquigarrow F_g}$$

$$\nabla_3 : (F_{15})\dfrac{}{\underbrace{(k, F_g) \oplus \mathrm{Env}_3}_{\mathrm{Env}_5} \vdash \mathsf{fun}\ x \mapsto (k\, x) \rightsquigarrow F}$$

with $F = \langle x, (k\, x), \mathrm{Env}_5 \rangle$ and:

$$\nabla_4 : (F_{11})\dfrac{}{\mathrm{Env}_6 \vdash k \rightsquigarrow F_g} \qquad \nabla_5 : (F_{11})\dfrac{}{\mathrm{Env}_6 \vdash x \rightsquigarrow 2}$$

$$\nabla_6 : (F_{11})\dfrac{}{\underbrace{(x, 2) \oplus (y, 3) \oplus \mathrm{Env}_2}_{\mathrm{Env}_7} \vdash y \rightsquigarrow 3}$$

Solution to exercise 3.4

Let e_1' be the subexpression $\mathsf{let}\ f = e_1\ \mathsf{in}\ e$.

1) Evaluation tree for e_0 in environment Env_0:

$$(F_{13}')\dfrac{(F_{13}')\dfrac{()\dfrac{\nabla}{\underbrace{(f, \lfloor e_1, \mathrm{Env}_1 \rfloor) \oplus \mathrm{Env}_1}_{\mathrm{Env}_2} \vdash e \rightsquigarrow v}}{\underbrace{(x, \lfloor 2, \mathrm{Env}_0 \rfloor) \oplus \mathrm{Env}_0}_{\mathrm{Env}_1} \vdash e_1' \rightsquigarrow v}}{\mathrm{Env}_0 \vdash e_0 \rightsquigarrow v}$$

2) Evaluation tree for e in environment Env_2:

$$(F_{13}')\dfrac{(F_{16}')\dfrac{(F_{11}')\dfrac{(F_{14}')\dfrac{}{\mathrm{Env}_1 \vdash e_1 \rightsquigarrow F_f}}{\mathrm{Env}_3 \vdash f \rightsquigarrow F_f} \quad (F_4)\dfrac{(F_{11}')\dfrac{(F_{11}')\dfrac{(F_1)\dfrac{}{\mathrm{Env}_2 \vdash 3 \rightsquigarrow 3}}{\mathrm{Env}_3 \vdash x \rightsquigarrow 3}}{\mathrm{Env}_4 \vdash y \rightsquigarrow 3} \quad (F_{11}')\dfrac{(F_1)\dfrac{}{\mathrm{Env}_0 \vdash 2 \rightsquigarrow 2}}{\mathrm{Env}_4 \vdash x \rightsquigarrow 2}}{\underbrace{(y, \lfloor x, \mathrm{Env}_3 \rfloor) \oplus \mathrm{Env}_1}_{\mathrm{Env}_4} \vdash y + x \rightsquigarrow 5}}{\underbrace{(x, \lfloor 3, \mathrm{Env}_2 \rfloor) \oplus \mathrm{Env}_2}_{\mathrm{Env}_3} \vdash (f\, x) \rightsquigarrow 5}}{\mathrm{Env}_2 \vdash e \rightsquigarrow 5}$$

with $F_f = \langle y, y + x, \mathrm{Env}_1 \rangle$.

3) Let e' be the expression $\mathsf{fun}\ y \mapsto y + x$. Evaluation tree for e in environment Env_2:

$$
(F'_{13})\dfrac{(F'_{16})\dfrac{(F'_{14})\dfrac{}{\mathrm{Env}_3 \vdash e' \rightsquigarrow \langle y, y + x, \mathrm{Env}_3\rangle} \quad (F_4)\dfrac{(F'_{11})\dfrac{(F'_{11})\dfrac{(F_1)\dfrac{}{\mathrm{Env}_2 \vdash 3 \rightsquigarrow 3}}{\mathrm{Env}_3 \vdash x \rightsquigarrow 3}}{\mathrm{Env}_4 \vdash y \rightsquigarrow 3} \quad (F'_{11})\dfrac{(F_1)\dfrac{}{\mathrm{Env}_2 \vdash 3 \rightsquigarrow 3}}{\mathrm{Env}_4 \vdash x \rightsquigarrow 3}}{\underbrace{(y, \lfloor x, \mathrm{Env}_3\rfloor) \oplus \mathrm{Env}_3}_{\mathrm{Env}_4} \vdash y + x \rightsquigarrow 6}}{\underbrace{(x, \lfloor 3, \mathrm{Env}_2\rfloor) \oplus \mathrm{Env}_2}_{\mathrm{Env}_3} \vdash (e'\, x) \rightsquigarrow 6}}{\mathrm{Env}_2 \vdash e \rightsquigarrow 6}
$$

4) We note:

$$
\mathsf{rec}\ g\ z = \underbrace{\mathsf{fun}\ k \mapsto \underbrace{\mathsf{if}\ z = 0\ \mathsf{then}\ \underbrace{(k\, x)}_{e_5}\ \mathsf{else}\ \underbrace{((g\,(z - 1))\, k)}_{e_6}}_{e_4}}_{e_3}
$$

Evaluation tree for e in environment Env_2:

$$
(F'_{13})\dfrac{(F'_{16})\dfrac{\overset{\nabla}{1} \quad \overset{\nabla}{2}}{\underbrace{(g, \lfloor e_2, \mathrm{Env}_2\rfloor) \oplus \mathrm{Env}_2}_{\mathrm{Env}_3} \vdash ((g\,(x - 1))\, f) \rightsquigarrow 4}}{\mathrm{Env}_2 \vdash e \rightsquigarrow 4}
$$

Construction of tree ∇_1:

$$
(F'_{16})\dfrac{(F'_{11})\dfrac{(F'_{15})\dfrac{}{\mathrm{Env}_2 \vdash e_2 \rightsquigarrow F_g}}{\mathrm{Env}_3 \vdash g \rightsquigarrow F_g} \quad (F'_{14})\dfrac{}{\underbrace{(z, \lfloor x - 1, \mathrm{Env}_3\rfloor) \oplus \mathrm{Env}_g}_{\mathrm{Env}_4} \vdash e_3 \rightsquigarrow F_k}}{\mathrm{Env}_3 \vdash (g\,(x - 1)) \rightsquigarrow F_k}
$$

with:

$$
F_g = \langle z, e_3, \underbrace{(g, \lfloor e_2, \mathrm{Env}_2\rfloor) \oplus \mathrm{Env}_2}_{\mathrm{Env}_g}\rangle
$$

and:

$$
F_k = \langle k, e_4, \mathrm{Env}_4\rangle
$$

Construction of tree ∇_2:

$$
(F_8)\dfrac{(F_4)\dfrac{(F'_{11})\dfrac{(F_4)\dfrac{(F'_{11})\dfrac{(F_1)\dfrac{}{\mathrm{Env}_0 \vdash 2 \rightsquigarrow 2}}{\mathrm{Env}_3 \vdash x \rightsquigarrow 2} \quad (F_1)\dfrac{}{\mathrm{Env}_3 \vdash 1 \rightsquigarrow 1}}{\mathrm{Env}_3 \vdash x - 1 \rightsquigarrow 1}}{\mathrm{Env}_5 \vdash z \rightsquigarrow 1} \quad (F_1)\dfrac{}{\mathrm{Env}_5 \vdash 0 \rightsquigarrow 0}}{\mathrm{Env}_5 \vdash z = 0 \rightsquigarrow \mathit{false}} \quad \nabla_3}{\underbrace{(k, \lfloor f, \mathrm{Env}_3 \rfloor) \oplus \mathrm{Env}_4}_{\mathrm{Env}_5} \vdash e_4 \rightsquigarrow 4}
$$

Construction of tree ∇_3:

$$
(F'_{16})\dfrac{(F'_{16})\dfrac{(F'_{11})\dfrac{(F'_{15})\dfrac{}{\mathrm{Env}_2 \vdash e_2 \rightsquigarrow F_g}}{\mathrm{Env}_5 \vdash g \rightsquigarrow F_g} \quad (F'_{14})\dfrac{}{\underbrace{(z, \lfloor z - 1, \mathrm{Env}_5 \rfloor) \oplus \mathrm{Env}_g}_{\mathrm{Env}_6} \vdash e_3 \rightsquigarrow F'_k}}{\mathrm{Env}_5 \vdash (g\,(z - 1)) \rightsquigarrow F'_k} \quad \nabla_4}{\mathrm{Env}_5 \vdash e_6 \rightsquigarrow 4}
$$

with $F'_k = \langle k, e_4, \mathrm{Env}_6 \rangle$.

Construction of tree ∇_4:

$$
(F_7)\dfrac{\nabla_5 \quad (F'_{16})\dfrac{(F'_{11})\dfrac{(F'_{11})\dfrac{(F'_{11})\dfrac{(F'_{14})\dfrac{}{\mathrm{Env}_1 \vdash e_1 \rightsquigarrow \underbrace{\langle y, y + x, \mathrm{Env}_1 \rangle}_{F_p}}}{\mathrm{Env}_3 \vdash f \rightsquigarrow F_p}}{\mathrm{Env}_5 \vdash k \rightsquigarrow F_p}}{\mathrm{Env}_7 \vdash k \rightsquigarrow F_p} \quad \nabla_6}{\mathrm{Env}_7 \vdash (k\,x) \rightsquigarrow 4}}{\underbrace{(k, \lfloor k, \mathrm{Env}_5 \rfloor) \oplus \mathrm{Env}_6}_{\mathrm{Env}_7} \vdash e_4 \rightsquigarrow 4}
$$

Construction of tree ∇_5:

$$
(F_4)\dfrac{(F'_{11})\dfrac{(F_4)\dfrac{(F'_{11})\dfrac{(F_4)\dfrac{(F'_{11})\dfrac{(F_1)\dfrac{}{\mathrm{Env}_0 \vdash 2 \rightsquigarrow 2}}{\mathrm{Env}_3 \vdash x \rightsquigarrow 2} \quad (F_1)\dfrac{}{\mathrm{Env}_3 \vdash 1 \rightsquigarrow 1}}{\mathrm{Env}_3 \vdash x - 1 \rightsquigarrow 1}}{\mathrm{Env}_5 \vdash z \rightsquigarrow 1} \quad (F_1)\dfrac{}{\mathrm{Env}_5 \vdash 1 \rightsquigarrow 1}}{\mathrm{Env}_5 \vdash z - 1 \rightsquigarrow 0}}{\mathrm{Env}_7 \vdash z \rightsquigarrow 0} \quad (F_1)\dfrac{}{\mathrm{Env}_7 \vdash 0 \rightsquigarrow 0}}{\mathrm{Env}_7 \vdash z = 0 \rightsquigarrow \mathit{true}}
$$

Construction of tree ∇_6:

$$
(F_4)\dfrac{(F'_{11})\dfrac{(F'_{11})\dfrac{(F_1)\dfrac{}{\text{Env}_0 \vdash 2 \rightsquigarrow 2}}{\text{Env}_7 \vdash x \rightsquigarrow 2}}{\text{Env}_8 \vdash y \rightsquigarrow 2} \quad (F'_{11})\dfrac{(F_1)\dfrac{}{\text{Env}_0 \vdash 2 \rightsquigarrow 2}}{\text{Env}_8 \vdash x \rightsquigarrow 2}}{\underbrace{(y, \lfloor x, \text{Env}_7 \rfloor) \oplus \text{Env}_1}_{\text{Env}_8} \vdash y + x \rightsquigarrow 4}
$$

Solution to exercise 3.5

We proceed by induction over a (proving the implication in only one direction, since the relation $=_X$ is clearly symmetrical).

– If $a = k \in \mathbb{Z}$, then $\mathcal{F}(a) = \emptyset$, the evaluation tree for $\text{Env}_1 \vdash k \rightsquigarrow n$ must have been obtained using rule F_1 with $k = n$ and, using the same rule, we can clearly construct an evaluation tree for $\text{Env}_2 \vdash k \rightsquigarrow n$.

– If $a = x \in \mathbf{X}$, and thus $\mathcal{F}(a) = \{x\}$, the evaluation tree for $\text{Env}_1 \vdash x \rightsquigarrow n$ must have been obtained using rule F_{11} with $\text{Env}_1(x) = n$. Since, by hypothesis, $\text{Env}_1 =_{\mathcal{F}(a)} \text{Env}_2$, we obtain $\text{Env}_2(x) = n$ and, by applying rule F_{11}, we obtain an evaluation tree for $\text{Env}_2 \vdash x \rightsquigarrow n$.

– If $a = a_1 \,\text{op}\, a_2$ ($\text{op} \in \{+, -, *\}$), the evaluation tree for $\text{Env}_1 \vdash a_1 \,\text{op}\, a_2 \rightsquigarrow n$ must have been obtained using rule F_4:

$$
(F_4)\dfrac{(F_i)\dfrac{\vdots}{\text{Env}_1 \vdash a_1 \rightsquigarrow n_1} \quad (F_j)\dfrac{\vdots}{\text{Env}_1 \vdash a_2 \rightsquigarrow n_2}}{\text{Env}_1 \vdash a_1 \,\text{op}\, a_2 \rightsquigarrow n}
$$

with $n = n_1 \,\text{op}_{\mathbf{V}}\, n_2$. Since $\mathcal{F}(a_1) \subseteq \mathcal{F}(a)$ and $\mathcal{F}(a_2) \subseteq \mathcal{F}(a)$ (giving us $\text{Env}_1 =_{\mathcal{F}(a_1)} \text{Env}_2$ and $\text{Env}_1 =_{\mathcal{F}(a_2)} \text{Env}_2$), by induction hypothesis, there exists an evaluation tree for $\text{Env}_2 \vdash a_1 \rightsquigarrow n_1$ and for $\text{Env}_2 \vdash a_2 \rightsquigarrow n_2$. From these two trees, we can then build an evaluation tree for $\text{Env}_2 \vdash a_1 \,\text{op}\, a_2 \rightsquigarrow n$ by applying rule F_4.

– If $a =$ ($\mathsf{let}\ x = a_1\ \mathsf{in}\ a_2$), the evaluation tree for $\text{Env}_1 \vdash \mathsf{let}\ x = a_1\ \mathsf{in}\ a_2 \rightsquigarrow n$, must have been obtained using rule F_{13}:

$$
(F_{13})\dfrac{(F_i)\dfrac{\vdots}{\text{Env}_1 \vdash a_1 \rightsquigarrow n_1} \quad (F_j)\dfrac{\vdots}{(x, n_1) \oplus \text{Env}_1 \vdash a_2 \rightsquigarrow n_2}}{\text{Env}_1 \vdash \mathsf{let}\ x = a_1\ \mathsf{in}\ a_2 \rightsquigarrow n_2}
$$

Since, by definition, $\mathcal{F}(a) = \mathcal{F}(a_1) \cup (\mathcal{F}(a_2)\backslash\{x\})$, we have $\mathcal{F}(a_1) \subseteq \mathcal{F}(a)$ and thus $\text{Env}_1 =_{\mathcal{F}(a_1)} \text{Env}_2$, which allows us, by induction hypothesis over a_1, to obtain the evaluation tree for $\text{Env}_2 \vdash a_1 \rightsquigarrow n_1$. Furthermore, since $\text{Env}_1 =_{\mathcal{F}(a)} \text{Env}_2$, we obtain $(x, n_1) \oplus \text{Env}_1 =_{\{x\}\cup\mathcal{F}(a)} (x, n_1) \oplus \text{Env}_2$ and since $\mathcal{F}(a_2) \subseteq \mathcal{F}(a) \cup \{x\}$, we have $(x, n_1) \oplus \text{Env}_1 =_{\mathcal{F}(a_2)} (x, n_1) \oplus \text{Env}_2$, and the induction hypothesis can be used over a_2 to obtain an evaluation tree for $(x, n_1) \oplus \text{Env}_2 \vdash a_2 \rightsquigarrow n_2$. Applying rule F_{13} from the two evaluation trees constructed in this way, we obtain an evaluation tree for $\text{Env}_2 \vdash \textsf{let}\ x = a_1\ \textsf{in}\ a_2 \rightsquigarrow n_2$, concluding our proof.

Solution to exercise 3.6

We proceed by induction over e.

– If $e = k \in \mathbb{Z}$, then $[\![k[x \leftarrow e']]\!]_{\text{Env}} = [\![k]\!]_{\text{Env}} = k = [\![k]\!]_{(x,[\![e']\!]_{\text{Env}})\oplus\text{Env}}$

– If $e = z \in \mathbf{X}$, then if $z = x$ we obtain:

$$[\![x[x \leftarrow e']]\!]_{\text{Env}} = [\![e']\!]_{\text{Env}} = [\![x]\!]_{(x,[\![e']\!]_{\text{Env}})\oplus\text{Env}}$$

otherwise, if $z \neq x$ we obtain

$$[\![z[x \leftarrow e']]\!]_{\text{Env}} = [\![z]\!]_{\text{Env}} = \text{Env}(z) = [\![z]\!]_{(x,[\![e']\!]_{\text{Env}})\oplus\text{Env}}$$

– If $e = e_1 \,\text{op}\, e_2$, then:

$$\begin{array}{ll} [\![(e_1 \,\text{op}\, e_2)[x \leftarrow e']]\!]_{\text{Env}} & \\ = [\![e_1[x \leftarrow e'] \,\text{op}\, e_2[x \leftarrow e']]\!]_{\text{Env}} & \text{(substitution)} \\ = [\![e_1[x \leftarrow e']]\!]_{\text{Env}} \,\text{op}_{\mathbf{V}}\, [\![e_2[x \leftarrow e']]\!]_{\text{Env}} & \text{(evaluation function)} \\ = [\![e_1]\!]_{(x,[\![e']\!]_{\text{Env}})\oplus\text{Env}} \,\text{op}_{\mathbf{V}}\, [\![e_2]\!]_{(x,[\![e']\!]_{\text{Env}})\oplus\text{Env}} & \text{(induction hypothesis)} \\ = [\![e_1 \,\text{op}\, e_2]\!]_{(x,[\![e']\!]_{\text{Env}})\oplus\text{Env}} & \text{(evaluation function)} \end{array}$$

Solution to exercise 3.7

1) The transition sequence is as follows:

$$\begin{array}{l} \langle (z + (x - y)) * (x + 7), \text{Env}\rangle \hookrightarrow \langle (2 + (x - y)) * (x + 7), \text{Env}\rangle \\ \hookrightarrow \langle (2 + (1 - y)) * (x + 7), \text{Env}\rangle \hookrightarrow \langle (2 + (1 - 3)) * (x + 7), \text{Env}\rangle \\ \hookrightarrow \langle (2 + -2) * (x + 7), \text{Env}\rangle \hookrightarrow \langle 0 * (x + 7), \text{Env}\rangle \hookrightarrow \langle 0 * (1 + 7), \text{Env}\rangle \\ \hookrightarrow \langle 0 * 8, \text{Env}\rangle \hookrightarrow \langle 0, \text{Env}\rangle \end{array}$$

2) a) We consider the lexicographic order $\preceq$ over $\mathbb{N} \times \mathbb{N}$, defined from the order relation $\leq$ over $\mathbb{N}$:

$$(n_1, n_2) \preceq (n'_1, n'_2) \text{ if } (n_1 < n'_1) \text{ or } (n_1 = n'_1 \text{ and } n_2 \leq n'_2)$$

b) Let $P(\langle e, \text{Env}\rangle \hookrightarrow \langle e', \text{Env}\rangle)$ be the property:

$$(\mathcal{N}_V(e'), \mathcal{N}_O(e')) \prec (\mathcal{N}_V(e), \mathcal{N}_O(e))$$

and let us show, by induction, that this property is verified considering each rule in the inference system presented in Box 3.5.

– Rule AS_1: Since $\text{Env}(x) \in \mathbb{Z}$, we have $\mathcal{N}_V(x) = 1$ and $\mathcal{N}_V(\text{Env}(x)) = 0$ which concludes our proof.

– Rule AS_2: On the one hand, $\mathcal{N}_V(n_1 + n_2) = \mathcal{N}_V(n) = 0$ (since $n, n_1, n_2 \in \mathbb{Z}$) and, on the other hand, $\mathcal{N}_O(n) = \mathcal{N}_O(n_1 + n_2) - 1$, which concludes our proof.

– Rules AS_3 and AS_4: Similar reasoning to that used for rule AS_2.

– Rule AS_5: By induction hypothesis, we have:

$$(\mathcal{N}_V(e_1'), \mathcal{N}_O(e_1')) \prec (\mathcal{N}_V(e_1), \mathcal{N}_O(e_1))$$

Additionally, either $\mathcal{N}_V(e_1') < \mathcal{N}_V(e_1)$ and we obtain:

$$\mathcal{N}_V(e_1' + e_2) = \mathcal{N}_V(e_1') + \mathcal{N}_V(e_2) < \mathcal{N}_V(e_1) + \mathcal{N}_V(e_2) = \mathcal{N}_V(e_1 + e_2)$$

which concludes our proof, or $\mathcal{N}_V(e_1') = \mathcal{N}_V(e_1)$ and $\mathcal{N}_O(e_1') < \mathcal{N}_O(e_1)$ and we obtain:

$$\begin{aligned}\mathcal{N}_V(e_1' + e_2) &= \mathcal{N}_V(e_1') + \mathcal{N}_V(e_2) = \mathcal{N}_V(e_1) + \mathcal{N}_V(e_2) = \mathcal{N}_V(e_1 + e_2) \text{ and}\\ \mathcal{N}_O(e_1' + e_2) &= \mathcal{N}_O(e_1') + \mathcal{N}_O(e_2) + 1 < \mathcal{N}_O(e_1) + \mathcal{N}_O(e_2) + 1\\ &= \mathcal{N}_O(e_1 + e_2)\end{aligned}$$

which also brings us to our conclusion.

– Rules AS_7 and AS_9: Similar reasoning to that used for rule AS_5.

– Rule AS_6: By induction hypothesis, we have:

$$(\mathcal{N}_V(e_2'), \mathcal{N}_O(e_2')) \prec (\mathcal{N}_V(e_2), \mathcal{N}_O(e_2))$$

Additionally, either $\mathcal{N}_V(e_2') < \mathcal{N}_V(e_2)$ and we obtain:

$$\mathcal{N}_V(n_1 + e_2') = \mathcal{N}_V(e_2') < \mathcal{N}_V(e_2) = \mathcal{N}_V(n_1 + e_2)$$

which concludes our proof, or $\mathcal{N}_V(e_2') = \mathcal{N}_V(e_2)$ and $\mathcal{N}_O(e_2') < \mathcal{N}_O(e_2)$ and we obtain:

$$\begin{aligned}\mathcal{N}_V(n_1 + e_2') &= \mathcal{N}_V(e_2') = \mathcal{N}_V(e_2) = \mathcal{N}_V(n_1 + e_2)\\ \mathcal{N}_O(n_1 + e_2') &= \mathcal{N}_O(e_2') + 1 < \mathcal{N}_O(e_2) + 1 = \mathcal{N}_O(n_1 + e_2)\end{aligned}$$

which also brings us to our conclusion.

– Rules AS_8 and AS_{10}: Similar reasoning to that used for rule AS_6.

c) For each step $\langle e_i, \text{Env}\rangle \hookrightarrow \langle e_{i+1}, \text{Env}\rangle$, we have:

$$(\mathcal{N}_V(e_{i+1}), \mathcal{N}_O(e_{i+1})) \prec (\mathcal{N}_V(e_i), \mathcal{N}_O(e_i))$$

and since $\preceq$ is a well-founded order relation over $\mathbb{N}\times\mathbb{N}$, there is no strictly decreasing infinite sequence:

$$(n_0, n_0') \succ (n_1, n_1') \succ (n_2, n_2') \succ \cdots$$

so we can conclude that any transition sequence is of finite length.

d) This property is no longer verified if we add rule AS_{11} since:

$$(\mathcal{N}_V(e_1 + e_2), \mathcal{N}_O(e_1 + e_2)) = (\mathcal{N}_V(e_2 + e_1), \mathcal{N}_O(e_2 + e_1))$$

For example, this would give us the following infinite transition sequence:

$$\langle e_1 + e_2, \text{Env}\rangle \hookrightarrow \langle e_2 + e_1, \text{Env}\rangle \hookrightarrow \langle e_1 + e_2, \text{Env}\rangle \hookrightarrow \langle e_2 + e_1, \text{Env}\rangle \hookrightarrow \cdots$$

3) We prove that there is a transition sequence $\langle e, \text{Env}\rangle \overset{*}{\hookrightarrow} n$ by induction over the evaluation tree for $\text{Env} \vdash e \rightsquigarrow n$.

– If the evaluation tree for $\text{Env} \vdash e \rightsquigarrow n$ has been obtained using rule F_1, then $e = n \in \mathbb{Z}$ and we clearly have $\langle n, \text{Env}\rangle \overset{*}{\hookrightarrow} n$.

– If the evaluation tree for $\text{Env} \vdash e \rightsquigarrow n$ has been obtained using rule F_{11}, then $e = x \in \mathbf{X}$ and $n = \text{Env}(x) \in \mathbb{Z}$. Rule AS_1 can then be used to construct a tree for $\langle x, \text{Env}\rangle \hookrightarrow \langle \text{Env}(x), \text{Env}\rangle$ and we can establish that $\langle x, \text{Env}\rangle \overset{*}{\hookrightarrow} \text{Env}(x)$.

– If the evaluation tree for $\text{Env} \vdash e \rightsquigarrow n$ has been obtained using rule F_4 and the binary operator $+$:

$$(F_4)\dfrac{(F_i)\dfrac{\vdots}{\text{Env} \vdash e_1 \rightsquigarrow n_1} \quad (F_j)\dfrac{\vdots}{\text{Env} \vdash e_2 \rightsquigarrow n_2}}{\text{Env} \vdash e_1 + e_2 \rightsquigarrow n_1 + n_2}$$

then, by induction hypothesis, we have:

$$\langle e_1, \text{Env}\rangle \overset{*}{\hookrightarrow} \text{Env}(n_1) \text{ and } \langle e_2, \text{Env}\rangle \overset{*}{\hookrightarrow} \text{Env}(n_2)$$

and there exist two transition sequences:

$$\begin{array}{l}\langle e_1, \text{Env}\rangle = \langle e_1^0, \text{Env}\rangle \hookrightarrow \langle e_1^1, \text{Env}\rangle \hookrightarrow \langle e_1^2, \text{Env}\rangle \hookrightarrow \cdots \\ \cdots \hookrightarrow \langle e_1^{k_1}, \text{Env}\rangle = \langle n_1, \text{Env}\rangle\end{array} \qquad [A.1]$$

$$\begin{array}{l}\langle e_2, \text{Env}\rangle = \langle e_2^0, \text{Env}\rangle \hookrightarrow \langle e_2^1, \text{Env}\rangle \hookrightarrow \langle e_2^2, \text{Env}\rangle \hookrightarrow \cdots \\ \cdots \hookrightarrow \langle e_2^{k_2}, \text{Env}\rangle = \langle n_2, \text{Env}\rangle\end{array} \qquad [A.2]$$

Rule AS_5 allows us to transform each transition $\langle e_1^i, \text{Env}\rangle \hookrightarrow \langle e_1^{i+1}, \text{Env}\rangle$ in sequence [A.1] into a transition $\langle e_1^i + e_2, \text{Env}\rangle \hookrightarrow \langle e_1^{i+1} + e_2, \text{Env}\rangle$. We can thus obtain the following transition sequence:

$$\begin{array}{l}\langle e_1 + e_2, \text{Env}\rangle = \langle e_1^0 + e_2, \text{Env}\rangle \hookrightarrow \langle e_1^1 + e_2, \text{Env}\rangle \hookrightarrow \langle e_1^2 + e_2, \text{Env}\rangle \hookrightarrow \cdots \\ \cdots \hookrightarrow \langle e_1^{k_1} + e_2, \text{Env}\rangle = \langle n_1 + a_2, \text{Env}\rangle\end{array} \qquad [A.3]$$

Rule AS_6 allows us to transform each transition $\langle e_2^i, \text{Env}\rangle \hookrightarrow \langle e_2^{i+1}, \text{Env}\rangle$ in sequence [A.2] into a transition $\langle n_1 + e_2^i, \text{Env}\rangle \hookrightarrow \langle n_1 + e_2^{i+1}, \text{Env}\rangle$. We thus obtain the sequence:

$$\begin{array}{l}\langle n_1 + e_2, \text{Env}\rangle = \langle n_1 + e_2^0, \text{Env}\rangle \hookrightarrow \langle n_1 + e_2^1, \text{Env}\rangle \hookrightarrow \langle n_1 + e_2^2, \text{Env}\rangle \hookrightarrow \cdots \\ \cdots \hookrightarrow \langle n_1 + e_2^{k_2}, \text{Env}\rangle = \langle n_1 + n_2, \text{Env}\rangle\end{array} \qquad [A.4]$$

Finally, rule AS_2 allows us to consider the transition $\langle n_1 + n_2, \text{Env}\rangle \hookrightarrow \langle n, \text{Env}\rangle$ where n is the value of $n_1 + n_2$, and by concatenating sequences [A.3] and [A.4] with this transition, we finally obtain $\langle e_1 + e_2, \text{Env}\rangle \overset{*}{\hookrightarrow} n_1 + n_2$.

– For the binary operators $*$ and $-$, a similar reasoning to that shown above is used.

4) We proceed by induction over expression e_1.

– If $e_1 = n \in \mathbb{Z}$, then hypothesis $\langle e_1, \text{Env}\rangle \hookrightarrow \langle e_2, \text{Env}\rangle$ results in a contradiction since $\langle e_1, \text{Env}\rangle$ is a terminal configuration.

– If $e_1 = x \in \mathbf{X}$, then we must have $e_2 = \text{Env}(x) \in \mathbb{Z}$ and $v = \text{Env}(x)$. It is thus possible to conclude our proof by constructing the tree:

$$(F_{11})\frac{}{\text{Env} \vdash x \rightsquigarrow v}$$

– If $e_1 = a_1 \,\text{op}\, a_2$ (with op $\in \{+, *, -\}$), three cases are possible.

Case 1: If $a_1, a_2 \in \mathbb{Z}$, then $e_2 = v$ with $v = a_1 \,\text{op}_{\mathbf{V}}\, a_2 \in \mathbb{Z}$ and we can conclude our proof by constructing the tree:

$$(F_4)\frac{(F_1)\dfrac{}{\text{Env} \vdash a_1 \rightsquigarrow a_1} \qquad (F_1)\dfrac{}{\text{Env} \vdash a_2 \rightsquigarrow a_2}}{\text{Env} \vdash a_1 \,\text{op}\, a_2 \rightsquigarrow v}$$

Case 2: If $a_1 \in E_A \backslash \mathbb{Z}$, then $e_2 = a'_1 \,\mathrm{op}\, a_2$ with $\langle a_1, \mathrm{Env}\rangle \hookrightarrow \langle a'_1, \mathrm{Env}\rangle$. The evaluation tree for $\mathrm{Env} \vdash a'_1 \,\mathrm{op}\, a_2 \rightsquigarrow v$ must contain the subtrees:

$$(F_j)\frac{\vdots}{\mathrm{Env} \vdash a'_1 \rightsquigarrow v'_1} \qquad (F_k)\frac{\vdots}{\mathrm{Env} \vdash a_2 \rightsquigarrow v'_2}$$

and thus, by induction hypothesis, we obtain $\mathrm{Env} \vdash a_1 \rightsquigarrow v'_1$ and, from the evaluation tree for $\mathrm{Env} \vdash a_2 \rightsquigarrow v'_2$, rule F_4 allows us to obtain the evaluation tree for $\mathrm{Env} \vdash a_1 \,\mathrm{op}\, a_2 \rightsquigarrow v$.

Case 3: If $a_1 \in \mathbb{Z}$ and $a_2 \in E_A \backslash \mathbb{Z}$, then a similar reasoning to that used in the previous case brings us to our conclusion.

5) Since $\langle e, \mathrm{Env}\rangle \overset{\star}{\hookrightarrow} v$, there exists a transition sequence:

$$\langle e, \mathrm{Env}\rangle = \langle e^0, \mathrm{Env}\rangle \hookrightarrow \langle e^1, \mathrm{Env}\rangle \hookrightarrow \cdots \hookrightarrow \langle e^k, \mathrm{Env}\rangle$$

with $e^k = v \in \mathbb{Z}$. The proof is obtained by induction over the length k of this sequence. Let us prove this property for $k = 0$, for $k = 1$ and for $k > 1$, which we assume to be true for $k - 1$ and $k - 2$.

– If $k = 0$, then $e = v \in \mathbb{Z}$ and rule F_1 concludes our proof. Else, if $k = k_0 + 1$, then we proceed by case over the first transition in the sequence.

– If $k = 1$, then two cases are possible.

Case 1: If the transition has been obtained using rule AS_1: $\langle x, \mathrm{Env}\rangle \hookrightarrow \langle \mathrm{Env}(x), \mathrm{Env}\rangle$, then $e = x$ and $v = \mathrm{Env}(x)$ and rule F_{11} concludes our proof.

Case 2: If the transition has been obtained using rule AS_2, AS_3 or AS_4:

$$\langle n_1 \,\mathrm{op}\, n_2, \mathrm{Env}\rangle \hookrightarrow \langle n, \mathrm{Env}\rangle$$

with $\mathrm{op} \in \{+, -, *\}$, then $n = n_1 \,\mathrm{op}_{\mathbf{V}}\, n_2$ and rule F_4 concludes our proof:

$$(F_4)\frac{(F_1)\dfrac{}{\mathrm{Env} \vdash n_1 \rightsquigarrow n_1} \qquad (F_1)\dfrac{}{\mathrm{Env} \vdash n_2 \rightsquigarrow n_2}}{\mathrm{Env} \vdash n_1 \,\mathrm{op}\, n_2 \rightsquigarrow n}$$

– If $k \geq 2$, then two cases are possible.

Case 1: If $\langle e_1 \,\mathrm{op}\, e_2, \mathrm{Env}\rangle \hookrightarrow \langle e'_1 \,\mathrm{op}\, e_2, \mathrm{Env}\rangle \hookrightarrow \cdots \hookrightarrow \langle e^k, \mathrm{Env}\rangle$ is the considered sequence, with $\langle e_1, \mathrm{Env}\rangle \hookrightarrow \langle e'_1, \mathrm{Env}\rangle$ (the first transition is obtained using rule AS_5, AS_7 or AS_9), then, by induction hypothesis over a transition sequence of length $k - 1$, we have $\mathrm{Env} \vdash e'_1 \,\mathrm{op}\, e_2 \rightsquigarrow v$ and, from the previous question, we obtain $\mathrm{Env} \vdash e_1 \,\mathrm{op}\, e_2 \rightsquigarrow v$.

Case 2: If $\langle e_1 \,\text{op}\, e_2, \text{Env}\rangle \hookrightarrow \langle e_1 \,\text{op}\, e_2', \text{Env}\rangle \hookrightarrow \cdots \hookrightarrow \langle e^k, \text{Env}\rangle$ is the considered sequence, with $\langle e_2, \text{Env}\rangle \hookrightarrow \langle e_2', \text{Env}\rangle$ and $e_1 \in \mathbb{Z}$ (the first transition is obtained using rule AS_6, AS_8 or AS_{10}), then this sequence may be decomposed as follows:

$$\langle e_1 \,\text{op}\, e_2, \text{Env}\rangle \hookrightarrow \underbrace{\langle e_1 \,\text{op}\, e_2', \text{Env}\rangle \hookrightarrow \cdots \langle e_1 \,\text{op}\, v_2', \text{Env}\rangle}_{\text{length } k-2} \hookrightarrow \langle v, \text{Env}\rangle$$

where the last transition is obtained with rule AS_2, AS_3 or AS_4 and thus $v_2' \in \mathbb{Z}$ and $v = e_1 \,\text{op}_{\mathbb{V}}\, v_2'$. The sequence of length $k-2$ corresponds to a simplification sequence for expression e_2, from which we can construct the sequence $\langle e_2', \text{Env}\rangle \hookrightarrow \cdots \hookrightarrow \langle v_2', \text{Env}\rangle$ of length $k-2$. Since $v_2' \in \mathbb{Z}$, by induction hypothesis we have $\text{Env} \vdash e_2' \rightsquigarrow v_2'$ and since $e_1 \in \mathbb{Z}$, rule F_1 allows us to obtain $\text{Env} \vdash e_1 \rightsquigarrow e_1$. Following rule F_4, we thus have $\text{Env} \vdash e_1 \,\text{op}\, e_2' \rightsquigarrow v$. From the previous question, since $\langle e_1 \,\text{op}\, e_2, \text{Env}\rangle \hookrightarrow \langle e_1 \,\text{op}\, e_2', \text{Env}\rangle$, we obtain $\text{Env} \vdash e_1 \,\text{op}\, e_2 \rightsquigarrow v$.

6) Rules AS_1, AS_2, AS_3 and AS_4 remain unchanged. The other rules are replaced by the following rules ($n_2 \in \mathbb{Z}$):

$$(AS_5)\frac{\langle e_2, \text{Env}\rangle \hookrightarrow \langle e_2', \text{Env}\rangle}{\langle e_1 + e_2, \text{Env}\rangle \hookrightarrow \langle e_1 + e_2', \text{Env}\rangle} \quad (AS_6)\frac{\langle e_1, \text{Env}\rangle \hookrightarrow \langle e_1', \text{Env}\rangle}{\langle e_1 + n_2, \text{Env}\rangle \hookrightarrow \langle e_1' + n_2, \text{Env}\rangle}$$

$$(AS_7)\frac{\langle e_2, \text{Env}\rangle \hookrightarrow \langle e_2', \text{Env}\rangle}{\langle e_1 * e_2, \text{Env}\rangle \hookrightarrow \langle e_1 * e_2', \text{Env}\rangle} \quad (AS_8)\frac{\langle e_1, \text{Env}\rangle \hookrightarrow \langle e_1', \text{Env}\rangle}{\langle e_1 * n_2, \text{Env}\rangle \hookrightarrow \langle e_1' * n_2, \text{Env}\rangle}$$

$$(AS_9)\frac{\langle e_2, \text{Env}\rangle \hookrightarrow \langle e_2', \text{Env}\rangle}{\langle e_1 - e_2, \text{Env}\rangle \hookrightarrow \langle e_1 - e_2', \text{Env}\rangle} \quad (AS_{10})\frac{\langle e_1, \text{Env}\rangle \hookrightarrow \langle e_1', \text{Env}\rangle}{\langle e_1 - n_2, \text{Env}\rangle \hookrightarrow \langle e_1' - n_2, \text{Env}\rangle}$$

A.3. Semantics of imperative features

Solution to exercise 4.1

1) Let Mem and Mem_3 be two memory states. We have to prove that:

$$\langle \text{Mem}, c_1 \,;\, (c_2 \,;\, c_3)\rangle \twoheadrightarrow_{\text{Env}} \text{Mem}_3 \Leftrightarrow \langle \text{Mem}, (c_1 \,;\, c_2) \,;\, c_3\rangle \twoheadrightarrow_{\text{Env}} \text{Mem}_3$$

An inference tree for $\langle \text{Mem}, (c_1;\, c_2);\, c_3\rangle \twoheadrightarrow_{\text{Env}} \text{Mem}_3$ can be obtained from an inference tree for $\langle \text{Mem}, c_1;\, (c_2;\, c_3)\rangle \twoheadrightarrow_{\text{Env}} \text{Mem}_3$, and vice versa. Indeed, from the tree:

$$\begin{array}{c}(C_3)\dfrac{(C_i)\dfrac{\overset{\nabla}{1}}{\langle \text{Mem}, c_1\rangle \twoheadrightarrow_{\text{Env}} \text{Mem}_1} \quad (C_3)\dfrac{(C_j)\dfrac{\overset{\nabla}{2}}{\langle \text{Mem}_1, c_2\rangle \twoheadrightarrow_{\text{Env}} \text{Mem}_2} \quad (C_k)\dfrac{\overset{\nabla}{3}}{\langle \text{Mem}_2, c_3\rangle \twoheadrightarrow_{\text{Env}} \text{Mem}_3}}{\langle \text{Mem}_1, c_2 \,;\, c_3\rangle \twoheadrightarrow_{\text{Env}} \text{Mem}_3}}{\langle \text{Mem}, c_1 \,;\, (c_2 \,;\, c_3)\rangle \twoheadrightarrow_{\text{Env}} \text{Mem}_3}\end{array}$$

we can obtain the tree:

$$
(C_3)\frac{(C_3)\dfrac{(C_i)\dfrac{\nabla_1}{\langle \text{Mem}, c_1\rangle \twoheadrightarrow_{\text{Env}} \text{Mem}_1} \quad (C_j)\dfrac{\nabla_2}{\langle \text{Mem}_1, c_2\rangle \twoheadrightarrow_{\text{Env}} \text{Mem}_2}}{\langle \text{Mem}, c_1 \,;\, c_2\rangle \twoheadrightarrow_{\text{Env}} \text{Mem}_2} \quad (C_k)\dfrac{\nabla_3}{\langle \text{Mem}_2, c_3\rangle \twoheadrightarrow_{\text{Env}} \text{Mem}_3}}{\langle \text{Mem}, (c_1 \,;\, c_2) \,;\, c_3\rangle \twoheadrightarrow_{\text{Env}} \text{Mem}_3}
$$

and vice versa.

2) Let c_1 be the instruction while e do c and c_2 the instruction if e then $(c;\ c_1)$ else skip. Let Mem_1 and Mem_2 be two memory states. We have to prove that:

$$
\langle \text{Mem}_1, c_1\rangle \twoheadrightarrow_{\text{Env}} \text{Mem}_2 \Leftrightarrow \langle \text{Mem}_1, c_2\rangle \twoheadrightarrow_{\text{Env}} \text{Mem}_2
$$

We have to construct an inference tree for $\langle \text{Mem}_1, c_2\rangle \twoheadrightarrow_{\text{Env}} \text{Mem}_2$ from an inference tree for $\langle \text{Mem}_1, c_1\rangle \twoheadrightarrow_{\text{Env}} \text{Mem}_2$, and vice versa.

($\Rightarrow$) If $\langle \text{Mem}_1, c_1\rangle \twoheadrightarrow_{\text{Env}} \text{Mem}_2$, then two cases are possible.

Case 1: The last applied rule is C_6:

$$
(C_6)\frac{\begin{array}{l}[\![e]\!]_{\text{Env}}^{\text{Mem}_1}\\ = (\mathit{true}, \text{Mem}'_1)\end{array} \quad (C_j)\dfrac{\nabla_1}{\langle \text{Mem}'_1, c\rangle \twoheadrightarrow_{\text{Env}} \text{Mem}''_1} \quad (C_k)\dfrac{\nabla_2}{\langle \text{Mem}''_1, c_1\rangle \twoheadrightarrow_{\text{Env}} \text{Mem}_2}}{\langle \text{Mem}_1, c_1\rangle \twoheadrightarrow_{\text{Env}} \text{Mem}_2}
\qquad [A.5]
$$

and, the following inference tree can be constructed:

$$
(C_4)\frac{\begin{array}{l}[\![e]\!]_{\text{Env}}^{\text{Mem}_1}\\ = (\mathit{true}, \text{Mem}'_1)\end{array} \quad (C_3)\dfrac{(C_j)\dfrac{\nabla_1}{\langle \text{Mem}'_1, c\rangle \twoheadrightarrow_{\text{Env}} \text{Mem}''_1} \quad (C_k)\dfrac{\nabla_2}{\langle \text{Mem}''_1, c_1\rangle \twoheadrightarrow_{\text{Env}} \text{Mem}_2}}{\langle \text{Mem}'_1, c;\ c_1\rangle \twoheadrightarrow_{\text{Env}} \text{Mem}_2}}{\langle \text{Mem}_1, c_2\rangle \twoheadrightarrow_{\text{Env}} \text{Mem}_2}
\qquad [A.6]
$$

Case 2: The last applied rule is C_7:

$$
(C_7)\frac{[\![e]\!]_{\text{Env}}^{\text{Mem}_1} = (\mathit{false}, \text{Mem}_2)}{\langle \text{Mem}_1, c_1\rangle \twoheadrightarrow_{\text{Env}} \text{Mem}_2}
\qquad [A.7]
$$

and the following inference tree can be constructed:

$$(C_5)\frac{[\![e]\!]_{\mathrm{Env}}^{\mathrm{Mem}_1} = (false, \mathrm{Mem}_2) \quad (C_2)\dfrac{}{\langle \mathrm{Mem}_2, \mathsf{skip}\rangle \twoheadrightarrow_{\mathrm{Env}} \mathrm{Mem}_2}}{\langle \mathrm{Mem}_1, c_2\rangle \twoheadrightarrow_{\mathrm{Env}} \mathrm{Mem}_2} \qquad [A.8]$$

($\Leftarrow$) If $\langle \mathrm{Mem}_1, c_2\rangle \twoheadrightarrow_{\mathrm{Env}} \mathrm{Mem}_2$ then two cases are possible. Either the last applied rule is C_4, and the tree [A.5] can be constructed from the tree [A.6], or the last applied rule is C_5, and the tree [A.7] can be constructed from the tree [A.8]. Both cases conclude the proof.

Solution to exercise 4.2

Throughout this exercise, we assume that for any expression e, $[\![e]\!]_{\mathrm{Env}}^{\mathrm{Mem}_1} = (true, \mathrm{Mem}_2)$ (respectively, $[\![e]\!]_{\mathrm{Env}}^{\mathrm{Mem}_1} = (false, \mathrm{Mem}_2)$) if and only if $[\![\mathrm{not}\ e]\!]_{\mathrm{Env}}^{\mathrm{Mem}_1} = (false, \mathrm{Mem}_2)$ (respectively, $[\![\mathrm{not}\ e]\!]_{\mathrm{Env}}^{\mathrm{Mem}_1} = (true, \mathrm{Mem}_2)$).

1) The semantics of the repeat until loop can be defined as follows:

$$(C_8)\frac{\langle \mathrm{Mem}_1, c\rangle \twoheadrightarrow_{\mathrm{Env}} \mathrm{Mem}_2 \quad [\![e]\!]_{\mathrm{Env}}^{\mathrm{Mem}_2} = (true, \mathrm{Mem}_3)}{\langle \mathrm{Mem}_1, \mathsf{repeat}\ c\ \mathsf{until}\ e\rangle \twoheadrightarrow_{\mathrm{Env}} \mathrm{Mem}_3}$$

$$(C_9)\frac{\langle \mathrm{Mem}_1, c\rangle \twoheadrightarrow_{\mathrm{Env}} \mathrm{Mem}_2 \quad \begin{array}{l}[\![e]\!]_{\mathrm{Env}}^{\mathrm{Mem}_2}\\ = (false, \mathrm{Mem}_3)\end{array} \quad \langle \mathrm{Mem}_3, \mathsf{repeat}\ c\ \mathsf{until}\ e\rangle \twoheadrightarrow_{\mathrm{Env}} \mathrm{Mem}_4}{\langle \mathrm{Mem}_1, \mathsf{repeat}\ c\ \mathsf{until}\ e\rangle \twoheadrightarrow_{\mathrm{Env}} \mathrm{Mem}_4}$$

2) Execution tree for the program c:

$$\mathsf{repeat}\ (\underbrace{z :=\downarrow z * \downarrow x}_{c_1}\ ;\ \underbrace{x :=\downarrow x - 1}_{c_2})\ \mathsf{until}\ \underbrace{\downarrow x = 0}_{e}$$

$$(C_9)\frac{\underset{1}{\nabla} \quad \begin{array}{l}[\![e]\!]_{\mathrm{Env}}^{\mathrm{Mem}_2}\\ = (false, \mathrm{Mem}_2)\end{array} \quad (C_8)\dfrac{\underset{2}{\nabla} \quad [\![e]\!]_{\mathrm{Env}}^{\mathrm{Mem}_3} = (true, \mathrm{Mem}_3)}{\langle \mathrm{Mem}_2, \mathsf{repeat}\ c_1\ ;\ c_2\ \mathsf{until}\ e\rangle \twoheadrightarrow_{\mathrm{Env}} \mathrm{Mem}_3}}{\langle \mathrm{Mem}, \mathsf{repeat}\ c_1\ ;\ c_2\ \mathsf{until}\ e\rangle \twoheadrightarrow_{\mathrm{Env}} \mathrm{Mem}_3}$$

where ∇_1 is the tree:

$$(C_3)\frac{(C_1)\dfrac{\begin{array}{l}\mathrm{Env}(z) = r_z\\ [\![\downarrow z * \downarrow x]\!]_{\mathrm{Env}}^{\mathrm{Mem}} = (6, \mathrm{Mem})\end{array}}{\langle \mathrm{Mem}, c_1\rangle \twoheadrightarrow_{\mathrm{Env}} \underbrace{\mathrm{Mem}[r_z := 6]}_{\mathrm{Mem}_1}} \quad (C_1)\dfrac{\begin{array}{l}\mathrm{Env}(x) = r_x\\ [\![\downarrow x - 1]\!]_{\mathrm{Env}}^{\mathrm{Mem}_1} = (1, \mathrm{Mem}_1)\end{array}}{\langle \mathrm{Mem}_1, c_2\rangle \twoheadrightarrow_{\mathrm{Env}} \underbrace{\mathrm{Mem}_1[r_x := 1]}_{\mathrm{Mem}_2}}}{\langle \mathrm{Mem}, c_1\ ;\ c_2\rangle \twoheadrightarrow_{\mathrm{Env}} \mathrm{Mem}_2}$$

and ∇_2 is the tree:

$$(C_3)\dfrac{(C_1)\dfrac{\begin{array}{c}\mathrm{Env}(z) = r_z \\ [\![\downarrow z * \downarrow x]\!]^{\mathrm{Mem}_2}_{\mathrm{Env}} = (6, \mathrm{Mem}_2)\end{array}}{\langle \mathrm{Mem}_2, c_1\rangle \twoheadrightarrow_{\mathrm{Env}} \mathrm{Mem}_2} \quad (C_1)\dfrac{\begin{array}{c}\mathrm{Env}(x) = r_x \\ [\![\downarrow x - 1]\!]^{\mathrm{Mem}_2}_{\mathrm{Env}} = (0, \mathrm{Mem}_2)\end{array}}{\langle \mathrm{Mem}_2, c_2\rangle \twoheadrightarrow_{\mathrm{Env}} \underbrace{\mathrm{Mem}_2[r_x := 0]}_{\mathrm{Mem}_3}}}{\langle \mathrm{Mem}_2, c_1 \,;\, c_2\rangle \twoheadrightarrow_{\mathrm{Env}} \mathrm{Mem}_3}$$

3) Let c_1 be the instruction repeat c until e and c_2 the instruction c; if e then skip else c_1. An inference tree for $\langle \mathrm{Mem}, c_1\rangle \twoheadrightarrow_{\mathrm{Env}} \mathrm{Mem}'$ can be constructed from the inference tree for $\langle \mathrm{Mem}, c_2\rangle \twoheadrightarrow_{\mathrm{Env}} \mathrm{Mem}'$ and vice versa. It suffices to consider the possible rules, which may be used to construct these trees and to examine the two possible values of e. In both cases, we simply "combine" the inference trees as follows.

Case 1: When $[\![e]\!]^{\mathrm{Mem}_1}_{\mathrm{Env}} = (true, \mathrm{Mem}')$, starting from the tree:

$$(C_8)\dfrac{(C_i)\dfrac{\nabla}{\langle \mathrm{Mem}, c\rangle \twoheadrightarrow_{\mathrm{Env}} \mathrm{Mem}_1} \quad [\![e]\!]^{\mathrm{Mem}_1}_{\mathrm{Env}} = (true, \mathrm{Mem}')}{\langle \mathrm{Mem}, c_1\rangle \twoheadrightarrow_{\mathrm{Env}} \mathrm{Mem}'}$$

we can obtain the tree:

$$(C_3)\dfrac{(C_i)\dfrac{\nabla}{\langle \mathrm{Mem}, c\rangle \twoheadrightarrow_{\mathrm{Env}} \mathrm{Mem}_1} \quad (C_4)\dfrac{\begin{array}{c}[\![e]\!]^{\mathrm{Mem}_1}_{\mathrm{Env}} \\ = (true, \mathrm{Mem}')\end{array} \quad (C_2)\dfrac{}{\langle \mathrm{Mem}', \text{skip}\rangle \twoheadrightarrow_{\mathrm{Env}} \mathrm{Mem}'}}{\langle \mathrm{Mem}_1, \text{if } e \text{ then skip else } c_1\rangle \twoheadrightarrow_{\mathrm{Env}} \mathrm{Mem}'}}{\langle \mathrm{Mem}, c_2\rangle \twoheadrightarrow_{\mathrm{Env}} \mathrm{Mem}'}$$

and vice versa.

Case 2: When $[\![e]\!]^{\mathrm{Mem}_1}_{\mathrm{Env}} = (false, \mathrm{Mem}_2)$, starting from the tree:

$$(C_9)\dfrac{(C_i)\dfrac{\nabla_1}{\langle \mathrm{Mem}, c\rangle \twoheadrightarrow_{\mathrm{Env}} \mathrm{Mem}_1} \quad \begin{array}{c}[\![e]\!]^{\mathrm{Mem}_1}_{\mathrm{Env}} \\ = (false, \mathrm{Mem}_2)\end{array} \quad (C_j)\dfrac{\nabla_2}{\langle \mathrm{Mem}_2, c_1\rangle \twoheadrightarrow_{\mathrm{Env}} \mathrm{Mem}'}}{\langle \mathrm{Mem}, c_1\rangle \twoheadrightarrow_{\mathrm{Env}} \mathrm{Mem}'}$$

we can obtain the tree:

$$(C_3)\dfrac{(C_i)\dfrac{\nabla_1}{\langle \mathrm{Mem}, c\rangle \twoheadrightarrow_{\mathrm{Env}} \mathrm{Mem}_1} \quad (C_5)\dfrac{\begin{array}{c}[\![e]\!]^{\mathrm{Mem}_1}_{\mathrm{Env}} \\ = (false, \mathrm{Mem}_2)\end{array} \quad (C_j)\dfrac{\nabla_2}{\langle \mathrm{Mem}_2, c_1\rangle \twoheadrightarrow_{\mathrm{Env}} \mathrm{Mem}'}}{\langle \mathrm{Mem}_1, \text{if } e \text{ then skip else } c_1\rangle \twoheadrightarrow_{\mathrm{Env}} \mathrm{Mem}'}}{\langle \mathrm{Mem}, c_2\rangle \twoheadrightarrow_{\mathrm{Env}} \mathrm{Mem}'}$$

and vice versa.

4) Let c'_1 be the instruction while not e do c and c'_2 be the instruction repeat c until not e (see section 4.8.3).

For (P1), let us prove by structural induction over $\langle \mathrm{Mem}_H, c_H\rangle \twoheadrightarrow_{\mathrm{Env}} \mathrm{Mem}_H$ (using the induction scheme given in Box 4.4) that if:

$$\langle \mathrm{Mem}_H, c_H\rangle \twoheadrightarrow_{\mathrm{Env}} \mathrm{Mem}_H = \langle \mathrm{Mem}, c_1\rangle \twoheadrightarrow_{\mathrm{Env}} \mathrm{Mem}'$$

then $\langle \mathrm{Mem}, c_2\rangle \twoheadrightarrow_{\mathrm{Env}} \mathrm{Mem}'$. If $\langle \mathrm{Mem}_H, c_H\rangle \twoheadrightarrow_{\mathrm{Env}} \mathrm{Mem}_H$ was obtained using one of the rules C_1 to C_5 or C_8 or C_9, then the equality given above is false, which concludes the proof. Hence, we only need to consider rules C_6 and C_7.

Case 1: If $\langle \mathrm{Mem}, c_1\rangle \twoheadrightarrow_{\mathrm{Env}} \mathrm{Mem}'$ was obtained using rule C_6:

$$(C_6)\frac{[\![e]\!]^{\mathrm{Mem}}_{\mathrm{Env}} = (true, \mathrm{Mem}_1) \quad (C_i)\dfrac{\nabla_1}{\langle \mathrm{Mem}_1, c\rangle \twoheadrightarrow_{\mathrm{Env}} \mathrm{Mem}_2} \quad (C_j)\dfrac{\vdots}{\langle \mathrm{Mem}_2, c_1\rangle \twoheadrightarrow_{\mathrm{Env}} \mathrm{Mem}'}}{\langle \mathrm{Mem}, c_1\rangle \twoheadrightarrow_{\mathrm{Env}} \mathrm{Mem}'}$$

then, by induction hypothesis, there exists an inference tree for $\langle \mathrm{Mem}_2, c_2\rangle \twoheadrightarrow_{\mathrm{Env}} \mathrm{Mem}'$. There are two subcases.

Case 1.a: If this tree has been obtained using rule C_4:

$$(C_4)\frac{[\![e]\!]^{\mathrm{Mem}_2}_{\mathrm{Env}} = (true, \mathrm{Mem}_3) \quad (C_k)\dfrac{\nabla_2}{\langle \mathrm{Mem}_3, c'_2\rangle \twoheadrightarrow_{\mathrm{Env}} \mathrm{Mem}'}}{\langle \mathrm{Mem}_2, c_2\rangle \twoheadrightarrow_{\mathrm{Env}} \mathrm{Mem}'}$$

then we construct the tree ∇_3:

$$(C_9)\frac{(C_i)\dfrac{\nabla_1}{\langle \mathrm{Mem}_1, c\rangle \twoheadrightarrow_{\mathrm{Env}} \mathrm{Mem}_2} \quad [\![\mathsf{not}\ e]\!]^{\mathrm{Mem}_2}_{\mathrm{Env}} = (false, \mathrm{Mem}_3) \quad (C_k)\dfrac{\nabla_2}{\langle \mathrm{Mem}_3, c'_2\rangle \twoheadrightarrow_{\mathrm{Env}} \mathrm{Mem}'}}{\langle \mathrm{Mem}_1, c'_2\rangle \twoheadrightarrow_{\mathrm{Env}} \mathrm{Mem}'}$$

We get the conclusion, since we can construct the tree:

$$(C_4)\frac{[\![e]\!]^{\mathrm{Mem}}_{\mathrm{Env}} = (true, \mathrm{Mem}_1) \quad (C_9)\dfrac{\nabla_3}{\langle \mathrm{Mem}_1, c'_2\rangle \twoheadrightarrow_{\mathrm{Env}} \mathrm{Mem}'}}{\langle \mathrm{Mem}, c_2\rangle \twoheadrightarrow_{\mathrm{Env}} \mathrm{Mem}'}$$

Case 1.b: If this tree has been obtained using rule C_5:

$$(C_5)\frac{[\![e]\!]^{\mathrm{Mem}_2}_{\mathrm{Env}} = (false, \mathrm{Mem}') \quad (C_2)\dfrac{}{\langle \mathrm{Mem}', \mathsf{skip}\rangle \twoheadrightarrow_{\mathrm{Env}} \mathrm{Mem}'}}{\langle \mathrm{Mem}_2, c_2\rangle \twoheadrightarrow_{\mathrm{Env}} \mathrm{Mem}'}$$

then we get the conclusion, since we can construct the tree:

$$(C_4)\dfrac{[\![e]\!]^{\mathrm{Mem}}_{\mathrm{Env}} = (true, \mathrm{Mem}_1) \qquad (C_8)\dfrac{(C_i)\dfrac{\overset{\nabla}{1}}{\langle \mathrm{Mem}_1, c\rangle \twoheadrightarrow_{\mathrm{Env}} \mathrm{Mem}_2} \qquad [\![\mathrm{not}\ e]\!]^{\mathrm{Mem}_2}_{\mathrm{Env}} = (true, \mathrm{Mem}')}{\langle \mathrm{Mem}_1, c_2'\rangle \twoheadrightarrow_{\mathrm{Env}} \mathrm{Mem}'}}{\langle \mathrm{Mem}, c_2\rangle \twoheadrightarrow_{\mathrm{Env}} \mathrm{Mem}'}$$

Case 2: If $\langle \mathrm{Mem}, c_1\rangle \twoheadrightarrow_{\mathrm{Env}} \mathrm{Mem}'$ has been obtained using rule C_7:

$$(C_7)\dfrac{[\![e]\!]^{\mathrm{Mem}}_{\mathrm{Env}} = (false, \mathrm{Mem}')}{\langle \mathrm{Mem}, c_1\rangle \twoheadrightarrow_{\mathrm{Env}} \mathrm{Mem}'}$$

we conclude directly, constructing the tree:

$$(C_5)\dfrac{[\![e]\!]^{\mathrm{Mem}}_{\mathrm{Env}} = (false, \mathrm{Mem}') \qquad (C_2)\dfrac{}{\langle \mathrm{Mem}', \mathsf{skip}\rangle \twoheadrightarrow_{\mathrm{Env}} \mathrm{Mem}'}}{\langle \mathrm{Mem}, c_2\rangle \twoheadrightarrow_{\mathrm{Env}} \mathrm{Mem}'}$$

For (P3), as for (P1), we proceed by structural induction over $\langle \mathrm{Mem}, c_3\rangle \twoheadrightarrow_{\mathrm{Env}} \mathrm{Mem}'$ to prove $\langle \mathrm{Mem}, c_4\rangle \twoheadrightarrow_{\mathrm{Env}} \mathrm{Mem}'$. Thus, only rules C_8 and C_9 need to be considered.

Case 1: If $\langle \mathrm{Mem}, c_3\rangle \twoheadrightarrow_{\mathrm{Env}} \mathrm{Mem}'$ has been obtained using rule C_8:

$$(C_8)\dfrac{(C_i)\dfrac{\overset{\nabla}{1}}{\langle \mathrm{Mem}, c\rangle \twoheadrightarrow_{\mathrm{Env}} \mathrm{Mem}_1} \qquad [\![e]\!]^{\mathrm{Mem}_1}_{\mathrm{Env}} = (true, \mathrm{Mem}')}{\langle \mathrm{Mem}, c_3\rangle \twoheadrightarrow_{\mathrm{Env}} \mathrm{Mem}'}$$

we can conclude, since we can construct the tree:

$$(C_3)\dfrac{(C_i)\dfrac{\overset{\nabla}{1}}{\langle \mathrm{Mem}, c\rangle \twoheadrightarrow_{\mathrm{Env}} \mathrm{Mem}_1} \qquad (C_7)\dfrac{[\![\mathrm{not}\ e]\!]^{\mathrm{Mem}_1}_{\mathrm{Env}} = (false, \mathrm{Mem}')}{\langle \mathrm{Mem}_1, c_1'\rangle \twoheadrightarrow_{\mathrm{Env}} \mathrm{Mem}'}}{\langle \mathrm{Mem}, c_4\rangle \twoheadrightarrow_{\mathrm{Env}} \mathrm{Mem}'}$$

Case 2: If $\langle \mathrm{Mem}, c_3\rangle \twoheadrightarrow_{\mathrm{Env}} \mathrm{Mem}'$ has been obtained using rule C_9:

$$(C_9)\dfrac{(C_i)\dfrac{\overset{\nabla}{1}}{\langle \mathrm{Mem}, c\rangle \twoheadrightarrow_{\mathrm{Env}} \mathrm{Mem}_1} \qquad [\![e]\!]^{\mathrm{Mem}_1}_{\mathrm{Env}} = (false, \mathrm{Mem}_2) \qquad (C_k)\dfrac{\vdots}{\langle \mathrm{Mem}_2, c_3\rangle \twoheadrightarrow_{\mathrm{Env}} \mathrm{Mem}'}}{\langle \mathrm{Mem}, c_3\rangle \twoheadrightarrow_{\mathrm{Env}} \mathrm{Mem}'}$$

then, by induction hypothesis, there exists an inference tree for $\langle \text{Mem}_2, c_4\rangle \twoheadrightarrow_{\text{Env}} \text{Mem}'$:

$$(C_3)\dfrac{(C_j)\dfrac{\overset{\nabla}{_2}}{\langle \text{Mem}_2, c\rangle \twoheadrightarrow_{\text{Env}} \text{Mem}_3} \quad (C_k)\dfrac{\overset{\nabla}{_3}}{\langle \text{Mem}_3, c_1'\rangle \twoheadrightarrow_{\text{Env}} \text{Mem}'}}{\langle \text{Mem}_2, c_4\rangle \twoheadrightarrow_{\text{Env}} \text{Mem}'}$$

and we can construct the tree ∇_4:

$$(C_6)\dfrac{\begin{array}{l}[\![\text{not } e]\!]^{\text{Mem}_1}_{\text{Env}}\\ = (true, \text{Mem}_2)\end{array} \quad (C_j)\dfrac{\overset{\nabla}{_2}}{\langle \text{Mem}_2, c\rangle \twoheadrightarrow_{\text{Env}} \text{Mem}_3} \quad (C_k)\dfrac{\overset{\nabla}{_3}}{\langle \text{Mem}_3, c_1'\rangle \twoheadrightarrow_{\text{Env}} \text{Mem}'}}{\langle \text{Mem}_1, c_1'\rangle \twoheadrightarrow_{\text{Env}} \text{Mem}'}$$

meaning that we can conclude by constructing the tree:

$$(C_3)\dfrac{(C_i)\dfrac{\overset{\nabla}{_1}}{\langle \text{Mem}, c\rangle \twoheadrightarrow_{\text{Env}} \text{Mem}_1} \quad (C_6)\dfrac{\overset{\nabla}{_4}}{\langle \text{Mem}_1, c_1'\rangle \twoheadrightarrow_{\text{Env}} \text{Mem}'}}{\langle \text{Mem}, c_4\rangle \twoheadrightarrow_{\text{Env}} \text{Mem}'}$$

For (P2), if $\langle \text{Mem}, c_2\rangle \twoheadrightarrow_{\text{Env}} \text{Mem}'$, then two cases are possible:

Case 1: If the tree for $\langle \text{Mem}, c_2\rangle \twoheadrightarrow_{\text{Env}} \text{Mem}'$ has been obtained using rule C_4:

$$(C_4)\dfrac{[\![e]\!]^{\text{Mem}}_{\text{Env}} = (true, \text{Mem}_1) \quad (C_i)\dfrac{\vdots}{\langle \text{Mem}_1, \texttt{repeat } c \texttt{ until not } e\rangle \twoheadrightarrow_{\text{Env}} \text{Mem}'}}{\langle \text{Mem}, c_2\rangle \twoheadrightarrow_{\text{Env}} \text{Mem}'}$$

then, according to (P3), there exists a tree for $\langle \text{Mem}_1, c; \texttt{while not not } e \texttt{ do } c\rangle \twoheadrightarrow_{\text{Env}} \text{Mem}'$, from which a tree for $\langle \text{Mem}_1, c; \texttt{while } e \texttt{ do } c\rangle \twoheadrightarrow_{\text{Env}} \text{Mem}'$ can be obtained:

$$(C_3)\dfrac{(C_i)\dfrac{\overset{\nabla}{_1}}{\langle \text{Mem}_1, c\rangle \twoheadrightarrow_{\text{Env}} \text{Mem}_2} \quad (C_j)\dfrac{\overset{\nabla}{_2}}{\langle \text{Mem}_2, \texttt{while } e \texttt{ do } c\rangle \twoheadrightarrow_{\text{Env}} \text{Mem}'}}{\langle \text{Mem}_1, c; \texttt{while } e \texttt{ do } c\rangle \twoheadrightarrow_{\text{Env}} \text{Mem}'}$$

We conclude by constructing the tree:

$$(C_6)\dfrac{\begin{array}{l}[\![e]\!]^{\text{Mem}}_{\text{Env}}\\ = (true, \text{Mem}_1)\end{array} \quad (C_i)\dfrac{\overset{\nabla}{_1}}{\langle \text{Mem}_1, c\rangle \twoheadrightarrow_{\text{Env}} \text{Mem}_2} \quad (C_j)\dfrac{\overset{\nabla}{_2}}{\langle \text{Mem}_2, c_1\rangle \twoheadrightarrow_{\text{Env}} \text{Mem}'}}{\langle \text{Mem}, c_1\rangle \twoheadrightarrow_{\text{Env}} \text{Mem}'}$$

Case 2: If the tree for $\langle \text{Mem}, c_2 \rangle \twoheadrightarrow_{\text{Env}} \text{Mem}'$ was obtained using rule C_5:

$$(C_5)\frac{[\![e]\!]_{\text{Env}}^{\text{Mem}} = (false, \text{Mem}') \qquad \langle \text{Mem}', \textsf{skip} \rangle \twoheadrightarrow_{\text{Env}} \text{Mem}'}{\langle \text{Mem}, c_2 \rangle \twoheadrightarrow_{\text{Env}} \text{Mem}'}$$

then we conclude by constructing the tree:

$$(C_7)\frac{[\![e]\!]_{\text{Env}}^{\text{Mem}} = (false, \text{Mem}')}{\langle \text{Mem}, c_1 \rangle \twoheadrightarrow_{\text{Env}} \text{Mem}'}$$

For (P4), if $\langle \text{Mem}, c_4 \rangle \twoheadrightarrow_{\text{Env}} \text{Mem}'$, we have the tree:

$$(C_3)\frac{(C_i)\dfrac{\overset{\nabla}{1}}{\langle \text{Mem}, c \rangle \twoheadrightarrow_{\text{Env}} \text{Mem}_1} \quad (C_j)\dfrac{\vdots}{\langle \text{Mem}_1, \textsf{while not } e \textsf{ do } c \rangle \twoheadrightarrow_{\text{Env}} \text{Mem}'}}{\langle \text{Mem}, c_4 \rangle \twoheadrightarrow_{\text{Env}} \text{Mem}'}$$

Following (P4), there exists a tree for:

$$\langle \text{Mem}_1, \textsf{if not } e \textsf{ then repeat } c \textsf{ until not not } e \textsf{ else skip} \rangle \twoheadrightarrow_{\text{Env}} \text{Mem}'$$

from which we can obtain a tree for:

$$\langle \text{Mem}_1, \textsf{if not } e \textsf{ then repeat } c \textsf{ until } e \textsf{ else skip} \rangle \twoheadrightarrow_{\text{Env}} \text{Mem}'$$

by using rule C_4 or rule C_5.

Case 1: If it has been obtained using rule C_4:

$$(C_4)\frac{[\![\textsf{not } e]\!]_{\text{Env}}^{\text{Mem}_1} = (true, \text{Mem}_2) \quad (C_k)\dfrac{\overset{\nabla}{2}}{\langle \text{Mem}_2, \textsf{repeat } c \textsf{ until } e \rangle \twoheadrightarrow_{\text{Env}} \text{Mem}'}}{\langle \text{Mem}_1, \textsf{if not } e \textsf{ then repeat } c \textsf{ until } e \textsf{ else skip} \rangle \twoheadrightarrow_{\text{Env}} \text{Mem}'}$$

then we conclude by constructing the tree:

$$(C_9)\frac{(C_i)\dfrac{\overset{\nabla}{1}}{\langle \text{Mem}, c \rangle \twoheadrightarrow_{\text{Env}} \text{Mem}_1} \quad [\![e]\!]_{\text{Env}}^{\text{Mem}_1} = (false, \text{Mem}_2) \quad (C_k)\dfrac{\overset{\nabla}{2}}{\langle \text{Mem}_2, c_3 \rangle \twoheadrightarrow_{\text{Env}} \text{Mem}'}}{\langle \text{Mem}, c_3 \rangle \twoheadrightarrow_{\text{Env}} \text{Mem}'}$$

Case 2: If it has been obtained using rule C_5:

$$(C_5)\frac{[\![\textsf{not } e]\!]_{\text{Env}}^{\text{Mem}_1} = (false, \text{Mem}') \quad (C_2)\dfrac{}{\langle \text{Mem}', \textsf{skip} \rangle \twoheadrightarrow_{\text{Env}} \text{Mem}'}}{\langle \text{Mem}_1, \textsf{if not } e \textsf{ then repeat } c \textsf{ until } e \textsf{ else skip} \rangle \twoheadrightarrow_{\text{Env}} \text{Mem}'}$$

then we conclude by constructing the tree:

$$(C_8)\frac{(C_i)\dfrac{\overset{\nabla}{1}}{\langle \text{Mem}, c \rangle \twoheadrightarrow_{\text{Env}} \text{Mem}_1} \quad [\![e]\!]_{\text{Env}}^{\text{Mem}_1} = (true, \text{Mem}')}{\langle \text{Mem}, c_3 \rangle \twoheadrightarrow_{\text{Env}} \text{Mem}'}$$

Solution to exercise 4.3

To ease proof reading, we introduce the following notation:

$$\underbrace{(\underbrace{a := 0}_{c_1}\,;\,\underbrace{b := x}_{c_2})\,;\,\underbrace{\textsf{while}\downarrow\underbrace{b \geq y}_{e_w}\,\textsf{do}\,\underbrace{b :=\downarrow b - y}_{c_3}\,;\,\underbrace{a :=\downarrow a + 1}_{c_4}}_{c_w}}_{c_d}$$

1) In what follows, we assume that $\text{Env}(a)$ and $\text{Env}(b)$ are elements in $\mathbf{R}$, denoted as r_a and r_b, respectively. Let v_x and v_y be the values $\text{Env}(x)$ and $\text{Env}(y)$. By definition, the functions associated with each of the components of this program are as follows:

$$\begin{array}{l}
[\![c_1]\!]_{\text{Env}} : \text{Mem} \mapsto \text{Mem}[r_a := 0] \\
[\![c_2]\!]_{\text{Env}} : \text{Mem} \mapsto \text{Mem}[r_b := v_x] \\
[\![c_1; c_2]\!]_{\text{Env}} : \text{Mem} \mapsto \text{Mem}[r_a := 0][r_b := v_x] \\
[\![c_3]\!]_{\text{Env}} : \text{Mem} \mapsto \text{Mem}[r_b := \text{Mem}(r_b) - v_y] \\
[\![c_4]\!]_{\text{Env}} : \text{Mem} \mapsto \text{Mem}[r_a := \text{Mem}(r_a) + 1] \\
[\![c_3; c_4]\!]_{\text{Env}} : \text{Mem} \mapsto \text{Mem}[r_b := \text{Mem}(r_b) - v_y][r_a := \text{Mem}(r_a) + 1] \\
[\![c_w]\!]_{\text{Env}} = \text{lfp}(F) \text{ where } F : f \mapsto \textsf{cond}([\![e_w]\!]_{\text{Env}}, f \circ [\![c_3; c_4]\!]_{\text{Env}}, id) \\
F(f) : \text{Mem} \mapsto \begin{cases} f\left([\![c_3; c_4]\!]_{\text{Env}}(\text{Mem})\right) & \text{if } \text{Mem}(r_b) \geq v_y \\ \text{Mem} & \text{if } \text{Mem}(r_b) < v_y \end{cases}
\end{array}$$

The function $[\![c_3; c_4]\!]_{\text{Env}}$ describes the state obtained after executing one iteration of the loop. When the body of the loop is executed twice, the function $[\![c_3; c_4]\!]^2_{\text{Env}} = [\![c_3; c_4]\!]_{\text{Env}} \circ [\![c_3; c_4]\!]_{\text{Env}}$ is defined by:

$$\begin{array}{l}
\text{Mem} \mapsto \left(\begin{array}{l} \text{Mem } [r_b := \text{Mem}(r_b) - v_y][r_a := \text{Mem}(r_a) + 1] \\ \qquad [r_b := \text{Mem}(r_b) - v_y - v_y][r_a := \text{Mem}(r_a) + 1 + 1] \end{array} \right) \\
\qquad = \text{Mem}[r_b := \text{Mem}(r_b) - 2v_y][r_a := \text{Mem}(r_a) + 2]
\end{array}$$

More generally, when the body of the loop is executed n times, we have (by an easy induction on n):

$$[\![c_3; c_4]\!]^n_{\text{Env}} : \text{Mem} \mapsto \text{Mem}[r_b := \text{Mem}(r_b) - nv_y][r_a := \text{Mem}(r_a) + n]$$

Using theorem 4.2, let us compute $\text{lfp}(F) = \sqcup(\{F^n(\bot) \mid n \geq 0\})$ from $F^0(\bot) = \bot : \text{Mem} \mapsto \text{undef}$. By definition, we have:

$$F^1(\bot) = F(\bot) : \text{Mem} \mapsto \begin{cases} \text{undef if } \text{Mem}(r_b) \geq v_y \\ \text{Mem} \;\; \text{if } \text{Mem}(r_b) < v_y \end{cases}$$

Similarly:

$$F^2(\bot) = F(F^1(\bot)) : \text{Mem} \mapsto \begin{cases} F^1(\bot)([\![c_3; c_4]\!]_{\text{Env}}(\text{Mem})) & \text{if } \text{Mem}(r_b) \geq v_y \\ \text{Mem} & \text{if } \text{Mem}(r_b) < v_y \end{cases}$$

and since $[\![c_3; c_4]\!]_{\text{Env}}(\texttt{Mem})(r_b) = \texttt{Mem}(r_b) - v_y$, we obtain:

$$F^2(\perp) : \texttt{Mem} \mapsto \begin{cases} \text{undef} & \text{if } \texttt{Mem}(r_b) \geq 2v_y \text{ and } \texttt{Mem}(r_b) \geq v_y \\ [\![c_3; c_4]\!]_{\text{Env}}(\texttt{Mem}) & \text{if } \texttt{Mem}(r_b) < 2v_y \text{ and } \texttt{Mem}(r_b) \geq v_y \\ \texttt{Mem} & \text{if } \texttt{Mem}(r_b) < v_y \end{cases}$$

With one more iteration, we obtain:

$$F^3(\perp) = F(F^2(\perp)) : \texttt{Mem} \mapsto \begin{cases} F^2(\perp)([\![c_3; c_4]\!]_{\text{Env}}(\texttt{Mem})) & \text{if } \texttt{Mem}(r_b) \geq v_y \\ \texttt{Mem} & \text{if } \texttt{Mem}(r_b) < v_y \end{cases}$$

and thus:

$$F^3(\perp) : \texttt{Mem} \mapsto \begin{cases} \text{undef} & \text{if } \texttt{Mem}(r_b) \geq 3v_y \text{ and } \texttt{Mem}(r_b) \geq 2v_y \\ & \text{and } \texttt{Mem}(r_b) \geq v_y \\ [\![c_3; c_4]\!]^2_{\text{Env}}(\texttt{Mem}) & \text{if } \texttt{Mem}(r_b) < 3v_y \text{ and } \texttt{Mem}(r_b) \geq 2v_y \\ & \text{and } \texttt{Mem}(r_b) \geq v_y \\ [\![c_3; c_4]\!]_{\text{Env}}(\texttt{Mem}) & \text{if } \texttt{Mem}(r_b) < 2v_y \text{ and } \texttt{Mem}(r_b) \geq v_y \\ \texttt{Mem} & \text{if } \texttt{Mem}(r_b) < v_y \end{cases}$$

By an easy induction on n, we finally obtain:

$$F^n(\perp) : \texttt{Mem} \mapsto \begin{cases} \text{undef} & \text{if } \texttt{Mem}(r_b) \geq nv_y \text{ and } \texttt{Mem}(r_b) \geq (n-1)v_y \\ & \text{and } \cdots \text{ and } \texttt{Mem}(r_b) \geq v_y \\ [\![c_3; c_4]\!]^{n-1}_{\text{Env}}(\texttt{Mem}) & \text{if } \texttt{Mem}(r_b) < nv_y \text{ and } \texttt{Mem}(r_b) \geq (n-1)v_y \\ & \text{and } \cdots \text{ and } \texttt{Mem}(r_b) \geq v_y \\ \ldots & \\ [\![c_3; c_4]\!]^1_{\text{Env}}(\texttt{Mem}) & \text{if } \texttt{Mem}(r_b) < 2v_y \text{ and } \texttt{Mem}(r_b) \geq v_y \\ \texttt{Mem} & \text{if } \texttt{Mem}(r_b) < v_y \end{cases}$$

which may be written in a simpler form:

$$F^n(\perp) : \texttt{Mem} \mapsto \begin{cases} \text{undef} & \text{if } \texttt{Mem}(r_b) \geq nv_y \\ [\![c_3; c_4]\!]^{p-1}_{\text{Env}}(\texttt{Mem}) & \text{if } (p-1)v_y \leq \texttt{Mem}(r_b) < pv_y \\ \texttt{Mem} & \text{if } \texttt{Mem}(r_b) < v_y \end{cases}$$

and the function $[\![c_w]\!]_{\text{Env}} = \text{lfp}(F) = \sqcup_{n\geq 0}\{F^n(\perp)\}$ is thus defined by:

$$[\![c_w]\!]_{\text{Env}} : \texttt{Mem} \mapsto \begin{cases} [\![c_3; c_4]\!]^{p-1}_{\text{Env}}(\texttt{Mem}) & \text{if } (p-1)v_y \leq \texttt{Mem}(r_b) < pv_y \\ \texttt{Mem} & \text{if } \texttt{Mem}(r_b) < v_y \end{cases}$$

The denotational semantics of program c_d is thus defined by:

$$[\![c_d]\!]_{\text{Env}} : \texttt{Mem} \mapsto [\![c_w]\!]_{\text{Env}}(\texttt{Mem}[r_a := 0][r_b := v_x])$$

and we finally obtain:

$$[\![c_d]\!]_{\text{Env}} : \texttt{Mem} \mapsto \begin{cases} \texttt{Mem}[r_b := v_x - (p-1)v_y][r_a := p-1] \\ \text{if } (p-1)v_y \leq v_x < pv_y \\ \texttt{Mem}[r_a := 0][r_b := v_x] \quad \text{if } v_x < v_y \end{cases}$$

When $v_x < v_y$, the quotient of the division of v_x by v_y is:

$$[\![c_d]\!]_{\text{Env}}(\text{Mem})(a) = 0,$$

while the remainder of this division is

$$[\![c_d]\!]_{\text{Env}}(\text{Mem})(b) = v_x.$$

When $v_x \geq v_y$, we have:

$$v_x = (v_y \times [\![c_d]\!]_{\text{Env}}(\text{Mem})(r_a)) + [\![c_d]\!]_{\text{Env}}(\text{Mem})(r_b)$$

with $[\![c_d]\!]_{\text{Env}}(\text{Mem}(r_b)) < v_y$ which, again, concludes the proof.

2) Let us prove the total correctness of the following Hoare triple:

$$\underbrace{\{x \geq 0 \text{ and } y > 0\}}_{P_1} c_d \underbrace{\{x = (\downarrow a \times y) + \downarrow b \text{ and } \downarrow b < y\}}_{P_2}$$

First, we use rules (Seq) and (Cons$_\text{D}$):

$$\text{(Seq)}\dfrac{\text{(Cons}_\text{G})\dfrac{\nabla_1}{\{P_1\}c_1\,;\,c_2\{P_I\}} \qquad \text{(Cons}_\text{D})\dfrac{\text{(W}_\text{T})\dfrac{\nabla_2}{\{P_I\}c_w\left\{\begin{array}{l}x = (\downarrow a \times y) + \downarrow b\\ \text{and } \downarrow b \geq 0 \text{ and } y > 0\\ \text{and not} \downarrow b \geq y\end{array}\right\}}}{\{P_I\}c_w\{P_2\}}}{\{P_1\}c_d\{P_2\}}$$

The application of rule (Cons$_\text{D}$) is based on the following implication:

$$(x = (\downarrow a \times y) + \downarrow b \ \text{ and } \ \downarrow b \geq 0 \ \text{ and } \ y > 0 \ \text{ and not } \downarrow b \geq y) \Rightarrow P_2$$

The loop invariant P_I used here is:

$$x = (\downarrow a \times y) + \downarrow b \ \text{ and } \ \downarrow b \geq 0 \ \text{ and } \ y > 0$$

and the variant is $V = \downarrow b$. Since $P_I \Rightarrow \downarrow b \geq 0$, the rule (W$_\text{T}$) can be applied. The inference tree ∇_2 is constructed as follows:

$$\text{(Seq)}\dfrac{\text{(Cons}_\text{G})\dfrac{\text{(Aff)}\dfrac{}{\left\{\begin{array}{l}P_I'' \text{ and}\\ \downarrow b - y < n\end{array}\right\} c_3 \left\{\begin{array}{l}P_I' \text{ and}\\ \downarrow b < n\end{array}\right\}}}{\left\{\begin{array}{l}P_I \text{ and}\\ \downarrow b \geq y \text{ and}\\ \downarrow b = n\end{array}\right\} c_3 \left\{\begin{array}{l}P_I' \text{ and}\\ \downarrow b < n\end{array}\right\}} \qquad \text{(Aff)}\dfrac{}{\left\{\begin{array}{l}P_I' \text{ and}\\ \downarrow b < n\end{array}\right\} c_4 \left\{\begin{array}{l}P_I \text{ and}\\ \downarrow b < n\end{array}\right\}}}{\{P_I \ \text{ and } \downarrow b \geq y \ \text{ and } \downarrow b = n\}c_3\,;\,c_4\{P_I \ \text{ and } \downarrow b < n\}}$$

where P'_I is the assertion P_I in which $\downarrow a$ has been substituted by $\downarrow a + 1$:

$$x = ((\downarrow a + 1) \times y) + \downarrow b \text{ and } \downarrow b \geq 0 \text{ and } y > 0$$

and P''_I is the assertion P'_I in which $\downarrow b$ has been substituted by $\downarrow b - y$:

$$x = ((\downarrow a + 1) \times y) + \downarrow b - y \text{ and } \downarrow b - y \geq 0 \text{ and } y > 0$$

The application of rule (Cons_{G}) is based on the following implication:

$$(P_I \text{ and } \downarrow b \geq y \text{ and } \downarrow b = n) \Rightarrow (P''_I \text{ and } \downarrow b - y < n)$$

which is easy to establish. Finally, the construction of tree ∇_1 is based on the implication $P_1 \Rightarrow (P_1 \text{ and } x = 0 \times y + x)$ and is obtained as follows:

$$(\text{Seq})\frac{(\text{Aff})\dfrac{}{\{P_1 \text{ and } x = 0 \times y + x\}c_2\{P''_1\}} \quad (\text{Aff})\dfrac{}{\{P''_1\}c_2\{P_I\}}}{\{P_1 \text{ and } x = 0 \times y + x\}c_1\,;\, c_2\{P_I\}}$$

where P''_1 is the assertion P_I in which $\downarrow b$ has been substituted by x:

$$x = (\downarrow a \times y) + x \text{ and } x \geq 0 \text{ and } y > 0$$

List of Notations

F, 120
Maj, 122
Min, 122
$[A \rightharpoonup B]$, 121
$\mathtt{Mem}[r := v]$, 19
$\mathtt{Env} \vdash e \leadsto v$, 55, 56, 62
$\twoheadrightarrow_{\mathtt{Env}}$, 89, 109, 115
$\langle \mathtt{Mem}, c \rangle \twoheadrightarrow_{\mathtt{Env}} \mathtt{Mem}'$, 89, 109, 115
$\perp$, 122
$[\![e]\!]^{\mathtt{Mem}}_{\mathtt{Env}}$, 24, 81, 131
codom(θ), 149
cond(f_b, f_1, f_2), 120
$\mathtt{Env} \vdash d : \mathtt{Env}'$, 169
$[\![c]\!]_{\mathtt{Env}}$, 119
dom($\mathtt{Env}$), 16
dom($\mathtt{Mem}$), 19
dom(θ), 149
$\mathtt{Env}$, 16, 42
$\equiv$, 69, 96
$\equiv_F$, 71
Err, 22
$\mathrm{Err_E}$, 42, 81
$\mathrm{Err_T}$, 42
$\mathrm{Err_U}$, 42, 81
$[\![e]\!]_{\mathtt{Env}}$, 42
$\mathtt{Env} \vdash e : \tau$, 169
$[\![\_]\!]_{\_}$, 81
$\forall \alpha_1 \ldots \alpha_n . \tau$, 145
$\mathcal{F}(e)$, 40
$\mathcal{F}(\sigma)$, 147
$\hookrightarrow^{\star}_{\mathtt{Env}}$, 94
$\mathtt{Gen}\,(\tau, \mathtt{Env})$, 147
undef, 119
$\mathtt{Env} \vdash e \triangleright (\tau, \theta)$, 154
$\mathtt{Env} \vdash c : WT$, 169
$\mathtt{Inst}\,(\sigma)$, 159
$\mathbf{Lang}_1$, 29
$\mathbf{Lang}_3$, 79, 109, 115, 118, 168
$\leq$, 210
lfp(f), 122
$\rightarrow_{\mathbf{Def}_3}$, 113
$\rightarrow_{\mathbf{Def}_1}$, 26
$\rightarrow_{\mathbf{Lang}_1}$, 29
$\rightarrow_{\mathbf{Def}_3}$, 86
$\mathtt{Mem}$, 18
$\mathtt{Mgu}\,(Q)$, 156
$\oplus$, 16
$\downarrow e$, 78
$\uparrow e$, 78
$\preceq$, 173
$\mathbb{A}_{\mathbb{S}}$, 65
$\mathbf{K}$, 35, 77
$\mathbf{Def}_1$, 26
$\mathbf{Def}_3$, 77, 112
$\mathbf{E}$, 16, 42
$\mathbb{E}$, 43, 81
$\mathbf{Exp}_1$, 21
$\mathbf{Exp}_2$, 35

$\mathbf{Exp}_3$, 77
$\mathbb{F}$, 43
$\mathbf{M}$, 18
$\mathbf{R}$, 18
$\mathbf{V}$, 16, 22, 43, 81
$\mathbb{V}$, 22, 42, 43, 81
$\mathbf{X}$, 16, 35
$\mathbf{X}^{\mathrm{cst}}_{\mathrm{Env}}$, 20
$\mathbf{X}^{\mathrm{ref}}_{\mathrm{Env}}$, 20
$\langle c_1, \mathrm{Mem}_1\rangle \hookrightarrow_{\mathrm{Env}} \langle c_2, \mathrm{Mem}_2\rangle$, 93
$\sqcup$, 122
$\sqsubseteq$, 121
$\{P_1\}c\{P_2\}$, 129
id, 163
$\mathrm{Env} \vdash e : \tau$, 151

Index of Programs

'a const1 (type), 23
'a def1 (type), 27
'a def4 (type), 79, 113
'a exp1 (type), 22
'a exp2 (type), 37
'a exp4, 79
'a option (type), 17
'a pgm4 (type), 80, 110, 115
'a val_err2 (type), 44
'a valeur2 (type), 44
'a valeurs1 (type), 23
'a valeurs4 (type), 82, 113
And2, 37
And4, 78
App2, 38
B_Op2, 38
B_Op4, 79
Bang1, 22
CBool2, 37
CInt1, 23
CInt2, 37
CRef1, 23
Clos2, 43
ClosRec2, 44
Cste1, 22
Cste2, 37
Cste4, 79
Cycle, 157
Deref4, 79
Div2, 37
Div4, 78
E_Exec2, 43
E_Typ2, 43
E_Undef2, 43
Eq2, 37
Eq4, 78
Err2, 44
Erreur1, 23
Eval_Error (exception), 83
Exec_Error (exception), 91
Ferm4, 113, 114
Fun2, 37
IAffect4, 80
IBloc4, 110
ICall4, 115
ICond4, 79
ISeq4, 80
ISkip4, 80
IWhile4, 80
If2, 38
Incompatible_tys, 157
Leq2, 37
Leq4, 78
Let_def1, 27
Let_def4, 79
Let_proc4, 112

Letin2, 37
Mgu, 156
Minus2, 37
Minus4, 78
Mult2, 37
Mult4, 78
Not2, 37
Not4, 78
Not_found, 159
Or2, 37
Or4, 78
Plus1, 22
Plus2, 37
Plus4, 78
Rec2, 38
Ref4, 79
Ref_Var1, 27
T_bool, 144
T_fun, 144
T_int, 144
T_var, 144
U_Op2, 37
U_Op4, 79
Unbound_var (type), 145
VBool2, 43
VCste1, 23
VCste4, 82
VInt2, 43
Val2, 44
Var1, 22
Var2, 37
Var4, 79
Var_def1, 27
a_fact, 38
add_mem, 83
ajout_liaison_env, 18
apply2_binary_op, 45
apply2_unary_op, 45
apply4_binary_op, 83
apply4_unary_op, 83
apply_one transf, 150
apply, 150
bin_op2 (type), 37
bin_op4 (type), 78
compose, 162
const2 (type), 37
const4 (type), 78
dom_env, 18
e25, 52
empty, 149
erreur2 (type), 44
eval2, 50
eval_exp1, 25
eval_exp4, 85
ex_e0, 29
ex_e1, 30
ex_exp1, 22, 38
ex_la0, 30
ex_ld0, 29
ex_ld4_1, 87
ex_m0, 29
ex_m1, 30
ex_pgm4_1, 80
ex_pgm4_2, 112
exec_pgm4, 91, 110, 117
find_id_ty (type), 145
find_vars_in_env, 148
find_vars_in_type, 148
free_var2, 41
generalize, 148
list_assoc, 159
list_mem_assoc, 162
new_ty_var, 144
occur_check, 157
refer (type), 27
scheme_t (type), 145
singleton, 149
specialize, 159
subst_env, 161
subst_t (type), 149
trans_def1_exec, 28
trans_def1, 28
trans_def4_exec, 86
trans_def4, 86, 114
trans_lang1_exec, 30
trans_lang1, 30

trivial_scheme, 146
ty_env_t (type), 145
type_expr, 165
type_t (type), 144
unary_op2 (type), 37
unary_op4 (type), 78
unify, 157
union_singleton, 18
valeur_ref, 19
var_name_t (type), 144
write_mem, 20

References

[ACC 92] ACCART HARDIN T., VIGUIÉ DONZEAU-GOUGE V., *Concepts et outils de programmation*, InterEditions, Paris, 1992.

[ARM 10] ARM, Cortex-M3 devices generic user guide: Revision A, available at: https://developer.arm.com/documentation/dui0552/a/the-cortex-m3-processor/programmers-model/core-registers, 2010.

[DAM 82] DAMAS L., MILNER R., "Principal type-schemes for functional programs", *Proceedings of the 9th Annual ACM Symposium on Principles of Programming Languages (POPL)*, pp. 207–212, ACM Press, 1982.

[DOL 93] DOLIGEZ D., LEROY X., "A concurrent, generational garbage collector for a multithreaded implementation of ML", *Proceedings of the 20th ACM Symposium on Principles of Programming Languages (POPL)*, pp. 113–123, ACM Press, 1993.

[DOL 94] DOLIGEZ D., GONTHIER G., "Portable, unobtrusive garbage collection for multiprocessor systems", *Proceedings of the 21st ACM Symposium on Principles of Programming Languages (POPL)*, pp. 70–83, ACM Press, 1994.

[DOW 09] DOWEK G., *Principles of Programming Languages*, Springer-Verlag, London, 2009.

[DOW 11] DOWEK G., LÉVY J.-J., *Introduction to the Theory of Programming Languages*, Springer-Verlag, London, 2011.

[DUB 18] DUBOIS C., GRINCHTEIN O., PEARSON J. *et al.*, "Exploring properties of a telecommunication protocol with message delay using interactive theorem prover", in JOHNSEN E., SCHAEFER I. (eds), *Proceedings of the 16th International Conference on Software Engineering and Formal Methods*, June 27–29, Toulouse, France, 2018.

[FRI 01] FRIEDMAN D.P., WAND M., HAYNES C.T., *Essentials of Programming Languages*, MIT Press, Cambridge, 2001.

[HOA 83] HOARE C.A.R., "An axiomatic basis for computer programming", *Communication of the ACM*, vol. 26, no.1, pp. 53–56, 1983.

[JAU 16] JAUME M., *Eléments de mathématiques discrètes – cours, exercices résolus, implémentations avec les langages Python et OCaml*, Ellipses, Paris, 2016.

[KLE 52] KLEENE S., *Introduction to Metamathematics*, North-Holland Publishing Company, Amsterdam, 1952.

[LER 91] LEROY X., WEIS P., "Polymorphic type inference and assignment", *Proceedings of the 18th ACM Symposium on Principles of Programming Languages (POPL)*, pp. 291–302, ACM Press, 1991.

[MIP 13] MIPS Technologies, *MIPS® Architecture for Programmers Volume II-A: The MIPS32® Instruction Set*, MIPS Tech, California, 2013.

[NIE 07] NIELSON H.R., NIELSON F., *Semantics with Applications: An Appetizer*, Springer-Verlag, London, 2007.

[PES 99] PESSAUX F., LEROY X., "Type-based analysis of uncaught exceptions", *Proceedings of the 26th ACM Symposium on Principles of Programming Languages (POPL)*, pp. 276–290, ACM Press, 1999.

[PIE 02] PIERCE B.C., *Types and Programming Languages*, MIT Press, Cambridge, 2002.

[PLO 04] PLOTKIN G.D., "A structural approach to operational semantics", *J. Logical and Algebraic Programming*, vols 60–61, pp. 17–139, 2004.

[STO 77] STOY J.E., *Denotational Semantics: The Scott-Strachey Approach to Programming Language Theory*, MIT Press, Cambridge, 1977.

[TUR 95] TURING A., GIRARD J.-Y., *La machine de Turing*, Le Seuil, Paris, 1995.

[WIN 93] WINSKEL G., *The Formal Semantics of Programming Languages: An Introduction*, MIT Press, Cambridge, 1993.

Index

A

algebra
  type, 143, 168, 194, 248
ALUs, 4
analysis
  lexical, 10
  semantic, 11
  syntactic, 11
array, 191, 225
assignment, 29, 89

B, C

binding, 16
  lexical, 57
  static, 57
block, 109
bytecode, 14
call
  -by-name, 61, 62
  -by-need, 61, 62
  -by-value, 56, 62
  by reference, 118, 226, 233
  by value, 115, 226
ccpo, 122
chain, 122
character, 186
  string, 194
closure, 43, 46, 60, 113
codomain of a substitution, 149
compiler, 12
continuous function, 124
contravariance, 174
correctness
  partial, 130
  total, 133
covariance, 174
CPU, 3

D, E

determinism, 65, 100
domain
  of a memory, 18, 19
  of a substitution, 149
  of an environment, 16, 17
endianness, 225

environment(s)
  equivalent, 71
  evaluation, 16
  execution, 16
  typing, 145
equality, 216
  between functions, 220
  of functions, 138
  physical, 217
  structural, 218
error
  execution, 42, 81, 139
  initialization, 19, 42, 81
  typing, 42
evaluation of an expression, 24, 42, 81
exception, 246
  handler, 246
expression(s)
  closed, 40
  equivalent, 69
  polymorphic, 143

F

fixpoint, 122
floating point, 187
frozen variable/expression, 61
function
  closed, 40
  compositional, 119
  logic, 2
  monotonic, 123
  partial, 119
  strict, 124

G

GC, 239
  conservative, 242
  copying, 240
  generational, 241
  mark-and-sweep, 240
  reference counting, 239
generality, 210
generalization, 147
generalized conditional, 205
guard, 213

H, I

Hoare triple, 129
identifier, 15
  definition, 26
in-out mode, 228
induction
  over an inference tree, 135
  over evaluation trees, 65
  structural, 39, 65
  well-founded, 98
inference, 154, 159
  type (algorithm), 163
input/output, 7
instance, 143, 151, 159
integer
  conversion (C), 185
  overflow, 182
  signed, 182
  size, 181
  two's complement, 181
interpreter, 12
interrupt, 8

L

lattice, 122
least upper bound, 122
linking, 13
loop
  invariant, 132
  variant, 133, 134
lower bound, 122

M

match/with, 208
memory, 2, 18
  allocation, 78, 234
    dynamic, 235
    static, 234
  EPROM, 3
  freeing, 237
  leak, 238
  RAM, 2, 3
  ROM, 3
minimum, 122

N, O

name binding
  masked, 16
object code, 12
order
  partial, 122
    over partial functions, 121
  well-founded, 75

P

parameter
  actual, 36, 46, 57
  formal, 36, 46, 57
pattern, 208
  disjunction, 214
  guard, 213
  matching, 208
pointer, 223
polymorphism, 143, 145, 171
postcondition, 129
precondition, 129
preorder types, 173
procedure, 112
programs
  equivalent, 96

R

record, 196, 214
reference, 18, 229
  accessible, 20
  counting, 239
register, 3
rule(s)
  evaluation, 55, 56, 62
  inference, 54
  subtyping, 174
  typing, 139, 151, 169

S

scheme
  field, 145
  trivial, 146
  type, 143, 145
scope
  dynamic, 46, 53, 113
  lexical, 46, 53, 113
  static, 46, 53, 113
semantics
  axiomatic, 129
  denotational, 118
  dynamic, 42
  operational, 54
    big-step, 60, 89
    small-step, 60, 74, 93
stack, 6
state, 20
string of characters, 194
structure, 196
substitution, 131, 149, 210
  application, 160
  composition, 161
  idempotent, 149
subtyping, 173
switch/case, 205

syntax
  abstract, 21, 35
  concrete, 21

T

term, 209
termination, 59, 63, 75, 98, 134
tokens, 11
tree
  abstract syntax, 11, 36, 80
  evaluation, 65
  inference, 65
  typing, 153
type
  abbreviation, 195
  compatible, 151
  enumerated, 200
  principal, 154
  record, 196
  sum, 202
typing
  completeness, 168
  correctness, 168
  dynamic, 139
  nominal, 195
  polymorphic, 143, 171
  safety, 167
  static, 140
  strong, 139
  structural, 195
  weak, 139

U

unification, 155
  algorithm, 156
unifier, 155
  principal, 156
`union`, 202
upper bound, 122

V

variable
  bound, 39, 146
  dummy, 39
  free, 40, 147
  fresh, 143
  generalized, 143, 145
  local, 35
  mutable, 15, 20, 87
  type, 141
virtual machine
  interpreter, 12–14

Other titles from

iSTE

in

Computer Engineering

2021

DELHAYE Jean-Loic
Inside the World of Computing: Technologies, Uses, Challenges

DUVAUT Patrick, DALLOZ Xavier, MENGA David, KOEHL François, CHRIQUI Vidal, BRILL Joerg
Internet of Augmented Me, I.AM: Empowering Innovation for a New Sustainable Future

2020

DARCHE Philippe
Microprocessor 1: Prolegomena – Calculation and Storage Functions – Models of Computation and Computer Architecture
Microprocessor 2: Core Concepts – Communication in a Digital System
Microprocessor 3: Core Concepts – Hardware Aspects
Microprocessor 4: Core Concepts – Software Aspects
Microprocessor 5: Software and Hardware Aspects of Development, Debugging and Testing – The Microcomputer

LAFFLY Dominique
TORUS 1 – Toward an Open Resource Using Services: Cloud Computing for Environmental Data
TORUS 2 – Toward an Open Resource Using Services: Cloud Computing for Environmental Data
TORUS 3 – Toward an Open Resource Using Services: Cloud Computing for Environmental Data

LAURENT Anne, LAURENT Dominique, MADERA Cédrine
Data Lakes
(Databases and Big Data Set – Volume 2)

OULHADJ Hamouche, DAACHI Boubaker, MENASRI Riad
Metaheuristics for Robotics
(Optimization Heuristics Set – Volume 2)

SADIQUI Ali
Computer Network Security

VENTRE Daniel
Artificial Intelligence, Cybersecurity and Cyber Defense

2019

BESBES Walid, DHOUIB Diala, WASSAN Niaz, MARREKCHI Emna
Solving Transport Problems: Towards Green Logistics

CLERC Maurice
Iterative Optimizers: Difficulty Measures and Benchmarks

GHLALA Riadh
Analytic SQL in SQL Server 2014/2016

TOUNSI Wiem
Cyber-Vigilance and Digital Trust: Cyber Security in the Era of Cloud Computing and IoT

2018

ANDRO Mathieu
Digital Libraries and Crowdsourcing
(Digital Tools and Uses Set – Volume 5)

ARNALDI Bruno, GUITTON Pascal, MOREAU Guillaume
Virtual Reality and Augmented Reality: Myths and Realities

BERTHIER Thierry, TEBOUL Bruno
From Digital Traces to Algorithmic Projections

CARDON Alain
Beyond Artificial Intelligence: From Human Consciousness to Artificial Consciousness

HOMAYOUNI S. Mahdi, FONTES Dalila B.M.M.
Metaheuristics for Maritime Operations
(Optimization Heuristics Set – Volume 1)

JEANSOULIN Robert
JavaScript and Open Data

PIVERT Olivier
NoSQL Data Models: Trends and Challenges
(Databases and Big Data Set – Volume 1)

SEDKAOUI Soraya
Data Analytics and Big Data

SALEH Imad, AMMI Mehdi, SZONIECKY Samuel
Challenges of the Internet of Things: Technology, Use, Ethics
(Digital Tools and Uses Set – Volume 7)

SZONIECKY Samuel
Ecosystems Knowledge: Modeling and Analysis Method for Information and Communication
(Digital Tools and Uses Set – Volume 6)

2017

BENMAMMAR Badr
Concurrent, Real-Time and Distributed Programming in Java

HÉLIODORE Frédéric, NAKIB Amir, ISMAIL Boussaad, OUCHRAA Salma, SCHMITT Laurent
Metaheuristics for Intelligent Electrical Networks
(Metaheuristics Set – Volume 10)

MA Haiping, SIMON Dan
Evolutionary Computation with Biogeography-based Optimization
(Metaheuristics Set – Volume 8)

PÉTROWSKI Alain, BEN-HAMIDA Sana
Evolutionary Algorithms
(Metaheuristics Set – Volume 9)

PAI G A Vijayalakshmi
Metaheuristics for Portfolio Optimization
(Metaheuristics Set – Volume 11)

2016

BLUM Christian, FESTA Paola
Metaheuristics for String Problems in Bio-informatics
(Metaheuristics Set – Volume 6)

DEROUSSI Laurent
Metaheuristics for Logistics
(Metaheuristics Set – Volume 4)

DHAENENS Clarisse and JOURDAN Laetitia
Metaheuristics for Big Data
(Metaheuristics Set – Volume 5)

LABADIE Nacima, PRINS Christian, PRODHON Caroline
Metaheuristics for Vehicle Routing Problems
(Metaheuristics Set – Volume 3)

LEROY Laure
Eyestrain Reduction in Stereoscopy

LUTTON Evelyne, PERROT Nathalie, TONDA Albert
Evolutionary Algorithms for Food Science and Technology
(Metaheuristics Set – Volume 7)

MAGOULÈS Frédéric, ZHAO Hai-Xiang
Data Mining and Machine Learning in Building Energy Analysis

RIGO Michel
Advanced Graph Theory and Combinatorics

2015

BARBIER Franck, RECOUSSINE Jean-Luc
COBOL Software Modernization: From Principles to Implementation with the BLU AGE® Method

CHEN Ken
Performance Evaluation by Simulation and Analysis with Applications to Computer Networks

CLERC Maurice
Guided Randomness in Optimization
(Metaheuristics Set – Volume 1)

DURAND Nicolas, GIANAZZA David, GOTTELAND Jean-Baptiste, ALLIOT Jean-Marc
Metaheuristics for Air Traffic Management
(Metaheuristics Set – Volume 2)

MAGOULÈS Frédéric, ROUX François-Xavier, HOUZEAUX Guillaume
Parallel Scientific Computing

MUNEESAWANG Paisarn, YAMMEN Suchart
Visual Inspection Technology in the Hard Disk Drive Industry

2014

BOULANGER Jean-Louis
Formal Methods Applied to Industrial Complex Systems

BOULANGER Jean-Louis
Formal Methods Applied to Complex Systems:Implementation of the B Method

GARDI Frédéric, BENOIST Thierry, DARLAY Julien, ESTELLON Bertrand, MEGEL Romain
Mathematical Programming Solver based on Local Search

KRICHEN Saoussen, CHAOUACHI Jouhaina
Graph-related Optimization and Decision Support Systems

LARRIEU Nicolas, VARET Antoine
Rapid Prototyping of Software for Avionics Systems: Model-oriented Approaches for Complex Systems Certification

OUSSALAH Mourad Chabane
Software Architecture 1
Software Architecture 2

PASCHOS Vangelis Th
Combinatorial Optimization – 3-volume series, 2nd Edition
Concepts of Combinatorial Optimization – Volume 1, 2nd Edition
Problems and New Approaches – Volume 2, 2nd Edition
Applications of Combinatorial Optimization – Volume 3, 2nd Edition

QUESNEL Flavien
Scheduling of Large-scale Virtualized Infrastructures: Toward Cooperative Management

RIGO Michel
Formal Languages, Automata and Numeration Systems 1: Introduction to Combinatorics on Words
Formal Languages, Automata and Numeration Systems 2: Applications to Recognizability and Decidability

SAINT-DIZIER Patrick
Musical Rhetoric: Foundations and Annotation Schemes

TOUATI Sid, DE DINECHIN Benoit
Advanced Backend Optimization

2013

ANDRÉ Etienne, SOULAT Romain
The Inverse Method: Parametric Verification of Real-time Embedded Systems

BOULANGER Jean-Louis
Safety Management for Software-based Equipment

DELAHAYE Daniel, PUECHMOREL Stéphane
Modeling and Optimization of Air Traffic

FRANCOPOULO Gil
LMF — Lexical Markup Framework

GHÉDIRA Khaled
Constraint Satisfaction Problems

ROCHANGE Christine, UHRIG Sascha, SAINRAT Pascal
Time-Predictable Architectures

WAHBI Mohamed
Algorithms and Ordering Heuristics for Distributed Constraint Satisfaction Problems

ZELM Martin *et al.*
Enterprise Interoperability

2012

ARBOLEDA Hugo, ROYER Jean-Claude
Model-Driven and Software Product Line Engineering

BLANCHET Gérard, DUPOUY Bertrand
Computer Architecture

BOULANGER Jean-Louis
Industrial Use of Formal Methods: Formal Verification

BOULANGER Jean-Louis
Formal Method: Industrial Use from Model to the Code

CALVARY Gaëlle, DELOT Thierry, SÈDES Florence, TIGLI Jean-Yves
Computer Science and Ambient Intelligence

MAHOUT Vincent
Assembly Language Programming: ARM Cortex-M3 2.0: Organization, Innovation and Territory

MARLET Renaud
Program Specialization

SOTO Maria, SEVAUX Marc, ROSSI André, LAURENT Johann
Memory Allocation Problems in Embedded Systems: Optimization Methods

2011

BICHOT Charles-Edmond, SIARRY Patrick
Graph Partitioning

BOULANGER Jean-Louis
Static Analysis of Software: The Abstract Interpretation

CAFERRA Ricardo
Logic for Computer Science and Artificial Intelligence

HOMES Bernard
Fundamentals of Software Testing

KORDON Fabrice, HADDAD Serge, PAUTET Laurent, PETRUCCI Laure
Distributed Systems: Design and Algorithms

KORDON Fabrice, HADDAD Serge, PAUTET Laurent, PETRUCCI Laure
Models and Analysis in Distributed Systems

LORCA Xavier
Tree-based Graph Partitioning Constraint

TRUCHET Charlotte, ASSAYAG Gerard
Constraint Programming in Music

VICAT-BLANC PRIMET Pascale *et al.*
Computing Networks: From Cluster to Cloud Computing

2010

AUDIBERT Pierre
Mathematics for Informatics and Computer Science

BABAU Jean-Philippe *et al.*
Model Driven Engineering for Distributed Real-Time Embedded Systems

BOULANGER Jean-Louis
Safety of Computer Architectures

MONMARCHE Nicolas *et al.*
Artificial Ants

PANETTO Hervé, BOUDJLIDA Nacer
Interoperability for Enterprise Software and Applications 2010

SIGAUD Olivier *et al.*
Markov Decision Processes in Artificial Intelligence

SOLNON Christine
Ant Colony Optimization and Constraint Programming

AUBRUN Christophe, SIMON Daniel, SONG Ye-Qiong *et al.*
Co-design Approaches for Dependable Networked Control Systems

2009

FOURNIER Jean-Claude
Graph Theory and Applications

GUEDON Jeanpierre
The Mojette Transform / Theory and Applications

JARD Claude, ROUX Olivier
Communicating Embedded Systems / Software and Design

LECOUTRE Christophe
Constraint Networks / Targeting Simplicity for Techniques and Algorithms

2008

BANÂTRE Michel, MARRÓN Pedro José, OLLERO Hannibal, WOLITZ Adam
Cooperating Embedded Systems and Wireless Sensor Networks

MERZ Stephan, NAVET Nicolas
Modeling and Verification of Real-time Systems

PASCHOS Vangelis Th
Combinatorial Optimization and Theoretical Computer Science: Interfaces and Perspectives

WALDNER Jean-Baptiste
Nanocomputers and Swarm Intelligence

2007

BENHAMOU Frédéric, JUSSIEN Narendra, O'SULLIVAN Barry
Trends in Constraint Programming

JUSSIEN Narendra
A TO Z OF SUDOKU

2006

BABAU Jean-Philippe *et al.*
From MDD Concepts to Experiments and Illustrations – DRES 2006

HABRIAS Henri, FRAPPIER Marc
Software Specification Methods

MURAT Cecile, PASCHOS Vangelis Th
Probabilistic Combinatorial Optimization on Graphs

PANETTO Hervé, BOUDJLIDA Nacer
Interoperability for Enterprise Software and Applications 2006 / IFAC-IFIP I-ESA'2006

2005

GÉRARD Sébastien *et al.*
Model Driven Engineering for Distributed Real Time Embedded Systems

PANETTO Hervé
Interoperability of Enterprise Software and Applications 2005